Current Topics in Behavioral Neurosciences

Volume 18

Series editors

Mark A. Geyer, La Jolla, CA, USA
Bart A. Ellenbroek, Wellington, New Zealand
Charles A. Marsden, Nottingham, UK

About this Series

Current Topics in Behavioral Neurosciences provides critical and comprehensive discussions of the most significant areas of behavioral neuroscience research, written by leading international authorities. Each volume offers an informative and contemporary account of its subject, making it an unrivalled reference source. Titles in this series are available in both print and electronic formats.

With the development of new methodologies for brain imaging, genetic and genomic analyses, molecular engineering of mutant animals, novel routes for drug delivery, and sophisticated cross-species behavioral assessments, it is now possible to study behavior relevant to psychiatric and neurological diseases and disorders on the physiological level. The *Behavioral Neurosciences* series focuses on "translational medicine" and cutting-edge technologies. Preclinical and clinical trials for the development of new diagnostics and therapeutics as well as prevention efforts are covered whenever possible.

More information about this series at http://www.springer.com/series/7854

Carmine M. Pariante · M. Danet Lapiz-Bluhm
Editors

Behavioral Neurobiology of Stress-related Disorders

Editors
Carmine M. Pariante
Department of Psychological Medicine
Institute of Psychiatry Kings College
  London
London
UK

M. Danet Lapiz-Bluhm
Department of Family and Community
  Health Systems
University of Texas Health Science Center
  at San Antonio
San Antonio, TX
USA

ISSN 1866-3370   ISSN 1866-3389 (electronic)
ISBN 978-3-662-45125-0   ISBN 978-3-662-45126-7 (eBook)
DOI 10.1007/978-3-662-45126-7

Library of Congress Control Number: 2014951150

Springer Heidelberg New York Dordrecht London

© Springer-Verlag Berlin Heidelberg 2014

This work is subject to copyright. All rights are reserved by the Publisher, whether the whole or part of the material is concerned, specifically the rights of translation, reprinting, reuse of illustrations, recitation, broadcasting, reproduction on microfilms or in any other physical way, and transmission or information storage and retrieval, electronic adaptation, computer software, or by similar or dissimilar methodology now known or hereafter developed. Exempted from this legal reservation are brief excerpts in connection with reviews or scholarly analysis or material supplied specifically for the purpose of being entered and executed on a computer system, for exclusive use by the purchaser of the work. Duplication of this publication or parts thereof is permitted only under the provisions of the Copyright Law of the Publisher's location, in its current version, and permission for use must always be obtained from Springer. Permissions for use may be obtained through RightsLink at the Copyright Clearance Center. Violations are liable to prosecution under the respective Copyright Law.

The use of general descriptive names, registered names, trademarks, service marks, etc. in this publication does not imply, even in the absence of a specific statement, that such names are exempt from the relevant protective laws and regulations and therefore free for general use.

While the advice and information in this book are believed to be true and accurate at the date of publication, neither the authors nor the editors nor the publisher can accept any legal responsibility for any errors or omissions that may be made. The publisher makes no warranty, express or implied, with respect to the material contained herein.

Printed on acid-free paper

Springer is part of Springer Science+Business Media (www.springer.com)

Preface

Since Selye defined stress in 1936, the field saw an exponential growth of research on the many facets of its effects. We are delighted to present this book that brings together some of the world leading experts on the neurobiology of stress at the preclinical and clinical level. Stress is such an over-used word that it is at times difficult to define its core features. When is an environment stressful? What does a stressful environment do to the brain and to the body? What are the biological mechanisms by which a stressor affects us? Why some environmental conditions are stressful for some individuals and not others? How does stress contributes to the onset and the progression of mental disorders? How do the effects of stress change over the lifetime of an individual? These are just some of the overarching questions that this book attempts to address, thanks to the contribution of 14 different chapters that cover a variety of topics.

In broad terms, the chapters can be grouped in three main streams. In the first stream, and over five chapters, this book present the biological pathways that are regulated by stress and that mediate the effects of stress on the brain and on the body, eventually affecting mental and physical health. These chapters cover brain-relevant mechanisms, ranging from "neurotransmitter systems" to "neuropeptides" to "neurogenesis and neuroplasticity". Moreover, they expand into mechanisms that are relevant to both the brain and the body, such as the "immune system" and the "hypothalamic–pituitary–adrenal axis" (HPA). In "Neuronal-Glial Mechanisms of Exercise-Evoked Stress Robustness", Dr. Monika Fleshner (University of Colorado, USA) and colleagues described an aspect of the interaction of stress, the immune system and behaviour. Their team presents a novel hypothesis on the role of exercise in promoting stress robustness through neuronal-glial mechanisms. Stress robustness incorporates resistance to stress and stress resilience. In "The Interface of Stress and the HPA Axis in Behavioural Phenotypes of Mental Illness", Dr. Carmine M. Pariante and Dr. David Baumeister from King's College London as well as Dr. Stafford Lightman (University of Bristol, UK) described the interface of stress and the HPA in behavioural phenotypes of mental illness. They elaborated on the clinical and molecular role of the neuroendocrine stress system in depressive, psychotic and post-traumatic stress disorders. In "Adult Hippocampal Neurogenesis in Depression: Behavioral Implications and Regulation by the Stress System", Dr. Chistoph Anacker (McGill University, Canada) addressed the interaction of stress, neurogenesis, neuroplasticity and behaviour. He reviewed some of the existing evidence for stress-and antidepressant-induced changes in adult hippocampal neurogenesis and their effects on depression and anxiety. In "Impact of Stress on Prefrontal Glutamatergic, Monoaminergic and Cannabinoid Systems", Dr. M. Danet Lapiz-Bluhm (University of Texas Health Science Center at San Antonio, USA) presented a concise review of the effects of stress and glucocorticoids on the glutamatergic, monoaminergic and cannabinoid signalling pathways modulating the prefrontal cortex. In "Interaction of Stress, Corticotropin-Releasing Factor, Arginine Vasopressin and Behaviour", Dr. Eleonore Beurel and Dr. Charles Nemeroff from the University of Miami (USA) focused on two peptidergic systems, i.e, corticotrophin releasing factor (CRF) and arginine vasopressin (AVP), on their roles in regulating stress response. Drugs that antagonize CRF and AVP receptors may have potential as a therapy for depression.

In the second stream, the emphasis is on psychological mechanisms that both mediate and modify the effects of stress, covering topics such as "cognition and emotional processing", the effects during "pregnancy and postnatal period" or "aging", and the important issue of "resilience". Across all chapters, the emphasis is on understanding the complex relationship between stress and behaviour, in all circumstances, leading sometimes to normal and sometimes to abnormal behavioural outcomes.

In "Long-lasting Consequences of Early Life Stress on Brain Structure, Emotion and Cognition", Dr. Harm Krugers (University of Amsterdam, Netherlands) and Dr. Marian Joëls (University Medical Center Utrecht, Netherlands) reviewed how early postnatal adversity determine the structure and function of the hippocampus, amygdala and the prefrontal cortex. These areas are crucial for the normal cognitive and emotional development. Along the same line, in "Mechanisms Linking In Utero Stress to Altered Offspring Behaviour", Dr. Theresia Mina and Dr. Rebecca Reynolds from the Queen's Medical Research Institute (Edinburgh, UK) highlighted the link between maternal *in utero* stressors on adverse behavioural outcomes of the offspring including poorer cognitive function as well as behavioural and emotional problems.

The succeeding chapters focus on the effects of stress on different psychiatric disorders. A team from Lundbeck Research USA and Denmark led by Dr. Connie Sanchez posed the question in "Does Stress Elicit Depression?", Dr. Helle Sickmann, Dr. Yan Li, Dr. Arne Mork, Dr. Connie Sanchez, and Dr. Maria Gulinello critically reviewed clinical and pre-clinical findings that may explain how stress can cause depression. Dr. M. Danet Lapiz-Bluhm and Dr. Alan Peterson, Director of STRONG STAR (South Texas Research Organizational Network Guiding Studies on Trauma and Resilience) PTSD Consortium, and both from the University of Texas Health Science Center at San Antonio (USA) reviewed the "Neurobehavioral Mechanisms of Traumatic Stress in Posttraumatic Stress Disorder". They reviewed the neurobiology of the effects of traumatic stress in the development of PTSD, specifically on mechanisms that are involved in fear conditioning and fear extinction. Dr. David

Baldwin and Dr. Hesham Yousry Elnazer from the University of Southampton (UK) addressed the role of stress in the development of anxiety disorders in "Investigation of Cortisol Levels in Patients with Anxiety Disorders: A Structured Review". Specifically, they reviewed HPA function across panic disorder, generalized anxiety disorder, specific phobias and social anxiety disorder.

In "Stress, Schizophrenia and Bipolar Disorder", a team from the University of New South Wales (Australia) and Schizophrenia Research Institute in Sydney (Australia) headed by Dr. Melissa Green reviewed the role of stress in the development of schizophrenia and bipolar disorder. They (Dr. Melissa Green, Dr. Leah Girshkin, Dr. Nina Teroganova and Dr. Yann Quidé) highlighted on how epigenetic studies of the effects of early life stress on gene expression may hold promise for unravelling the interaction between genes and environment to inform the 'stress-vulnerability' model of psychosis.

In "Stress, Substance Abuse, and Addiction", Drs. Charles Mathias and Dr. Donald Dougherty from the University of Texas Health Science Center at San Antonio (USA) in collaboration with Dr. Tiffany Duffing and Dr. Stefanie Greiner from Fielding Graduate University (USA) addressed the role of stress in substance abuse and addiction. They reviewed the developmental and biological processes involved in the relationship of stress exposure and substance use initiation, substance use maintenance and relapse, and response to substance abuse treatment. Special emphasis was given to describing the various stress-related mechanisms involved in substance use and abuse, highlighting the differences between each of these phases of drug use and drawing upon current research to make suggestions for treatments of substance use disorder (SUD) patients.

Dr. Mak Daulatzai (University of Melbourne, Australia) addressed the role of stress, depression and aging in cognitive decline and Alzheimer's disease in "Role of Stress, Depression, and Aging in Cognitive Decline and Alzheimer's Disease". He highlighted the role of gut systemic inflammation towards the development of neuroinflammation, which may subsequently upregulate hippocampal formation of amyloid beta and neurofibrillary tangles, synaptic and neuronal degeneration, gray matter volume atrophy, and progressive cognitive decline.

Last, but certainly not least, "Role of Stress, Depression, and Aging in Cognitive Decline and Alzheimer's Disease" addressed the issue on psychological resiliency to stress. Dr. Alan Peterson, Dr. Tabatha Blount and Dr. Donald McGreary from the University of Texas Health Science Center at San Antonio (USA) described how research on psychological resiliency is at its infancy and is limited by a number of factors including: (1) the broad use of the term resiliency; (2) the lack of standardized definitions of resiliency; (3) a primary focus on descriptive, assessment, and measurement studies; (4) relatively few randomized controlled trials to evaluate the efficacy of resiliency enhancement programs; and (5) methodological challenges inherent in conducting applied resiliency research. More studies are needed to better understand the behavioural neurobiology of stress and psychological resiliency.

To both the novice and the expert, this book will provide the reader "one-stop" resource on the most current body of knowledge and advances on the neurobiology of the pervasive effects of stress on various neurobiological systems and its role in the development of various stress-related disorder and resilience.

London, UK Carmine M. Pariante
San Antonio, USA M. Danet Lapiz-Bluhm

Contents

Neuronal-Glial Mechanisms of Exercise-Evoked Stress Robustness 1
Monika Fleshner, Benjamin N. Greenwood and Raz Yirmiya

The Interface of Stress and the HPA Axis in Behavioural Phenotypes of Mental Illness 13
David Baumeister, Staffor L. Lightman and Carmin M. Pariante

Adult Hippocampal Neurogenesis in Depression: Behavioral Implications and Regulation by the Stress System 25
Christoph Anacker

Impact of Stress on Prefrontal Glutamatergic, Monoaminergic and Cannabinoid Systems 45
M. Danet Lapiz-Bluhm

Interaction of Stress, Corticotropin-Releasing Factor, Arginine Vasopressin and Behaviour 67
Eléonore Beurel and Charle B. Nemeroff

Long-lasting Consequences of Early Life Stress on Brain Structure, Emotion and Cognition 81
Harm J. Krugers and Marian Joëls

Mechanisms Linking In Utero Stress to Altered Offspring Behaviour 93
Theresia H. Mina and Rebecc M. Reynolds

**Does Stress Elicit Depression? Evidence From Clinical
and Preclinical Studies**................................... 123
Helle M. Sickmann, Yan Li, Arne Mørk, Connie Sanchez
and Maria Gulinello

**Neurobehavioral Mechanisms of Traumatic Stress
in Post-traumatic Stress Disorder** 161
M. Danet Lapiz-Bluhm and Ala L. Peterson

**Investigation of Cortisol Levels in Patients with Anxiety
Disorders: A Structured Review**............................ 191
Hesham Yousry Elnazer and Davi S. Baldwin

Stress, Schizophrenia and Bipolar Disorder 217
Meliss J. Green, Leah Girshkin, Nina Teroganova and Yann Quidé

Stress, Substance Abuse, and Addiction 237
Tiffany M. Duffing, Stefanie G. Greiner, Charle W. Mathias
and Donal M. Dougherty

**Role of Stress, Depression, and Aging in Cognitive Decline
and Alzheimer's Disease** 265
Mak Adam Daulatzai

Stress and Psychological Resiliency........................ 297
Alan L. Peterson, Tabatha H. Blount and Donald D. McGeary

Index 313

Neuronal-Glial Mechanisms of Exercise-Evoked Stress Robustness

Monika Fleshner, Benjamin N. Greenwood and Raz Yirmiya

Abstract Stress robustness by definition, incorporates both stress resistance (organisms endure greater stressor intensity or duration before suffering negative consequences) and stress resilience (organisms recover faster after suffering negative consequences). Factors that influence stress robustness include the nature of the stressor, (i.e., controllability, intensity, chronicity) and features of the organism (i.e., age, genetics, sex, and physical activity status). Here we present a novel hypothesis for how physically active versus sedentary living promotes stress robustness in the face of intense uncontrollable stress. Advances in neurobiology have established microglia as an active player in the regulation of synaptic activity, and recent work has revealed mechanisms for modulating glial function, including cross talk between neurons and glia. This chapter presents supporting evidence that the physical activity status of an organism may modulate stress-evoked neuronal-glial responses by changing the CX3CL1-CX3CR1 axis. Specifically, we propose that sedentary animals respond to an intense acute uncontrollable stressor with excessive serotonin (5-HT) and noradrenergic (NE) activity and/or prolonged down-regulation of the CX3CL1-CX3CR1 axis resulting in activation and proliferation of hippocampal microglia in the absence of pathogenic signals and consequent hippocampal-dependent memory deficits and reduced neurogenesis. In contrast, physically active animals respond to the same stressor with constrained 5-HT and NE activity and rapidly recovering CX3CL1-CX3CR1 axis responses resulting in the quieting of microglia, and protection from negative cognitive and neurobiological effects of stress.

Keywords Stress resistance · Microglia · Fractalkine · Exercise

---

M. Fleshner (✉) · B. N. Greenwood
Department of Integrative Physiology and The Center for Neuroscience,
University of Colorado, Boulder, CO 80309, USA
e-mail: monika.fleshner@Colorado.EDU

R. Yirmiya
Department of Psychology, The Hebrew University of Jerusalem, Jerusalem, Israel

Abbreviations

ADR	Adrenergic
βADR	Beta Adrenergic Receptor
BDNF	Brain Derived Neurotrophic Factor
CX3CL1	CX3C Chemokine or Fractalkine
CX3CR1	CX3C Chemokine 1 Receptor or Fractalkine receptor
5-HT	Serotonin
IL-1β	Interleukin-1beta
NE	Norepinephrine
TNFα	Tumor Necrosis Factor Alpha
US	Uncontrollable stressor

Contents

1 Neuronal-Glia Consequences of Intense Uncontrollable Stressor (US) Exposure 2
  1.1 Mechanisms for Stress-Evoked Inflammatory/Destructive Microglia Activation and Hippocampal BDNF Downregulation: 5-HT and NE 3
2 Stress Robustness Produced by Physical Activity: Protection of Memory 4
  2.1 Mechanisms for Stress Robustness Produced by Physical Activity: 5-HT and NE Constraint 5
3 The CX3CL1-CX3CR1 Axis Promotes Microglia Quiescence and the Neuroprotective Phenotype 6
  3.1 Mechanism for Stress Robustness Produced by Physical Activity: CX3CL1-CX3CR1 Axis Modulation 7
4 Conclusions 8
References 9

1 Neuronal-Glia Consequences of Intense Uncontrollable Stressor (US) Exposure

The acute stress response is a highly adaptive cascade of physiological changes designed to facilitate behavioral escape and promote survival. If the stress response is excessive or prolonged, however, it can produce negative cognitive and affective consequences. Intense uncontrollable stressors (US) are especially capable of evoking excessive and prolonged stress responses and to result in negative consequences in some organisms. The neurobiological consequences of US in sedentary (stress vulnerable) organisms, for example, include sensitization of the brain raphe serotonergic (5-HT) system (Maier and Watkins 2005; Rozeske et al. 2011); excessive activation of locus coeruleus noradrenergic system (Greenwood et al. 2003; Weiss et al. 1994); increases in hippocampal inflammatory proteins (particularly interleukin-1beta, IL-1β, (Frank et al. 2007; Nguyen et al. 1998; O'Connor et al. 2003; Yirmiya and Goshen 2011); increases in hippocampal microglia proliferation (Bian et al. 2012) and activation (Frank et al. 2007); decreases in neuronal CX3C receptor R1 (CXC3R1) mRNA in dorsal raphe (Fig. 3); decreases in hippocampal neurogenesis (Bland et al. 2006); and decreases in plasticity-associated markers (i.e., brain derived neurotrophic factor, BDNF, (Greenwood et al. 2007; Zoladz et al. 2011). Importantly, these neural changes play a role in specific negative, cognitive, affective, and behavioral consequences of stress. For example, sensitized 5-HT produced by US interferes with the ability to escape subsequent threats (reviewed in (Greenwood and Fleshner 2008; Maier and Watkins 2005); whereas hippocampal IL-1β, perhaps via IL-1β-induced reductions in BDNF and neurogenesis, mediates US-evoked deficits in hippocampal-dependent learning and memory processes (Ekdahl 2012; Ekdahl et al. 2009; Kempermann and Neumann 2003; Leuner and Gould 2010; Maier et al. 1999). Dr. Raz Yirmiya and his research group have reported that exposure of mice to US activates microglia and elevates brain inflammatory cytokines, and these changes at least partially underlie the negative effects of stressor exposure on hippocampal dependent memory functioning and neurogenesis [reviewed in (Yirmiya and Goshen 2011)].

1.1 Mechanisms for Stress-Evoked Inflammatory/Destructive Microglia Activation and Hippocampal BDNF Downregulation: 5-HT and NE

Microglia both facilitate growth/repair and orchestrate inflammation/destruction (Tremblay et al. 2011). Recent evidence, for example, demonstrates that activated growth/repair microglia are essential during normal neural development (Ekdahl 2012; Paolicelli et al. 2011; Schafer et al. 2012) and may play a role in adult plasticity changes during learning, memory formation and neurogenesis (Ekdahl et al. 2009; Tremblay and Majewska 2011; Williamson et al. 2011; Yirmiya and Goshen 2011). In fact, there is some evidence to suggest that physical activity may promote the growth/repair microglia phenotype (Vukovic et al. 2012; Ziv et al. 2006) and that depletion of microglia can prevent some of the beneficial effects of exercise on learning and memory (Maggi et al. 2011). In contrast, after exposure to an acute uncontrollable stressor, inflammatory/destructive microglia can release inflammatory mediators that contribute to the negative, cognitive, affective, and neurogenic consequences of stress. Cross talk between neurons and microglia plays a dynamic role in microglia signaling, and are targets for how physical activity prevents damaging microglia activation and promotes stress robustness. This chapter focuses on two mechanisms of neuronal-microglia cross talk, stress-evoked 5-HT/NE, and the CX3CL1 (neuronal expressed)-CX3CR1 (glial expressed) axis.

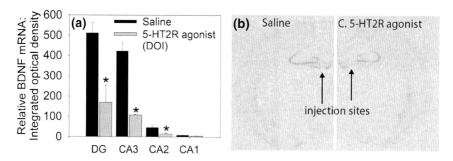

Fig. 1 Activation of 5-HT$_2$R reduces BDNF mRNA in the hippocampus. Adult, male, F344 rats (n = 3–4/group) were implanted with bilateral cannulae targeted to the dentate gyrus (DG) of the hippocampus. Following at least 7 days of recovery, rats received an intra-DG microinjection of saline into one hemisphere and the mixed 5-HT$_{2A/2C}$ receptor (5-HT$_2$R) agonist DOI (0.2 μl; 16 nmol) into the alternate hemisphere in a counterbalanced manner. Rats were sacrificed 2 h after injection and BDNF mRNA was quantified with in situ hybridization. **a** Relative BDNF mRNA levels throughout the hippocampal subfields. DOI reduced BDNF mRNA compared to saline injection. Representative coronal slices showing BDNF mRNA labeling after saline and DOI injections are shown in (**b**) and (**c**), respectively. Injection sites mark the location of the injector tips

Recent evidence suggests that both 5-HT and NE receptors are expressed on microglia (Pocock and Kettenmann 2007) and that these neurotransmitters can modulate microglia activity. For example, activation of 5-HT receptors on microglia, specifically the 5-HT2 subgroup, rapidly promotes microglia motility (Krabbe et al. 2012) and decreases hippocampal BDNF (Duman 1998; Vaidya et al. 1997, 1999); whereas activation of the beta adrenergic receptor (βADR) mediates hippocampal proinflammatory cytokine increases after US (Johnson et al. 2005). 5-HT and NE modulate glia function via transmitter "volume transmission" that allows neurotransmitters to diffuse into the extracellular space and interact with receptors on adjacent glia (Pocock and Kettenmann 2007), rather than via classical synaptic input. Figure 1 confirms that hippocampal injection of a drug that binds 5-HT2 receptors (DOI) reduces hippocampal BDNF mRNA.

2 Stress Robustness Produced by Physical Activity: Protection of Memory

Regular physical activity promotes stress robustness; whereas the lack of physical activity or a sedentary life style promotes stress vulnerability (the degree to which an organism succumbs to stressful influences) in humans and other mammals. The mechanisms for how differences in physical activity status contribute to changes in stress robustness remain unknown.

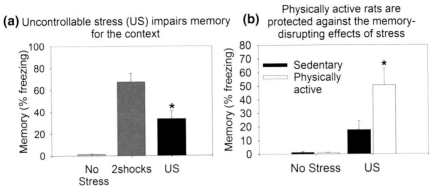

Fig. 2 Wheel running prevents memory impairment produced by uncontrollable stress (US). Adult, male F344 rats (N = 8/group) either remained sedentary or were allowed 6 weeks of voluntary access to running wheels prior to stressor exposure. Rats were handled (No Stress), or were exposed to a conditioning chamber for 5 min prior to exposure to either 2 foot shocks (0.6 mA, 1 s duration, 1 min ITI), or US consisting of 60 foot shocks (0.6 mA, 5 s duration, 1 min variable ITI). The next day, rats were placed back into the conditioning chambers and memory for the context where stress occurred was assessed using freezing. **a** Non stressed rats displayed no freezing behavior. Memory of the context where stress occurred was strong in sedentary rats following 2 foot shocks, as revealed by high levels of freezing. In contrast, rats exposed to US demonstrated 50 % less freezing, indicative of impaired memory for the context. Rats exposed to US have even greater freezing in a generalization context (not shown); suggesting the reduced freezing observed in the stressor context represents a memory failure and not a freezing impairment, per se. **b** Relative to sedentary rats, physically active rats exposed to US retain a strong memory of the stressor context

An example of stress robustness produced by physical activity is displayed in Fig. 2. Sedentary (stress vulnerable) rats exposed to US (60 uncontrollable foot-shocks) display a clear disruption in hippocampal-dependent contextual memory tested after 24 h (Fig. 2a). Rats allowed 6 weeks of prior wheel running are protected from the stress effect (Fig. 2b). Thus, physical activity promotes a stress robust phenotype such that US no longer produces memory disruptions.

2.1 Mechanisms for Stress Robustness Produced by Physical Activity: 5-HT and NE Constraint

Dr. Fleshner and her research group have been studying stress robustness produced by physical activity for more than a decade. Her group has evidence that 6 weeks of wheel running prevents sensitization of 5-HT neurons in the dorsal raphe nucleus produced by US [reviewed in (Greenwood and Fleshner 2008, 2011)], and constrains activation of locus coeruleus noradrenergic cells during exposure to US in rats (Fleshner 2005; Greenwood et al. 2003). Given that physically active rats respond to US with constrained 5-HT and NE responses, and that these neurotransmitters can stimulate proinflammatory cytokine increases and BDNF decreases, it is interesting to note that although we have not yet tested the hippocampus, we have preliminary data that exposure to US upregulates mRNA expression of the proinflammatory cytokines tumor necrosis factor alpha (TNFα) and IL-1β in the dorsal raphe of sedentary (stress vulnerable) organisms but not in physically active (stress robust) rats. In addition, we have also reported (Greenwood et al. 2007) that US reduces hippocampal BDNF in sedentary (stress vulnerable) rats but not physically active rats (stress robust). Given that, BDNF is required for consolidation of hippocampal-dependent memory (Alonso et al. 2002a, b; Tyler et al. 2002), a reduction in this plasticity-related molecule could be critical for the hippocampal-dependent memory disturbances produced by US. And finally, although the mechanisms for stressor- and pathogen-evoked cytokine release are *not* interchangeable (Campeau et al. 2010; Campisi et al. 2012; Maslanik et al. 2012), it is interesting to note that wheel running was recently reported to reduce hippocampal microglia activation, brain proinflammatory cytokine responses, and BDNF reductions after peripheral bacterial challenge in older rats (Barrientos et al. 2011) lending additional support to our hypothesis that regular physical activity may quiet stressor-evoked hippocampal microglia activation, and protect against BDNF down-regulation and reduced neurogenesis.

3 The CX3CL1-CX3CR1 Axis Promotes Microglia Quiescence and the Neuroprotective Phenotype

The chemokine CX3CL1 (fractalkine) is constitutively expressed in a transmembrane form on neurons and is released after metalloproteinase-mediated cleavage (Lauro et al. 2010). It has been suggested that CXCL1 helps to keep microglia in relative quiescence and to promote their neuroprotective phenotype (Mizuno et al. 2003). The effects of CXCL1 are mediated by binding to CX3CR1, which in the brain is exclusively expressed by microglia (Cardona et al. 2006; Jung et al. 2000). Homozygous mice with the CX3CR1-GFP mutation have a functional deletion of the microglial CX3CR1 (Jung et al. 2000). There is evidence that these CX3CR1$^{-/-}$ mice have microglial responses that are dysregulated, resulting in neurotoxicity and exacerbated neuronal cell loss in models of infection (peripheral LPS injections), Parkinson disease (Pabon et al. 2011), and Amyotrophic Lateral Sclerosis (Cardona et al. 2006). The CX3CL1-CX3CR1 axis is also involved in physiological processes, including modulation of hippocampal neurogenesis (Bachstetter et al. 2011; Maggi et al. 2011; Rogers et al. 2011). Interestingly, the aging brain responds to stressors and pathogens with exaggerated and prolonged microglia activation and inflammatory cytokines responses (Vukovic et al. 2012; Wynne et al. 2009); and recent evidence suggests that down-regulation of the

CX3CL1-CX3CR1 axis plays a major role in this effect (Bachstetter et al. 2011; Wynne et al. 2010).

3.1 Mechanism for Stress Robustness Produced by Physical Activity: CX3CL1-CX3CR1 Axis Modulation

In addition to constrained stress-evoked 5-HT and NE, physical activity may quiet stressor-evoked microglia activation by modulating the CX3CL1-CX3CR1 axis (D'Haese et al. 2012). We have preliminary data that exposure to US produces a *persistent* down-regulation of mRNA expression of CX3CR1 in the dorsal raphe of sedentary (stress vulnerable) organisms but not physically active (stress robust) rats. There is evidence that disruptions of the CX3CL1-CX3CR1 axis using genetic ($CX3CR^{+/-}$ or $CX3CR1^{-/-}$ mice) and pharmacological (CX3CR1 blocking Ig) approaches impairs in a graded fashion (with greater disruptions reported in knockouts than knockdowns) hippocampal neurogenesis and hippocampal-dependent (contextual fear conditioning) memory; and that exaggerated levels of IL-1β may directly contribute to these effects (Bachstetter et al. 2011; Rogers et al. 2011). Figure 3 reveals that exposure to US reduced CX3CR1 mRNA immediately after stressor exposure in all rats ($p < 0.001$), and mRNA levels remained reduced 2 h after stressor termination in the sedentary ($p < 0.05$) but not physical active rats. Interestingly, CX3CL1 mRNA expression was not changed by stress (not shown); however, this does not rule out a change in ligand concentrations. A lack of an effect on CX3CL1 mRNA is a common finding since secreted CXCL1 protein concentrations are primarily modulated post-transcriptionally by protein kinase C (Hatori et al. 2002). There were also no changes in other neuronalglia regulatory molecules (i.e., CD200, CD200R1), suggesting that physical activity may *selectively* modulate the CX3CL1-CX3CR1 axis. This improved rate of recovery could be a molecular signature of stress resilience.

Thus, we hypothesize that exposure to US in sedentary animals down regulates the CX3CL1-CX3CR1 axis, increases microglia motility (mediated by 5-HT; (Krabbe et al. 2012)), decreases BDNF (mediated by 5-HT (Duman 1998; Vaidya et al. 1997, 1999), Fig. 1), and stimulates excessive pro-inflammatory cytokine release (mediated by NE, (Johnson et al. 2005)) in the hippocampus and other areas of the stress circuit (i.e., dorsal raphe). These microglia responses can disrupt learning and memory processes and inhibit adaptive organismal behavioral performance to stress such as active coping in response to future stressful events and hippocampal-dependent learning and memory processes, by impairing neurogenesis and the expression of plasticity-associated molecules (e.g., BDNF). In contrast, physically active animals exposed to US are protected from excessive 5-HT and NE responses, proinflammatory cytokines responses/release, and more rapidly restore the CX3CL1-CX3CR1 axis and thus resist the negative neurobiological and behavioral effects.

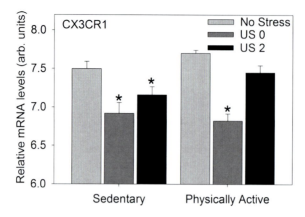

Fig. 3 Uncontrollable stress (US) reduces mRNA expression of CX3CR1. Rats (n = 8/group) lived with mobile (Physically Active) or locked (Sedentary) running wheels. After 6 weeks, rats were exposed to US and sacrificed immediately or 2 h after stressor termination. The dorsal raphe nucleus (DRN) was laser captured and mRNA expression was measured using Affymatrix gene array. The results were that US reduced CX3CR1 mRNA expression and that physically active rats recovered faster than sedentary rats

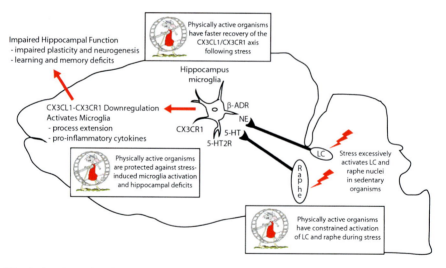

Fig. 4 A schematic of our working hypothesis for a neuronal-glal mechanisms for stress resistance produced by exercise

4 Conclusions

Clearly, regular physical activity changes the way our brain responds to stressors. The types of changes produced by regular physical activity are widespread and include adaptations in neurogenesis, growth factors, and neurotransmitter systems.

Only in the past few years have neurobiologists fully appreciated the dynamic interactions between neuron and microglia. This work suggests that the changes in neuronal-glia cross talk should be added to the growing list of brain adaptations produced by exercise. Figure 4 depicts our current hypothesis. We propose that sedentary animals respond to an intense acute uncontrollable stressor with excessive serotonin (5-HT) and noradrenergic (NE) activity and/or prolonged downregulation of the CX3CL1-CX3CR1 axis resulting in activation and proliferation of hippocampal microglia in the absence of pathogenic signals and consequent hippocampal-dependent memory deficits and reduced neurogenesis. In contrast, physically active animals respond to the same stressor with constrained 5-HT and NE activity and rapidly recovering CX3CL1-CX3CR1 axis responses resulting in the quieting of microglia and protection from negative cognitive and neurobiological effects of stress.

References

Alonso M, Vianna MR, Depino AM, Mello e Souza T, Pereira P, Szapiro G, Viola H, Pitossi F, Izquierdo I, Medina JH (2002a) BDNF-triggered events in the rat hippocampus are required for both short- and long-term memory formation. Hippocampus 12:551–560. doi:10.1002/hipo.10035

Alonso M, Vianna MR, Izquierdo I, Medina JH (2002b) Signaling mechanisms mediating BDNF modulation of memory formation in vivo in the hippocampus. Cell Mol Neurobiol 22:663–674

Bachstetter AD, Morganti JM, Jernberg J, Schlunk A, Mitchell SH, Brewster KW, Hudson CE, Cole MJ, Harrison JK, Bickford PC, Gemma C (2011) Fractalkine and CX 3 CR1 regulate hippocampal neurogenesis in adult and aged rats. Neurobiol Aging 32:2030–2044. doi:10.1016/j.neurobiolaging.2009.11.022

Barrientos RM, Frank MG, Crysdale NY, Chapman TR, Ahrendsen JT, Day HE, Campeau S, Watkins LR, Patterson SL, Maier SF (2011) Little exercise, big effects: reversing aging and infection-induced memory deficits, and underlying processes. J Neurosci Official J Soc Neurosci 31:11578–11586. doi:10.1523/JNEUROSCI.2266-11.2011

Bian Y, Pan Z, Hou Z, Huang C, Li W, Zhao B (2012) Learning, memory, and glial cell changes following recovery from chronic unpredictable stress. Brain Res Bull. doi:10.1016/j.brainresbull.2012.04.008

Bland ST, Schmid MJ, Greenwood BN, Watkins LR, Maier SF (2006) Behavioral control of the stressor modulates stress-induced changes in neurogenesis and fibroblast growth factor-2. NeuroReport 17:593–597

Campeau S, Nyhuis TJ, Kryskow EM, Masini CV, Babb JA, Sasse SK, Greenwood BN, Fleshner M, Day HE (2010) Stress rapidly increases alpha 1d adrenergic receptor mRNA in the rat dentate gyrus. Brain Res 1323: 109–118. doi:10.1016/j.brainres.2010.01.084 (S0006-8993 (10)00238-6[pii])

Campisi J, Sharkey C, Johnson JD, Asea A, Maslanik T, Bernstein-Hanley I, Fleshner M (2012) Stress-induced facilitation of host response to bacterial challenge in F344 rats is dependent on extracellular heat shock protein 72 and independent of alpha beta T cells. Stress 15:637–646. doi:10.3109/10253890.2011.653596

Cardona AE, Pioro EP, Sasse ME, Kostenko V, Cardona SM, Dijkstra IM, Huang D, Kidd G, Dombrowski S, Dutta R, Lee JC, Cook DN, Jung S, Lira SA, Littman DR, Ransohoff RM (2006) Control of microglial neurotoxicity by the fractalkine receptor. Nat Neurosci 9:917–924. doi:10.1038/nn1715

D'Haese JG, Friess H, Ceyhan GO (2012) Therapeutic potential of the chemokine-receptor duo fractalkine/CX3CR1: an update. Expert Opin Ther Targets 16:613–618. doi:10.1517/14728222.2012.682574

Duman RS (1998) Novel therapeutic approaches beyond the serotonin receptor. Biol Psychiatry 44:324–335

Ekdahl CT (2012) Microglial activation—tuning and pruning adult neurogenesis. Front Pharmacol 3:41. doi:10.3389/fphar.2012.00041

Ekdahl CT, Kokaia Z, Lindvall O (2009) Brain inflammation and adult neurogenesis: the dual role of microglia. Neuroscience 158:1021–1029. doi:10.1016/j.neuroscience.2008.06.052

Fleshner M (2005) Physical activity and stress resistance: sympathetic nervous system adaptations prevent stress-induced immunosuppression. Exerc Sport Sci Rev 33:120–126

Frank MG, Baratta MV, Sprunger DB, Watkins LR, Maier SF (2007) Microglia serve as a neuroimmune substrate for stress-induced potentiation of CNS pro-inflammatory cytokine responses. Brain Behav Immun 21:47–59. doi:10.1016/j.bbi.2006.03.005

Greenwood BN, Fleshner M (2008) Exercise, learned helplessness, and the stress-resistant brain. Neuromol Med 10:81–98

Greenwood BN, Fleshner M (2011) Exercise, stress resistance, and central serotonergic systems. Exerc Sport Sci Rev 39:140–149. doi:10.1097/JES.0b013e31821f7e45

Greenwood BN, Kennedy S, Smith TP, Campeau S, Day HE, Fleshner M (2003) Voluntary freewheel running selectively modulates catecholamine content in peripheral tissue and c-Fos expression in the central sympathetic circuit following exposure to uncontrollable stress in rats. Neuroscience 120:269–281

Greenwood BN, Strong PV, Foley TE, Thompson RS, Fleshner M (2007) Learned helplessness is independent of levels of brain-derived neurotrophic factor in the hippocampus. Neuroscience 144:1193–1208

Hatori K, Nagai A, Heisel R, Ryu JK, Kim SU (2002) Fractalkine and fractalkine receptors in human neurons and glial cells. J Neurosci Res 69:418–426. doi:10.1002/jnr.10304

Johnson JD, Campisi J, Sharkey CM, Kennedy SL, Nickerson M, Greenwood BN, Fleshner M (2005) Catecholamines mediate stress-induced increases in peripheral and central inflammatory cytokines. Neuroscience 135:1295–1307

Jung S, Aliberti J, Graemmel P, Sunshine MJ, Kreutzberg GW, Sher A, Littman DR (2000) Analysis of fractalkine receptor CX(3)CR1 function by targeted deletion and green fluorescent protein reporter gene insertion. Mol Cell Biol 20:4106–4114

Kempermann G, Neumann H (2003) Neuroscience. Microglia: the enemy within? Science 302:1689–1690. doi:10.1126/science.1092864

Krabbe G, Matyash V, Pannasch U, Mamer L, Boddeke HW, Kettenmann H (2012) Activation of serotonin receptors promotes microglial injury-induced motility but attenuates phagocytic activity. Brain Behav Immun 26:419–428. doi:10.1016/j.bbi.2011.12.002

Lauro C, Cipriani R, Catalano M, Trettel F, Chece G, Brusadin V, Antonilli L, van Rooijen N, Eusebi F, Fredholm BB, Limatola C (2010) Adenosine A1 receptors and microglial cells mediate CX3CL1-induced protection of hippocampal neurons against Glu-induced death. Neuropsychopharmacol Official Publ Am Coll Neuropsychopharmacol 35:1550–1559. doi:10.1038/npp.2010.26

Leuner B, Gould E (2010) Structural plasticity and hippocampal function. Annu Rev Psychol 61(111–40):C1–C3. doi:10.1146/annurev.psych.093008.100359

Maggi L, Scianni M, Branchi I, D'Andrea I, Lauro C, Limatola C (2011) CX(3)CR1 deficiency alters hippocampal-dependent plasticity phenomena blunting the effects of enriched environment. Front Cell Neurosci 5:22. doi:10.3389/fncel.2011.00022

Maier SF, Nguyen KT, Deak T, Milligan ED, Watkins LR (1999) Stress, learned helplessness, and brain interleukin-1 beta. Adv Exp Med Biol 461:235–249

Maier SF, Watkins LR (2005) Stressor controllability and learned helplessness: the roles of the dorsal raphe nucleus, serotonin, and corticotropin-releasing factor. Neurosci Biobehav Rev 29:829–841. doi:10.1016/j.neubiorev.2005.03.021

Maslanik T, Bernstein-Hanley I, Helwig B, Fleshner M (2012) The impact of acute-stressor exposure on splenic innate immunity: a gene expression analysis. Brain Behav Immun 26:142–149. doi:10.1016/j.bbi.2011.08.006

Mizuno T, Kawanokuchi J, Numata K, Suzumura A (2003) Production and neuroprotective functions of fractalkine in the central nervous system. Brain Res 979:65–70

Nguyen KT, Deak T, Owens SM, Kohno T, Fleshner M, Watkins LR, Maier SF (1998) Exposure to acute stress induces brain interleukin-1beta protein in the rat. J Neurosci 18:2239–2246

O'Connor KA, Johnson JD, Hansen MK, Wieseler Frank JL, Maksimova E, Watkins LR, Maier SF (2003) Peripheral and central proinflammatory cytokine response to a severe acute stressor. Brain Res 991:123–132

Pabon MM, Bachstetter AD, Hudson CE, Gemma C, Bickford PC (2011) CX3CL1 reduces neurotoxicity and microglial activation in a rat model of Parkinson's disease. J Neuroinflammation 8:9. doi:10.1186/1742-2094-8-9

Paolicelli RC, Bolasco G, Pagani F, Maggi L, Scianni M, Panzanelli P, Giustetto M, Ferreira TA, Guiducci E, Dumas L, Ragozzino D, Gross CT (2011) Synaptic pruning by microglia is necessary for normal brain development. Science 333:1456–1458. doi:10.1126/science. 1202529

Pocock JM, Kettenmann H (2007) Neurotransmitter receptors on microglia. Trends Neurosci 30:527–535. doi:10.1016/j.tins.2007.07.007

Rogers JT, Morganti JM, Bachstetter AD, Hudson CE, Peters MM, Grimmig BA, Weeber EJ, Bickford PC, Gemma C (2011) CX3CR1 deficiency leads to impairment of hippocampal cognitive function and synaptic plasticity. J Neurosci Official J Soc Neurosci 31:16241–16250. doi:10.1523/JNEUROSCI.3667-11.2011

Rozeske RR, Evans AK, Frank MG, Watkins LR, Lowry CA, Maier SF (2011) Uncontrollable, but not controllable, stress desensitizes 5-HT1A receptors in the dorsal raphe nucleus. J Neurosci Official J Soc Neurosci 31:14107–14115. doi:10.1523/JNEUROSCI.3095-11.2011

Schafer DP, Lehrman EK, Kautzman AG, Koyama R, Mardinly AR, Yamasaki R, Ransohoff RM, Greenberg ME, Barres BA, Stevens B (2012) Microglia sculpt postnatal neural circuits in an activity and complement-dependent manner. Neuron 74:691–705. doi:10.1016/j.neuron.2012. 03.026

Tremblay ME, Majewska AK (2011) A role for microglia in synaptic plasticity? Communicative Integr Biol 4:220–222. doi:10.4161/cib.4.2.14506

Tremblay ME, Stevens B, Sierra A, Wake H, Bessis A, Nimmerjahn A (2011) The role of microglia in the healthy brain. J Neurosci Official J Soc Neurosci 31:16064–16069. doi:10. 1523/JNEUROSCI.4158-11.2011

Tyler WJ, Alonso M, Bramham CR, Pozzo-Miller LD (2002) From acquisition to consolidation: on the role of brain-derived neurotrophic factor signaling in hippocampal-dependent learning. Learn Mem 9:224–237. doi:10.1101/lm.51202

Vaidya VA, Marek GJ, Aghajanian GK, Duman RS (1997) 5-HT2A receptor-mediated regulation of brain-derived neurotrophic factor mRNA in the hippocampus and the neocortex. J Neurosci Official J Soc Neurosci 17:2785–2795

Vaidya VA, Terwilliger RM, Duman RS (1999) Role of 5-HT2A receptors in the stress-induced down-regulation of brain-derived neurotrophic factor expression in rat hippocampus. Neurosci Lett 262:1–4

Vukovic J, Colditz MJ, Blackmore DG, Ruitenberg MJ, Bartlett PF (2012) Microglia modulate hippocampal neural precursor activity in response to exercise and aging. J Neurosci Official J Soc Neurosci 32:6435–6443. doi:10.1523/JNEUROSCI.5925-11.2012

Weiss JM, Stout JC, Aaron MF, Quan N, Owens MJ, Butler PD, Nemeroff CB (1994) Depression and anxiety: role of the locus coeruleus and corticotropin-releasing factor. Brain Res Bull 35:561–572

Williamson LL, Sholar PW, Mistry RS, Smith SH, Bilbo SD (2011) Microglia and memory: modulation by early-life infection. J Neurosci Official J Soc Neurosci 31:15511–15521. doi:10.1523/JNEUROSCI.3688-11.2011

Wynne AM, Henry CJ, Godbout JP (2009) Immune and behavioral consequences of microglial reactivity in the aged brain. Integr Comp Biol 49:254–266. doi:10.1093/icb/icp009

Wynne AM, Henry CJ, Huang Y, Cleland A, Godbout JP (2010) Protracted downregulation of CX3CR1 on microglia of aged mice after lipopolysaccharide challenge. Brain Behav Immun 24:1190–1201. doi:10.1016/j.bbi.2010.05.011

Yirmiya R, Goshen I (2011) Immune modulation of learning, memory, neural plasticity and neurogenesis. Brain Behav Immun 25:181–213. doi:10.1016/j.bbi.2010.10.015

Ziv Y, Ron N, Butovsky O, Landa G, Sudai E, Greenberg N, Cohen H, Kipnis J, Schwartz M (2006) Immune cells contribute to the maintenance of neurogenesis and spatial learning abilities in adulthood. Nat Neurosci 9:268–275. doi:10.1038/nn1629

Zoladz PR, Park CR, Halonen JD, Salim S, Alzoubi KH, Srivareerat M, Fleshner M, Alkadhi KA, Diamond DM (2011) Differential expression of molecular markers of synaptic plasticity in the hippocampus, prefrontal cortex, and amygdala in response to spatial learning, predator exposure, and stress-induced amnesia. Hippocampus. doi:10.1002/hipo.20922

The Interface of Stress and the HPA Axis in Behavioural Phenotypes of Mental Illness

David Baumeister, Stafford L. Lightman and Carmine M. Pariante

Abstract Abnormalities of hypothalamic-pituitary-adrenal (HPA) axis function are one of the most consistent biological findings across several mental disorders, but many of the mechanisms underlying this abnormality as well as the potential contribution to behavioural phenotypes remain only partially understood. Interestingly, evidence suggests a U-curve, with dysregulation of the HPA axis towards both hyper- or hypoactivity manifesting as a risk to mental wellbeing. This review will elaborate on both the clinical and molecular role of the neuroendocrine stress system in depressive, psychotic and post-traumatic stress disorders and present some of the most recent findings that have shed light on the complex interface between environmental stressors, molecular mechanisms and clinical presentation. Crucially, plasticity of the HPA axis confers both vulnerability to adverse events, particularly so in early developmental stages, as well as hope for the treatment of mental disorder, as evidenced by changes in HPA functioning associated with remission of symptoms.

Keywords Stress · Cortisol · HPA axis · Depression · Psychosis · PTSD

Contents

1 Overview of HPA Axis Functioning 14
2 HPA Function in Unipolar and Bipolar Depression 15
3 HPA Function in Psychosis 16

D. Baumeister · C. M. Pariante (✉)
Department of Psychological Medicine, Institute of Psychiatry, Kings College London, Room 2-055, The James Black Centre 125 Coldharbour Lane, London SE5 9NU, UK
e-mail: carmine.pariante@kcl.ac.uk

S. L. Lightman
Henry Wellcome Laboratories for Integrative Neuroscience and Endocrinology, University of Bristol, Dorothy Hodgkin Building, Whitson Street, Bristol BS1 3NY, UK

4	HPA Function in PTSD	17
5	The Impact of Stress in Early Life	18
6	The HPA-Stress Interface in Late Adolescence and Adulthood	19
7	Conclusion	20
References		21

1 Overview of HPA Axis Functioning

Activity of the HPA axis, the neuroendocrine stress system, is governed by the secretion of corticotrophin-releasing factor (CRF) and vasopressin (AVP) from the hypothalamus, which in turn activate the secretion of adrenocorticotrophic hormone (ACTH) from the pituitary. This finally stimulates the secretion of glucocorticoids (cortisol in humans and corticosterone in rodents) from the adrenal cortex, which then interact with their cognate receptors in multiple target tissues. Glucocorticoids have widespread regulatory roles as part of the stress response, both in peripheral functions such as immunity and metabolism as well as in the central nervous system (CNS). In the CNS, glucocorticoids moderate neuronal survival, neurogenesis, long-term potentiation and dendritic growth as well as atrophy in complex anatomical structures extensively implicated in psychopathology, particularly the hippocampus and amygdala [reviewed in Herbert et al. (2006)]. Notably, the HPA axis is embedded in bidirectional relationships to other allosteric systems that have been implicated in psychopathology, such as the inflammatory (Dantzer et al. 2008) and monoaminergic systems (Gotlib et al. 2008), and thus some of the behavioural effects of HPA functioning may be mediated by interaction with these systems.

In the HPA axis, glucocorticoids are responsible for feedback inhibition both on CRF and AVP from the hypothalamus and directly on secretion of ACTH from pituitary corticotropes. Endogenous glucocorticoids regulate release of CRF in the paraventricular nucleus and ACTH in the pituitary via activation of their cognate receptors—the glucocorticoid receptor (GR) and the mineralocorticoid receptor (MR). MR has a high affinity for endogenous glucocorticoids, whilst the GR has a lower affinity, suggesting the GR is more important in the regulation of the stress response, i.e. an acute elevation in glucocorticoids, whereas the high affinity of the MR tends to be tonically activated at most times of the day. The assertion that the GR modulates HPA function during stress is supported by research utilising the GR-selective synthetic glucocorticoid dexamethasone as a pharmacological challenge—which, in healthy individuals, is associated with a reduction in cortisol levels for up to 24 h, demonstrating GR mediated negative feedback within the HPA axis. Recent research also suggests MR can regulate fast feedback inhibition (Atkinson et al. 2008).

2 HPA Function in Unipolar and Bipolar Depression

Considering its role at the interface between stress and brain function, it is perhaps not surprising that the HPA axis has been found abnormal in many psychiatric disorders, albeit with idiosyncratic presentation. In depression, it appears clinical abnormalities of HPA function are, at least in part, related to reduced feedback inhibition by endogenous glucocorticoids, leading to hyperactivity of the axis [reviewed in Pariante (2006)]. Indicative of this hyperactivity, a significant percentage of depressed patients have increased levels of cortisol in the saliva, plasma and urine, and increased size (as well as activity) of the pituitary and adrenal glands [Reviewed in Nemeroff and Vale (2005)]. Recent research developments have utilised hair cortisol as a long-term measure of HPA functioning, and have confirmed elevated cortisol hair levels in depression, suggesting persistent HPA hyperactivity [reviewed in Staufenbiel et al. (2013)]. Depressed patients also show an increased HPA response to psychosocial stressors (Pariante and Lightman 2008) and are more likely to report daily events as stressful (Bylsma et al. 2011). Depression is also associated with an elevated cortisol response to awakening, a phenomenon that persists even after recovery (Bhagwagar et al. 2003, 2005; Vreeburg et al. 2009). Further, recent evidence suggests that unaffected individuals with a parental history of depression show a similarly augmented cortisol awakening response (Vreeburg et al. 2010). Interestingly, individuals at risk for depression show elevated waking cortisol levels similar to depressed patients, but their HPA axes recover more rapidly from psychosocial stress exposure (Dienes et al. 2012). This hyperactivity is likely to relate to impaired functioning of GR, reducing the ability of the HPA axis to feedback and inhibit its own activity. Studies have shown changes in both function and expression of GR in patients in major depression: non-suppression of cortisol secretion following administration of dexamethasone; impaired GR function in peripheral blood mononuclear cells isolated and cultivated in vitro, or in peripheral cells examined in vivo using metabolic or vascular indices; and reduced GR expression in neuropathological studies of post-mortem human brains [reviewed in (Pariante 2006; Pariante and Lightman 2008; Pariante and Miller 2001)].

Interestingly, evidence not only suggests idiosyncratic HPA activity depending on clinical status, but also variations depending on clinical subtypes. A recent review spanning four decades of HPA research found that whilst depression was generally associated with increased cortisol and ACTH but not CRH levels, individual subtypes differed: atypical depression was associated with a third of a standard deviation (SD) lower, melancholic depression a quarter of an SD higher, and psychotic depression nearly half an SD higher cortisol levels (Stetler and Miller 2011). Interestingly, it appears that melancholic depression is associated with greater cortisol awakening response and diurnal cortisol slope, whereas atypical depression appears to be more closely linked to elevation of inflammatory as well as metabolic markers (Lamers et al. 2012).

Similar to depression, bipolar disorder is associated with blunted response to dexamethasone challenge (Daban et al. 2005) and there is evidence of elevated baseline cortisol levels both during manic and depressive phases (Duffy et al. 2012). Bipolar patients also show an enhanced cortisol awakening response (Deshauer et al. 2003), and there is some evidence suggesting altered HPA function in the offspring of bipolar parents (Ellenbogen et al. 2010). Elevated hair cortisol has also been reported bipolar patients, but only when age of onset was older than 30 (Manenschijn et al. 2012). Moreover, manic episodes appear to be preceded by elevations of both cortisol and ACTH, suggesting relevance to the pathogenesis of bipolar disorder, rather than altered HPA functioning being a relict of depressive symptoms. In line with this, evidence obtained from post-mortem investigation of GR expression showed increased expression in both amygdalar neurons and astrocytes for unipolar, but not bipolar depressed patients or healthy controls (Wang et al. 2013).

3 HPA Function in Psychosis

The HPA axis has also been shown to be functionally altered in psychosis, with a high degree of similarities to depressive disorders: first episode psychosis patients show elevated baseline levels of HPA activity as well as blunted response to dexamethasone challenge in the context of elevated diurnal cortisol levels, the latter of which appears to be normalised by antipsychotic medications, as well as potentially enlarged size of the pituitary gland (Borges et al. 2013). A recent meta-analysis showed an increased pituitary volume of non-significant magnitude in first episode psychosis as well as a significant increase in individuals at ultra-high risk of psychosis who transitioned (Nordholm et al. 2013) and further evidence has shown similar pituitary volume elevations in non-affected relatives of patients with schizophrenia (Mondelli et al. 2008). Interestingly, patients who received medication had significantly larger pituitary glands compared to drug-naïve patients, possibly due to the effects of antipsychotics on prolactin production. Clinical high risk for psychosis in medication-free individuals is also associated with elevated basal salivary cortisol (Sugranyes et al. 2012).

Unlike depressed patients however, first episode patients show a significantly lower awakening cortisol response when compared to healthy controls (Mondelli et al. 2010). Interestingly, blunted cortisol response to awakening in schizophrenia patients predicts worse cognitive functioning (Aas et al. 2010), and is positively correlated with and predicted by the severity of positive symptoms in schizophrenia patients (Belvederi Murri et al. 2011). Schizophrenia patients further show a tendency towards attenuated cortisol response to psychosocial stress, however in the context of increased activity of the sympathetic nervous system as indicated by elevated heart rate and blood pressure (Brenner et al. 2009) and individuals at ultra-high risk for psychosis exhibit a significantly attenuated cortisol response to psychosocial stress compared to healthy controls (Brenner et al. 2009; Pruessner et al. 2013). Patients with psychosis also show a greater emotional reactivity to daily life stress (Myin-Germeys et al. 2005). Interestingly however, Pruessner et al. (2013) found that lower cortisol output in response to psychosocial stress in patients with psychosis is correlated with higher levels of self-reported stress during the preceding year.

4 HPA Function in PTSD

Perhaps the most mixed findings on HPA function have been obtained in individuals with post-traumatic stress disorder (PTSD). A meta-analysis on both basal as well as dynamic HPA functioning found no differences in PTSD patients, trauma-exposed (TE) and non-exposed (NE) individuals in terms of basal cortisol levels, consistent across saliva, urine and plasma sampling, although reductions in baseline cortisol have been reported in individual studies (Klaassens et al. 2011). Whilst exposure to trauma in adulthood had no significant overall impact on basal cortisol levels it was associated with enhanced cortisol suppression in response to dexamethasone. Conversely, some studies on hair cortisol in PTSD have reported elevated cortisol levels (Steudte et al. 2011; Luo et al. 2012), whilst others have not (Steudte et al. 2013), potentially due to the different kinds of trauma exposure in the respective samples.

Interestingly, evidence reviewed by de Kloet et al. (2006) showed that whilst PTSD is associated with enhanced inhibitory feedback in response to dexamethasone challenge, indicative of increased functioning of the GR, individuals with PTSD show augmented cortisol responses to psychosocial stress tests. In line with these findings, de Kloet et al. (2012) recently reported that cognitive challenge was rated as more stressful by and led to elevated ACTH but not noradrenaline responses in PTSD patients, but research utilising dexamethasone challenge found opposing effects, i.e., enhanced suppression of ACTH (Yehuda et al. 2004; Golier et al. 2006).

A recent meta-analysis comparing PTSD to PTSD comorbid with depression (PTSD + MDD) showed further interesting subtleties in differential HPA function: PTSD, PTSD + MDD and TE groups exhibited attenuated morning cortisol compared to NE groups, but whilst PTSD and TE groups showed similar patterns in afternoon cortisol levels, comorbid depression was associated with significant elevations compared to NE controls (Morris et al. 2012). Furthermore, PTSD, PTSD + MDD and TE groups all showed augmented cortisol suppression in response to dexamethasone, with no significant effect size differences between the groups. Interestingly, these findings were observed in the context of overall diminished daily output of cortisol in PTSD and PTSD + MDD patients but not TE individuals. However, some of this evidence remains mixed, as there have also been reports of elevated afternoon cortisol levels in patients with PTSD in another meta-analysis (Miller et al. 2007) (Table 1).

Table 1 Cortisol characteristics associated with disorders and adversity-exposure

Cortisol measure	Depression	Bipolar Disorder	Psychosis	PTSD	Childhood adversity	Adulthood adversity
Awakening response	↑	↑	↓	↓	Mixed evidence	↑[a]
Afternoon	↑	↑	↑	=	–	↑
Daily output	↑	↑	↑	↓[b]	↑	↑
Hair	↑	↑[c]	–	Mixed evidence	↓	–
Post-DST	↑	↑	↑	↓	↑	↓
Post-psychosocial stress test	↑	?	↓[b]	↑	↓	–

[a] Context-dependent increases or decreases
[b] Tendency
[c] Only when onset-age <30

5 The Impact of Stress in Early Life

Evidence over the last decades has provided evidence suggesting that HPA axis dysfunction is not a simple consequence or an epiphenomenon of mental disorder, but on the contrary it is a risk factor predisposing to the development of psychopathological behaviour, brought about by early life experiences programming molecular changes as well as by biological vulnerability to stress. Perhaps the most striking development in this field has been the realisation that abnormal functioning of the HPA axis may reflect a susceptibility that can be programmed through early life events—starting even as early as in prenatal development [reviewed in (Cottrell and Seckl 2009)]. Clinical studies have shown that women who are sexually or physically abused in childhood exhibit a markedly enhanced activation of the HPA axis as adults. Even if not currently depressed they exhibit enhanced ACTH and heart rate responses when exposed to psychosocial stress; and if they are currently depressed they exhibit the largest increase in ACTH secretion and heart rate, as well as a very large increases in cortisol secretion (Heim and Nemeroff 2002). Moreover, research using dexamethasone has also found persistent HPA axis hyperactivity in men with early life trauma (Heim et al. 2008). Notably however, evidence on associations of childhood trauma with awakening cortisol response has been inconsistent, with reports of both augmentation and attenuation of the awakening cortisol response (Lu et al. 2013; Mangold et al. 2010).

Research attempting to establish associations of HPA functioning profiles with psychopathological behaviours needs to control for the mediating effects of childhood trauma, as childhood trauma itself has been shown to be associated with a variety of adult mental disorder, including depression, bipolar disorder, psychosis and PTSD (Putnam et al. 2013; Varese et al. 2012; Subica 2013; Edwards et al. 2003). Interestingly, a recent study demonstrated that when participants meeting criteria for MDD were matched to healthy controls with no lifetime history of depression based on age, sex and experience of childhood adversity, using the dexamethasone/CRH test failed to distinguish depressed from non-depressed participants (Carpenter et al. 2009). Conversely, depressed individuals with a history of childhood trauma, as opposed to depressed patients without this type of early experience, show decreased cortisol hair levels (Hinkelmann et al. 2013).

One of the most frequently proposed mechanisms through which early life experiences may impact on the HPA axis is epigenetic programming. Indeed, there is evidence for greater methylation of the GR in hippocampal regions of suicide completers who had been subjected to childhood abuse compared to suicide completers without a history of childhood trauma (McGowan et al. 2009). Similarly, Tyrka et al. (2012) recently reported that a history of childhood adversity in healthy adults was associated with increased methylation of a promoter region of the GR gene in leukocyte DNA. Moreover, this methylation was associated with an attenuated response to the dexamethasone/CRH test. In line with this evidence, childhood trauma induces demethylation of glucocorticoid response elements of the gene coding for the GR-associated heat shock protein FKBP5, which normally inhibits the ability of the ligand to bind cytosolic GR and subsequently translocate to the cell nucleus, where it can increase FKBP5 transcription which in turn reduces GR activity. Individuals with a functional polymorphism of this gene are at greater risk of PTSD, depression and suicide (Klengel et al. 2012). Interestingly, polymorphisms of the FKBP5 gene associated with greater expression of the chaperone protein are also associated with prolonged elevation of cortisol levels following psychosocial stress exposure (Ising et al. 2008). Although most research has focused on the effects of early life events on programming changes in the HPA axis itself concentrating on epigenetic modifications of glucocorticoid receptor genes, it is important to emphasise that there are many other closely related systems that may be susceptible to programming. For example, a recent study showed increased methylation of the serotonin transporter gene (SERT) in bullied children when compared to their discordant mono-zygotic co-twins, which was associated with blunted cortisol response to the TSST (Ouellet-Morin et al. 2012).

6 The HPA-Stress Interface in Late Adolescence and Adulthood

The 3-hit model of vulnerability and resilience recently proposed by Daskalakis et al. (2013) suggests that the interaction of genetic predisposition with early life experience sets the course of neuroendocrine alterations in neural development via epigenetic programming towards an adult phenotype vulnerable to environmental stressors. Indeed, stressful life events in adulthood such as trauma or exposure to chronic stress may precipitate the onset of a range of disorders and can facilitate relapse in existing disorders (Melchior et al. 2007; Stilo et al. 2012; Bebbington et al. 1993; Francis et al. 2012; Lethbridge and Allen 2008). Interestingly, the type of stress one is exposed to appears to be associated with differential patterns of HPA response and subsequent vulnerability to specific psychopathological syndromes. There is evidence that the immediate response of the HPA axis to trauma (i.e. within 24 h of exposure) can predict the development of PTSD: several studies suggest that lower cortisol levels in the peritraumatic period are associated with a higher risk of subsequent PTSD, and there is evidence of enhanced cortisol suppression in response to dexamethasone in trauma-exposed individuals who go on to develop PTSD (Morris and Rao 2013). Furthermore, PTSD-specific HPA functioning in the form of diminished morning but elevated afternoon cortisol levels within a week of exposure are also linked with the subsequent development of the disorder (Aardal-Eriksson et al. 2001).

Marin et al. (2007) assessed life stress and HPA functioning in healthy women between the ages of 15 and 19, and found that exposure to episodic stressors in the context of high chronic stress led to increased cortisol release, both upon awakening and overall daily output, as well as reduced GR mRNA. However, exposure to episodic stressors in the context of low chronic stress led to decreased cortisol release and enhanced GR mRNA. In the context of medium chronic stress, the level of exposure to episodic events had no impact on either cortisol or GR levels. A meta-analysis by Miller et al. (2007) further showed that idiosyncratic stress signatures can differentially impact on HPA function: for example, the awakening cortisol response increases in response to significant stressors that pose a threat to the social self, but decreases when the stressor poses threat to physical integrity, involves a loss and/or is perceived as uncontrollable. Similarly, some stressor types only impact on certain HPA measures but not others, e.g. whilst stressors that pose a threat to the social self appear to increase afternoon cortisol levels, they do not impact on overall daily cortisol output or response to dexamethasone challenge.

7 Conclusion

The findings discussed in the present review show that specific HPA axis profiles appear to be characteristic of different disorders and syndromes. The high degree of neuroplasticity during early developmental stages acts as a window of sensitivity, allowing childhood adversity to convey vulnerability to mental illness in later life, which, even in the absence of the development of psychopathological behaviours, is associated with highly complex effects on measures of HPA function. Taken together, HPA axis dysfunction in mental disorders as described above may not be the consequence of these ailments per se, but rather the manifestation of persistent neurobiological abnormalities that predispose to their development dependent on specific combinations and characteristics of idiosyncratic stress exposure. As such, on-going disruption of HPA homeostasis, be it towards hyper- or hypoactivity, can have adverse impacts on mental and physical wellbeing. Due to its unique position at the interface between biological systems and adversity, the HPA axis not only presents as one of the most interesting examples of molecular interplay of the individual with their environment over the course of a lifetime, but also as one of the most challenging areas of mental health research.

Acknowledgments This work was supported by the grant "Persistent Fatigue Induced by Interferon-alpha: A New Immunological Model for Chronic Fatigue Syndrome" from the Medical Research Council (UK) MR/J002739/1. Additional support has been offered by the Commission of European Communities Seventh Framework Programme (Collaborative Project Grant Agreement no. 22963, Mood Inflame); by the National Institute for Health Research Mental Health Biomedical Research Centre in Mental Health at South London and Maudsley NHS Foundation Trust and King's College London; by a grant from the Psychiatry Research Trust, UK (McGregor 97); by Janssen Parmaceutica NV/Janssen Pharmaceutical Companies of Johnson & Johnson; and by the Institute of Psychiatry at Kings College London.

References

Aardal-Eriksson E, Eriksson TE, Thorell LH (2001) Salivary cortisol, posttraumatic stress symptoms, and general health in the acute phase and during 9-month follow-up. Biol Psychiatry 50(12):986–993

Aas M, Dazzan P, Mondelli V, Toulopoulou T, Reichenberg A, Di Forti M et al (2010) Abnormal cortisol awakening response predicts worse cognitive function in patients with first-episode psychosis. Psychol Med 41(3):463–476

Atkinson HC, Wood SA, Castrique ES, Kershaw YM, Wiles CC, Lightman SL (2008) Corticosteroids mediate fast feedback of the rat hypothalamic-pituitary-adrenal axis via the mineralocorticoid receptor. Am J Physiol Endocrinol Metab 294(6):E1011–E1022

Bebbington P, Wilkins S, Jones P, Foerster A, Murray R, Toone B et al (1993) Life events and psychosis. Initial results from the Camberwell collaborative psychosis study. Br J Psychiatry 162:72–79

Belvederi Murri M, Pariante CM, Dazzan P, Hepgul N, Papadopoulos AS, Zunszain P et al (2011) Hypothalamic-pituitary-adrenal axis and clinical symptoms in first-episode psychosis. Psychoneuroendocrinology 37(5):629–644

Bhagwagar Z, Hafizi S, Cowen PJ (2003) Increase in concentration of waking salivary cortisol in recovered patients with depression. Am J Psychiatry 160(10):1890–1891

Bhagwagar Z, Hafizi S, Cowen PJ (2005) Increased salivary cortisol after waking in depression. Psychopharmacology 182(1):54–57

Borges S, Gayer-Anderson C, Mondelli V (2013) A systematic review of the activity of the hypothalamic-pituitary-adrenal axis in first episode psychosis. Psychoneuroendocrinology 38(5):603–611

Brenner K, Liu A, Laplante DP, Lupien S, Pruessner JC, Ciampi A et al (2009) Cortisol response to a psychosocial stressor in schizophrenia: blunted, delayed, or normal? Psychoneuroendocrinology 34(6):859–868

Bylsma LM, Taylor-Clift A, Rottenberg J (2011) Emotional reactivity to daily events in major and minor depression. J Abnorm Psychol 120(1):155–167

Carpenter LL, Ross NS, Tyrka AR, Anderson GM, Kelly M, Price LH (2009) Dex/CRH test cortisol response in outpatients with major depression and matched healthy controls. Psychoneuroendocrinology 34(8):1208–1213

Cottrell EC, Seckl JR (2009) Prenatal stress, glucocorticoids and the programming of adult disease. Front Behav Neurosci 3:19

Daban C, Vieta E, Mackin P, Young AH (2005) Hypothalamic-pituitary-adrenal axis and bipolar disorder. Psychiatr Clin North Am 28(2):469–480

Dantzer R, O'Connor JC, Freund GG, Johnson RW, Kelley KW (2008) From inflammation to sickness and depression: when the immune system subjugates the brain. Nat Rev Neurosci 9(1):46–56

Daskalakis NP, Bagot RC, Parker KJ, Vinkers CH, de Kloet ER (2013) The three-hit concept of vulnerability and resilience: toward understanding adaptation to early-life adversity outcome. Psychoneuroendocrinology 38(9):1858–1873

de Kloet CS, Vermetten E, Geuze E, Kavelaars A, Heijnen CJ, Westenberg HG (2006) Assessment of HPA axis function in posttraumatic stress disorder: pharmacological and non-pharmacological challenge tests, a review. J Psychiatr Res 40(6):550–567

de Kloet CS, Vermetten E, Rademaker AR, Geuze E, Westenberg HG (2012) Neuroendocrine and immune responses to a cognitive stress challenge in veterans with and without PTSD. Eur J Psychotraumatol 3

Deshauer D, Duffy A, Alda M, Grof E, Albuquerque J, Grof P (2003) The cortisol awakening response in bipolar illness: a pilot study. Can J Psychiatry 48(7):462–466

Dienes KA, Hazel NA, Hammen CL (2012) Cortisol secretion in depressed, and at-risk adults. Psychoneuroendocrinology 38(6):927–940

Duffy A, Lewitzka U, Doucette S, Andreazza A, Grof P (2012) Biological indicators of illness risk in offspring of bipolar parents: targeting the hypothalamic-pituitary-adrenal axis and immune system. Early Interv Psychiatry 6(2):128–137

Edwards VJ, Holden GW, Felitti VJ et al (2003) Relationship between multiple forms of childhood maltreatment and adult mental health in community respondents: results from the adverse childhood experiences study. Am J Psychiatry 160:1453–1460

Ellenbogen MA, Santo JB, Linnen AM, Walker CD, Hodgins S (2010) High cortisol levels in the offspring of parents with bipolar disorder during two weeks of daily sampling. Bipolar Disord 12(1):77–86

Francis JL, Moitra E, Dyck I, Keller MB (2012) The impact of stressful life events on relapse of generalized anxiety disorder. Depress Anxiety 29(5):386–391

Golier JA, Legge J, Yehuda R (2006) The ACTH response to dexamethasone in Persian Gulf War veterans. Ann NY Acad Sci 1071:448–453

Gotlib IH, Joorman J, Minor KL, Hallmayer J (2008) HPA axis reactivity: a mechanism underlying the associations among. 5-HTTLPR, stress, and depression. Biol Psychiatry 63(9):847–851

Heim C, Nemeroff CB (2002) Neurobiology of early life stress: clinical studies. Semin Clin Neuropsychiatry 7(2):147–159

Heim C, Mletzko T, Purselle D, Musselman DL, Nemeroff CB (2008) The dexamethasone/corticotropin-releasing factor test in men with major depression: role of childhood trauma. Biol Psychiatry 63(4):398–405

Herbert J, Goodyer IM, Grossman AB, Hastings MH, de Kloet ER, Lightman SL et al (2006) Do corticosteroids damage the brain? J Neuroendocrinol 18(6):393–411

Hinkelmann K, Muhtz C, Dettenborn L, Agorastos A, Wingenfeld K, Spitzer C et al (2013) Association between childhood trauma and low hair cortisol in depressed patients and healthy control subjects. Biol Psychiatry 74(9):e15–e17

Ising M, Depping AM, Siebertz A, Lucae S, Unschuld PG, Kloiber S et al (2008) Polymorphisms in the FKBP5 gene region modulate recovery from psychosocial stress in healthy controls. Eur J Neurosci 28(2):389–398

Klaassens ER, Giltay EJ, Cuijpers P, van Veen T, Zitman FG (2011) Adulthood trauma and HPA-axis functioning in healthy subjects and PTSD patients: a meta-analysis. Psychoneuroendocrinology 37(3):317–331

Klengel T, Mehta D, Anacker C, Rex-Haffner M, Pruessner JC, Pariante CM et al (2012) Allele-specific FKBP5 DNA demethylation mediates gene-childhood trauma interactions. Nat Neurosci 16(1):33–41

Lamers F, Vogelzangs N, Merikangas KR, de Jonge P, Beekman AT, Penninx BW (2012) Evidence for a differential role of HPA-axis function, inflammation and metabolic syndrome in melancholic versus atypical depression. Mol Psychiatry 18(6):692–699

Lethbridge R, Allen NB (2008) Mood induced cognitive and emotional reactivity, life stress, and the prediction of depressive relapse. Behav Res Ther 46(10):1142–1150

Lu S, Gao W, Wei Z, Wu W, Liao M, Ding Y et al (2013) Reduced cingulate gyrus volume associated with enhanced cortisol awakening response in young healthy adults reporting childhood trauma. PLoS ONE 8(7):e69350

Luo H, Hu X, Liu X, Ma X, Guo W, Qiu C et al (2012) Hair cortisol level as a biomarker for altered hypothalamic-pituitary-adrenal activity in female adolescents with posttraumatic stress disorder after the 2008 Wenchuan earthquake. Biol Psychiatry 72(1):65–69

Manenschijn L, Spijker AT, Koper JW, Jetten AM, Giltay EJ, Haffmans J et al (2012) Long-term cortisol in bipolar disorder: associations with age of onset and psychiatric co-morbidity. Psychoneuroendocrinology 37(12):1960–1968

Mangold D, Wand G, Javors M, Mintz J (2010) Acculturation, childhood trauma and the cortisol awakening response in Mexican-American adults. Horm Behav 58(4):637–646

Marin TJ, Martin TM, Blackwell E, Stetler C, Miller GE (2007) Differentiating the impact of episodic and chronic stressors on hypothalamic-pituitary-adrenocortical axis regulation in young women. Health Psychol 26(4):447–455

McGowan PO, Sasaki A, D'Alessio AC, Dymov S, Labonte B, Szyf M et al (2009) Epigenetic regulation of the glucocorticoid receptor in human brain associates with childhood abuse. Nat Neurosci 12(3):342–348

Melchior M, Caspi A, Milne BJ, Danese A, Poulton R, Moffitt TE (2007) Work stress precipitates depression and anxiety in young, working women and men. Psychol Med 37(8):1119–1129

Miller GE, Chen E, Zhou ES (2007) If it goes up, must it come down? Chronic stress and the hypothalamic-pituitary-adrenocortical axis in humans. Psychol Bull 133(1):25–45

Mondelli V, Dazzan P, Gabilondo A, Tournikioti K, Walshe M, Marshall N et al (2008) Pituitary volume in unaffected relatives of patients with schizophrenia and bipolar disorder. Psychoneuroendocrinology 33(7):1004–1012

Mondelli V, Dazzan P, Hepgul N, Di Forti M, Aas M, D'Albenzio A et al (2010) Abnormal cortisol levels during the day and cortisol awakening response in first-episode psychosis: the role of stress and of antipsychotic treatment. Schizophr Res 116(2–3):234–242

Morris MC, Rao U (2013) Psychobiology of PTSD in the acute aftermath of trauma: integrating research on coping, HPA function and sympathetic nervous system activity. Asian J Psychiatr 6(1):3–21

Morris MC, Compas BE, Garber J (2012) Relations among posttraumatic stress disorder, comorbid major depression, and HPA function: a systematic review and meta-analysis. Clin Psychol Rev 32(4):301–315

Myin-Germeys I, Delespaul P, van Os J (2005) Behavioural sensitization to daily life stress in psychosis. Psychol Med 35(5):733–741

Nemeroff CB, Vale WW (2005) The neurobiology of depression: inroads to treatment and new drug discovery. J Clin Psychiatry 66(Suppl 7):5–13

Nordholm D, Krogh J, Mondelli V, Dazzan P, Pariante C, Nordentoft M (2013) Pituitary gland volume in patients with schizophrenia, subjects at ultra high-risk of developing psychosis and healthy controls: a systematic review and meta-analysis. Psychoneuroendocrinology 38(11):2394–2404

Ouellet-Morin I, Wong CC, Danese A, Pariante CM, Papadopoulos AS, Mill J et al (2012) Increased serotonin transporter gene (SERT) DNA methylation is associated with bullying victimization and blunted cortisol response to stress in childhood: a longitudinal study of discordant monozygotic twins. Psychol Med 43(9):1813–1823

Pariante CM (2006) The glucocorticoid receptor: part of the solution or part of the problem? J Psychopharmacol 20(4 Suppl):79–84

Pariante CM, Lightman SL (2008) The HPA axis in major depression: classical theories and new developments. Trends Neurosci 31(9):464–468

Pariante CM, Miller AH (2001) Glucocorticoid receptors in major depression: relevance to pathophysiology and treatment. Biol Psychiatry 49(5):391–404

Pruessner M, Bechard-Evans L, Boekestyn L, Iyer SN, Pruessner JC, Malla AK (2013) Attenuated cortisol response to acute psychosocial stress in individuals at ultra-high risk for psychosis. Schizophr Res 146(1–3):79–86

Putnam KT, Harris WW, Putnam FW (2013) Synergistic childhood adversities and complex adult psychopathology. J Trauma Stress 26(4):435–442

Staufenbiel SM, Penninx BW, Spijker AT, Elzinga BM, van Rossum EF (2013) Hair cortisol, stress exposure, and mental health in humans: a systematic review. Psychoneuroendocrinology 38(8):1220–1235

Stetler C, Miller GE (2011) Depression and hypothalamic-pituitary-adrenal activation: a quantitative summary of four decades of research. Psychosom Med 73(2):114–126

Steudte S, Kolassa IT, Stalder T, Pfeiffer A, Kirschbaum C, Elbert T (2011) Increased cortisol concentrations in hair of severely traumatized Ugandan individuals with PTSD. Psychoneuroendocrinology 36(8):1193–1200

Steudte S, Kirschbaum C, Gao W, Alexander N, Schonfeld S, Hoyer J et al (2013) Hair cortisol as a biomarker of traumatization in healthy individuals and posttraumatic stress disorder patients. Biol Psychiatry 74(9):639–646

Stilo SA, Di Forti M, Mondelli V, Falcone AM, Russo M, O'Connor J et al (2012) Social disadvantage: cause or consequence of impending psychosis? Schizophr Bull 39(6):1288–1295

Subica AM (2013) Psychiatric and physical sequelae of childhood physical and sexual abuse and forced sexual trauma among individuals with serious mental illness. J Trauma Stress 26(5):588–596

Sugranyes G, Thompson JL, Corcoran CM (2012) HPA-axis function, symptoms, and medication exposure in youths at clinical high risk for psychosis. J Psychiatr Res 46(11):1389–1393

Tyrka AR, Price LH, Marsit C, Walters OC, Carpenter LL (2012) Childhood adversity and epigenetic modulation of the leukocyte glucocorticoid receptor: preliminary findings in healthy adults. PLoS ONE 7(1):e30148

Varese F, Smeets F, Drukker M, Lieverse R, Lataster T, Viechtbauer W et al (2012) Childhood adversities increase the risk of psychosis: a meta-analysis of patient-control, prospective- and cross-sectional cohort studies. Schizophr Bull 38(4):661–671

Vreeburg SA, Hoogendijk WJ, van Pelt J, Derijk RH, Verhagen JC, van Dyck R et al (2009) Major depressive disorder and hypothalamic-pituitary-adrenal axis activity: results from a large cohort study. Arch Gen Psychiatry 66(6):617–626

Vreeburg SA, Hartman CA, Hoogendijk WJ, van Dyck R, Zitman FG, Ormel J et al (2010) Parental history of depression or anxiety and the cortisol awakening response. Br J Psychiatry 197(3):180–185

Wang Q, Verweij EW, Krugers HJ, Joels M, Swaab DF, Lucassen PJ (2013) Distribution of the glucocorticoid receptor in the human amygdala; changes in mood disorder patients. Brain Struct Funct. doi:10.1016/j.neuroscience.2013.05.043

Yehuda R, Golier JA, Halligan SL, Meaney M, Bierer LM (2004) The ACTH response to dexamethasone in PTSD. Am J Psychiatry 161(8):1397–1403

Adult Hippocampal Neurogenesis in Depression: Behavioral Implications and Regulation by the Stress System

Christoph Anacker

Abstract Adult hippocampal neurogenesis, the birth of new neurons in the dentate gyrus of the adult brain, can be regulated by stress and antidepressant treatment, and has consistently been implicated in the behavioral neurobiology of stress-related disorders, especially depression and anxiety. A reciprocal relationship between hippocampal neurogenesis and the hypothalamus–pituitary–adrenal (HPA) axis has recently been suggested, which may play a crucial role in the development and in the resolution of depressive symptoms. This chapter will review some of the existing evidence for stress- and antidepressant-induced changes in adult hippocampal neurogenesis, and critically evaluate the behavioral effects of these changes for depression and anxiety. The potential role of neurogenesis as a neurobiological mechanism for sustained remission from depressive symptoms will be discussed, integrating existing data from clinical studies, animal work, and cellular models. The effect of glucocorticoid hormones and the glucocorticoid receptor (GR) will thereby be evaluated as a central mechanism by which stress and antidepressant may exert their opposing effects on neurogenesis, and ultimately, on mood and behavior.

Keywords Neuroplasticity · Stem cells · Neuropsychiatric disorders · Proliferation · Fluoxetine

Contents

1 Introduction 26
2 Plasticity Through New Brain Cells: Neurogenesis in the Adult Hippocampus 26

C. Anacker (✉)
Departments of Psychiatry, Neurology and Neurosurgery,
Douglas Mental Health University Institute, McGill University,
6875 Boulevard La Salle, Montreal, QC H4H 1R3, Canada
e-mail: Christoph.Anacker@mail.mcgill.ca

3	Regulation of Adult Hippocampal Neurogenesis by Stress	28
4	The HPA Axis and Neurogenesis Regulation	29
5	Neurogenesis-Dependent Effects of Stress	32
6	Regulation of Adult Hippocampal Neurogenesis by Antidepressants	33
7	Neurogenesis-Dependent Effects of Antidepressants	34
8	Adult Hippocampal Neurogenesis in Depression: Evidence from Human Studies	36
9	Neurogenesis and Mood Regulation: Evolutionary Adaptation Gone Awry?	37
10	Conclusion and Outlook	38
References		39

1 Introduction

Already the psychologist William James (1842–1910) hypothesized that the ability of human behavior to adapt to the environment may require plastic psychological mechanisms that depend on "[...] the possession of a structure weak enough to yield to an influence, but strong enough not to yield all at once [...]" (reviewed in Pascual-Leone et al. 2005). Today we have a much clearer understanding of what comprises this "structure," and we have access to a multitude of neurobiological data, demonstrating the enormous capacity of the brain to constantly adapt, rewire, and develop in a tightly controlled manner from early life into adulthood. One of the perhaps most fascinating forms of neuroplasticity is the ability of the adult brain to not only modify connections of existing cells, but to also constantly generate completely new neurons in defined regions of the brain. Over recent years, our appreciation of this phenomenon of "neurogenesis" in the adult brain, and for its role in many important brain functions, has steadily grown. Central to this review, neurogenesis in the adult hippocampus has been extensively studied for its implications in the pathophysiology and treatment of neuropsychiatric disorders. This chapter will therefore discuss the interactions of stress and neurogenesis in the adult brain, and review existing evidence for the behavioral effects of adult hippocampal neurogenesis in stress-related psychopathologies, particularly depression and anxiety.

2 Plasticity Through New Brain Cells: Neurogenesis in the Adult Hippocampus

The long-held dogma that the formation of new neurons is confined to prepubertal development and absent in the adult brain was based on views of many distinguished scientists at the time, including Ramon y Cajal, who had observed that the anatomical structure of the mammalian brain was composed of individual neurons that remained stable in their microscopic appearance (Ramon y Cajal 1928). Technical advances, including the development of tritiated thymidine [H^3]-autoradiography and electron microscopy, later on enabled visualization and characterization of dividing cells and their progeny, which led to the discovery of newborn neurons in the subventricular zone (SVZ) of the lateral ventricles and in the subgranular zone (SGZ) of the hippocampal dentate gyrus (Altman and Das 1965; Kaplan and Hinds 1977).

In a pivotal study in 1998, Eriksson and colleagues discovered that adult neurogenesis is not only confined to the brain of lower species, but indeed also occurs in the adult brain of humans. Using postmortem brain tissue of cancer patients, who had received injections of the synthetic nucleotide 5-bromo-deoxyuridine (BrdU) in order to stage their tumor growth, the study revealed BrdU incorporation not only into the growing tumor tissue, but also into proliferating neural progenitor cells in the SVZ of the lateral ventricles and in the SGZ of the hippocampal dentate gyrus. Moreover, some of these BrdU labeled cells were shown to develop into neurons, demonstrating for the first time that neurogenesis indeed also occurs in the adult human brain (Eriksson et al. 1998).

Although several studies have now confirmed these findings and demonstrated continuous cell birth in the adult hippocampus, the functional relevance of neurogenesis in humans has been the subject of intense debate. Points of criticism have thereby mainly been the slow rate of neurogenesis and the small number of newborn neurons, as many of these cells die shortly after birth (Goritz and Frisen 2012). These findings had raised the question whether such few new neurons may at all be relevant to human brain function, or rather represent an evolutionary remnant that may only have functional implications in rodents (Rakic 1985). Spalding and colleagues recently addressed this issue, by making use of the radioactive carbon isotope, ^{14}C, to date the birth of new neurons in the human brain. They have taken advantage of the 1955–1963 aboveground nuclear bomb tests during the cold war, which had caused an elevation of atmospheric ^{14}C that got incorporated into plants and subsequently made its way through the food chain into the human body. The defined period of elevated atmospheric ^{14}C before the Partial Nuclear Test Ban Treaty in 1963 was therefore used to retrospectively birth-date proliferating cells that had incorporated ^{14}C into the DNA during mitosis. Using this method, the study demonstrated that in contrast to earlier estimations, the number of adult-born neurons in the human hippocampus is in fact much higher, and that indeed as many as 700 new neurons are being added to the human hippocampus every day (Spalding et al. 2013).

But if these newborn neurons are indeed so abundant in the adult hippocampus, what is their purpose for brain function and behavior? Although direct evidence for a behavioral role of neurogenesis in humans is still missing, the study by Spalding and colleagues suggests that the extent of neurogenesis in the human brain is indeed comparable to that in mice and rats, species in which essential effects of neurogenesis for the regulation of mood and behavior have consistently been demonstrated. Some of this evidence for the role of hippocampal neurogenesis on mood and behavior in rodents will therefore be discussed in the following section. I will first review the effects of stress on adult hippocampal neurogenesis and then delineate specifically which behaviors have been shown to be dependent on such changes in neurogenesis.

3 Regulation of Adult Hippocampal Neurogenesis by Stress

The capacity of the adult hippocampus to generate new neurons has spurred interest in how this process may be regulated by environmental influences. Considering evidence for hippocampal volume changes in depressed patients (Sheline et al. 1996) and atrophy of hippocampal neurons during stress (Watanabe et al. 1992; Woolley et al. 1990), neurogenesis gained significant attention as a potential mediator for stress effects on the hippocampus. One of the first studies reporting an effect of psychosocial stress on neurogenesis had demonstrated that tree shrews that are socially defeated once by a dominant male, develop a reduction in newborn hippocampal neurons as a result of this acute stress exposure (Gould et al. 1997). Moreover, a one-time exposure to the stressful stimulus of predator odors reduces cell proliferation in the developing dentate gyrus of newborn rat pups (Tanapat et al. 1998). Such *acute* stress-induced reductions in neurogenesis also extend to various paradigms of *chronic* stress. For example, unpredictable chronic mild stress (UCMS), an experimental paradigm that uses a sequence of alternating environmental stressors over a period of 4–8 weeks, causes a reduction in the number of proliferating hippocampal progenitor cells and newborn neurons in rodents (Santarelli et al. 2003; Surget et al. 2008, 2011; Tanti et al. 2012). At the same time, UCMS causes anhedonia, as well as despair- and anxiety-like behavior, which are commonly interpreted as a depression-like phenotype in rodents (Willner 2005).

In addition to UCMS, prenatal stress also causes lifelong reductions in hippocampal cell proliferation, in the total number of hippocampal granule neurons, and in the overall volume of the hippocampus (Coe et al. 2003; Lemaire et al. 2000). In line with these detrimental effects of prenatal stress on neurogenesis, early life adversity, modeled by daily maternal separation during the first 2 weeks of life, causes a reduction in hippocampal progenitor cell proliferation and neuronal development in the adult offspring (Mirescu et al. 2004). These findings are particularly relevant for the role of early mother-infant interactions in regulating neurogenesis and behavior later in adulthood. Using cross fostering of mice pups to dams of two different mouse strains with inherent differences in maternal care, Koehl and colleagues demonstrated that mice raised by dams with higher levels of maternal care show increased levels of newborn neurons as adults (Koehl et al. 2012). This finding is striking, considering that maternal care crucially determines neuroendocrine function and stress reactivity in the offspring, and poses the question whether changes in adult hippocampal neurogenesis may indeed mediate these effects of maternal care on stress susceptibility later in life (Caldji et al. 2000; Liu et al. 2000; Weaver et al. 2006).

Consistent with the effects of environmental and early-life stress models, chronic psychosocial stress also inhibits hippocampal neurogenesis and thereby induces depression-like behavior (Lehmann et al. 2013). These effects of social stress have also been demonstrated in nonhuman primates, in which social isolation causes a complex behavioral phenotype of anhedonia and subordination, which is accompanied by a reduction in neurogenesis and reversible by antidepressant treatment (Perera et al. 2011).

Interestingly, on an anatomical level, chronic stress-induced reductions in hippocampal neurogenesis are most prominent in the ventral pole of the hippocampus (Tanti et al. 2012; Lehmann et al. 2013). This finding is indeed striking, considering that the ventral hippocampus, as opposed to the dorsal hippocampus, forms neuronal connections to the hypothalamus and the amygdala, two important structures for neuroendocrine function and mood regulation. Accordingly, lesions of the ventral, but not the dorsal hippocampus, alter emotional behavior and stress responses (Henke 1990), and optogenetic activation of newborn neurons in the ventral hippocampus reduces anxiety, while their activation in the dorsal hippocampus regulates learning and memory (Kheirbek et al. 2013). Regulation of neurogenesis by stress specifically in the ventral hippocampus therefore further supports a role for neurogenesis in the control of neural networks that are crucial for mood regulation.

But what are the mechanisms that drive these above-described stress-induced changes in adult neurogenesis and behavior? Several molecular signaling systems have been proposed, including (but not limited to) stress-induced elevations in glutamate and subsequent NMDA receptor activation (Cameron et al. 1995; Gould et al. 1997), increased levels of proinflammatory cytokines (Zunszain et al. 2011, 2012), reduced neurotrophic factors, such as brain-derived neurotrophic factor, BDNF, (Duman 2004) and hyperactivity of the hypothalamus–pituitary–adrenal (HPA) axis, which causes increased glucocorticoid hormone levels and activation of the glucocorticoid receptor (GR) (Anacker et al. 2011a; Pariante and Lightman 2008). While all these signaling mechanisms likely play important parts in the trajectory to depression, the interest in glucocorticoid effects on neurogenesis has recently gained momentum, in light of evidence for a reciprocal interaction of HPA axis abnormalities and hippocampal neurogenesis that may potentially underlie the development and relapse of depression and anxiety. This chapter will therefore focus primarily on these interactions of the HPA axis and neurogenesis, and discuss how glucocorticoid hormones may regulate neurogenesis and thereby ultimately affect mood and behavior.

4 The HPA Axis and Neurogenesis Regulation

The HPA axis is a major part of the neuroendocrine system, which regulates the body's response to stress. The HPA axis is primarily regulated by the hippocampus, which controls the release of corticotrophin releasing hormone (CRH) and arginine-vasopressin (AVP) from the paraventricular nucleus (PVN) of the hypothalamus upon exposure to stress. CRH then induces the synthesis of adrenocorticotrophic hormone (ACTH) from the anterior pituitary gland, which in turn stimulates the production of glucocorticoids (cortisol in humans and corticosterone in rodents) in the adrenal cortex, and their release into the blood stream. Stress causes an impairment of this tightly regulated system, subsequently resulting in increased release of CRH from the PVN, increased synthesis of ACTH, and enhanced release of glucocorticoids (Anacker et al. 2011a; Holsboer et al. 1984).

Glucocorticoid hormones predominantly bind to two different steroid receptors: the type I, or mineralocorticoid receptor (MR), and the type II, or GR (Holsboer 2000). Glucocorticoids have multiple functions in almost every tissue of the human body, including regulation of energy metabolism, immune functions, sexuality, and mood. They can also exert negative feedback inhibition of the HPA axis, by activating both MR and GR in the hippocampus, the PVN and the anterior pituitary, thereby maintaining low glucocorticoid levels under normal physiological conditions (Anacker et al. 2011a; Jacobson and Sapolsky 1991; Sapolsky et al. 1985). However, upon exposure to chronic stress, this MR- and GR-mediated negative feedback loop is impaired, resulting in constant HPA axis hyperactivity, increased pituitary and adrenal gland volume, and chronically high levels of glucocorticoids (Meaney et al. 1995; Nemeroff et al. 1992; Pariante 2006; Pariante and Miller 2001; Sapolsky et al. 1985; Anacker and Pariante 2012b).

Around 80 % of severely depressed patients exhibit such chronic hyperactivity of the HPA axis (Anacker et al. 2011a; Juruena et al. 2006; Young et al. 1991), and this phenomenon can also be modeled by rodent studies, which have shown HPA axis abnormalities and increased glucocorticoid levels upon exposure to unpredictable chronic mild stress, early life stress or chronic social stress (Lehmann et al. 2013; Surget et al. 2008, 2011). Indeed, it is this persistent increase in glucocorticoid hormones that critically contributes to the reduction in neurogenesis and to the development of a depression-like phenotype. For example, chronic treatment with the rodent glucocorticoid, corticosterone, reduces neurogenesis and induces both depression- and anxiety-like behavior that is reversible by antidepressant treatment (David et al. 2009; Murray et al. 2008). These detrimental effects on neurogenesis are predominantly mediated by the GR, as brief treatment with the GR antagonist, RU486, counteracts the decrease in neurogenesis upon glucocorticoid treatment and stress in rodents (Hu et al. 2012; Mayer et al. 2006; Oomen et al. 2007). Moreover, treatment with high concentrations of the human glucocorticoid, cortisol, reduces cell proliferation, and neuronal differentiation of human hippocampal progenitor cells in vitro, an effect that is dependent on GR-induced expression of serum- and glucocorticoid-regulated kinase 1 (SGK1) (Anacker et al. 2013a, b).

This role of the GR as a crucial mediator for stress effects on neurogenesis and behavior is further supported by a number of transgenic mouse studies. Mice with a 50 % gene dose reduction of the GR in the entire body ($GR^{+/-}$ mice) exhibit depressive behavior and decreased hippocampal neurogenesis, most likely because glucocorticoid levels are increased in these mice as a result of impairments in GR-mediated feedback regulation of the HPA axis (Kronenberg et al. 2009). Accordingly, GR deletion in the entire central nervous system (GR^{NesCre} mice) causes HPA axis hyperactivity and increased glucocorticoid levels, again, likely caused by disrupted GR-mediated feedback inhibition on the hypothalamus and forebrain. Interestingly, elevated glucocorticoid levels do not impair hippocampal neurogenesis or precipitate depressive behavior in this model in which the GR is deleted in all cells of the entire central nervous system (Gass et al. 2000; Tronche et al. 1999). Although these findings may appear contradictory at first sight, it has also been shown that GR deletion only in mature forebrain neurons (using GR$^{\text{CamKII}\alpha\text{Cre}}$ mice) increases glucocorticoid levels and indeed impairs adult hippocampal neurogenesis and precipitates depressive behavior (Boyle et al. 2005, 2006). It is noteworthy, that the latter GR deletion in mature forebrain neurons spares the GR in stem cells and in newborn neurons of the hippocampus, suggesting that glucocorticoid-induced activation of the GR in hippocampal stem cells and their progeny may account for the reduction in neurogenesis and the subsequent development of depressive behavior in situations of stress and hypercortisolemia. This is supported by a recent study, which has shown that lentiviral-mediated knockdown of the GR, specifically in newborn cells of the hippocampus, accelerates their neuronal differentiation, dendritic arborization, and migration into the molecular layer of the dentate gyrus. While this GR knockdown in newborn neurons impairs hippocampus-dependent memory consolidation (Fitzsimons et al. 2013), the role of the GR in adult hippocampal stem cells and their progeny as a mediator for stress effects on depression remains elusive.

Taken together, the above-described studies suggest that GR activation in the hypothalamus, pituitary, and in mature neurons of the hippocampus, is important to regulate HPA axis feedback inhibition and glucocorticoid levels, while GR activation specifically in newborn neurons may be detrimental for neurogenesis and thereby possibly contribute to the development of depressive symptoms under chronic stress. However, the effects of glucocorticoids on neurogenesis and behavior are even more complex. For example, Lehman and colleagues have used two powerful paradigms that both stimulate glucocorticoid secretion, but have opposing effects on neurogenesis: chronic social defeat stress and environmental enrichment. While glucocorticoids released during social defeat precipitate depressive behavior by decreasing neurogenesis, glucocorticoids released during environmental enrichment are in fact responsible for counteracting depressive behavior by enhancing neurogenesis (Lehmann et al. 2013). These findings are striking, as they demonstrate that glucocorticoids can have differential effects, depending on which environmental stimulus precipitates their release. This is in line with the emerging view that glucocorticoids may increase neurogenesis when the underlying stress has hedonic value, as it is also the case after mating or after physical exercise (Brown et al. 2003; Leuner et al. 2010; van Praag et al. 1999). The concurrent regulation of protective factors, such as oxytocin or dopamine, may be part of the mechanism that increases neurogenesis in the presence of high glucocorticoid levels under eustress conditions (Hoglinger et al. 2004; Leuner et al. 2012). This is further supported by data showing that one and the same GR can exert opposing effects on gene transcription depending on which protein–protein interactions are induced by the combination of activating stimuli (Diamond et al. 1992; Kappeler and Meaney 2010). It will therefore be important for future studies to take into consideration that the function of the GR is indeed highly complex, and experimental investigations of stress-induced impairments in neurogenesis and behavior will need to thoroughly scrutinize the molecular regulation of the GR in defined brain regions and cell populations, in order to fully capture the molecular and neurobiological pathways that ultimately contribute to the pathophysiology of depression.

5 Neurogenesis-Dependent Effects of Stress

As discussed above, a multitude of studies have demonstrated consistent glucocorticoid-mediated effects of stress on hippocampal neurogenesis. However, the causal relationship between reduced neurogenesis and the development of anxiety and depression has long been controversial. Indeed, rodent studies investigating a functional role of neurogenesis for mood and behavior have shown that reducing neurogenesis by focal X-ray irradiation of the hippocampus or by treatment with the cytostatic drug, methylazoxymethanol (MAM), do not induce depression-like behaviors per se (Bessa et al. 2009; David et al. 2009; Santarelli et al. 2003; Surget et al. 2008). These findings had initially suggested that, a reduction in neurogenesis by itself might not primarily be involved in precipitating behavioral abnormalities. However, it is noteworthy that depleting neurogenesis with MAM treatment increases the latency to feed in the novelty suppressed feeding test. In this test, MAM-treated, neurogenesis-deficient rats show a greater latency to feed in a novel environment, such as an open field arena, indicating a higher level of anxiety and approach-avoidance behavior in rats in which neurogenesis in not intact (Bessa et al. 2009). These findings may therefore point toward a potential role for neurogenesis primarily in the behavioral domain of anxiety, rather than in depression. Accordingly, Revest and colleagues demonstrated that neurogenesis-deficient mice are characterized by increased fear. Using overexpression of the proapoptotic gene *Bax* in adult hippocampal neurons to ablate neurogenesis, the authors found that neurogenesis-deficient mice escaped faster into a protective cylinder when exposed to a predator, indicating a particular involvement of neurogenesis for fear-related behaviors (Revest et al. 2009).

Recent work has further expanded on these findings, by examining the role of neurogenesis specifically as a mediator for the behavioral response to stress. Snyder and colleagues used transgenic mice in which neurogenesis can be specifically ablated during adulthood (GFAP-TK mice). Using this transgenic approach, the authors demonstrated that acute restraint stress causes higher levels of anxiety in neurogenesis-deficient mice as compared to mice in which neurogenesis is still intact, pointing toward a potential role for neurogenesis as a "buffer" for the behavioral response to stress. Interestingly, in contrast to the data from the above-described earlier studies, the paper by Snyder et al. also showed that total ablation of neurogenesis did indeed precipitate behavioral despair and anhedonia, even under baseline conditions when mice were not exposed to any stress (Snyder et al. 2011).

Importantly, neurogenesis may be particularly relevant for achieving sustained spontaneous remission from depressive symptoms: While rats with intact neurogenesis recover from stress-induced depressive symptoms within 4 weeks after a 6-week UCMS period, neurogenesis-ablation by MAM treatment prevents this spontaneous recovery from depression-like behavior (Mateus-Pinheiro et al. 2013). When comparing these rodent findings with clinical studies, it is interesting to note that remission from depressive symptoms in patients has been associated with a normalization of HPA axis hyperactivity after antidepressant treatment. Indeed, depressed patients that continue to show HPA axis hyperactivity after treatment are less likely to achieve remission and are at higher risk for relapse (Appelhof et al. 2006; Zobel et al. 2001). This is particularly striking, considering the preclinical evidence for neurogenesis as a crucial component of HPA axis regulation. Specifically, neurogenesis-deficient mice show a more pronounced glucocorticoid surge after acute stress when compared with mice in which neurogenesis is intact (Schloesser et al. 2009; Snyder et al. 2011). However, whether a stress-induced, partial reduction in neurogenesis can be responsible for the development of depressive behavior and for impairments in spontaneous remission, to the same extent as the aforementioned complete ablation of neurogenesis affects these behaviors, still remains to be elucidated. Nevertheless, these findings may suggest that neurogenesis disturbances impair hippocampal inhibitory control over the HPA axis, subsequently contributing to persistent HPA axis hyperactivity, which in turn further reduces hippocampal neurogenesis through chronic elevations in glucocorticoid hormones. This may set into motion a vicious cycle of increased glucocorticoids and reduced neurogenesis, ultimately leading to sustained anxiety and depression-like behavior and a higher risk for relapse (Anacker and Pariante 2012a, b). Counteracting these deeply manifested impairments in adult neurogenesis may therefore be a promising strategy for future antidepressant treatments to restore plastic disturbances in the hippocampal-neuroendocrine circuitry, which may be particularly necessary to confer long-term remission from anxiety and depression.

6 Regulation of Adult Hippocampal Neurogenesis by Antidepressants

If, as outlined above, adult hippocampal neurogenesis mediates stress-induced impairments in HPA axis function and behavior, do antidepressants exert their behavioral effects by increasing neurogenesis? In the first study to ever examine the effects of chronic antidepressant treatment on hippocampal neurogenesis, Malberg and colleagues found that different pharmacological classes of antidepressant drugs, as well as electroconvulsive shocks, all increase progenitor cell proliferation in the rat dentate gyrus by $\sim$ 20–50 % (Malberg et al. 2000). The study also reported that 75 % of these adult-born cells develop into neurons, while only 13 % become glia (Malberg et al. 2000). A large number of rodent studies has now replicated these findings, and demonstrated that the stress- and glucocorticoid-induced reduction in neurogenesis is reversed by a broad range of pharmacologically different antidepressants and experimental conditions (e.g., (Banasr et al. 2006; Dagyte et al. 2010; David et al. 2009; Egeland et al. 2010; Surget et al. 2008, 2011). Furthermore, the effects of antidepressants to increase neurogenesis have also been observed in adult bonnet monkeys (Perera et al. 2007, 2011), in human postmortem brain tissue of depressed patients (Boldrini et al. 2009, 2012), and in vitro, in antidepressant-treated human hippocampal progenitor cells (Anacker et al. 2011b).

It is particularly noteworthy that, in line with neurogenesis-dependent regulation of HPA axis function, the antidepressant-induced increase in hippocampal neurogenesis is indeed crucial to restore hippocampal inhibitory control over the HPA axis and to normalize glucocorticoid levels after chronic stress in rodents (Surget et al. 2011). These findings therefore indicate that antidepressants may ameliorate the detrimental effects of stress on HPA axis function and mood by restoring adult hippocampal neurogenesis. Interestingly, while antidepressants counteract the stress- and glucocorticoid-induced reduction in neurogenesis, their pro-neurogenic effect is at the same time dependent on the presence of glucocorticoids: For example, SSRIs increase neurogenesis only upon cotreatment with glucocorticoids in mice and in human hippocampal progenitor cells (Anacker et al. 2011b; David et al. 2009). In addition, in adrenalectomized rats, in which corticosterone concentrations have been surgically clamped at low levels, fluoxetine does no longer increase hippocampal neurogenesis (Huang and Herbert 2006). Interestingly, this effect is also observed when corticosterone concentrations are clamped at constant high levels, suggesting that the circadian rhythm of the HPA axis may be relevant for the SSRI-induced stimulation of neurogenesis (Huang and Herbert 2006).

In line with neurogenesis and HPA axis interactions, antidepressants have been shown to regulate the function of the GR both in vivo and in vitro (Anacker et al. 2011a, b; Pariante et al. 1997, 2001). However, in contrast to glucocorticoids, antidepressants activate the GR by inducing cyclic adenosine monophosphate (cAMP) and protein kinase A (PKA)-dependent phosphorylation of the receptor, which causes GR binding to the DNA and activates a GR-dependent set of downstream target genes that is different from the gene expression profile induced by glucocorticoid hormones (Anacker et al. 2011b; Guidotti et al. 2013; Miller et al. 2002). Importantly, this cAMP/PKA-dependent activation of the GR is a crucial mechanism for the antidepressant-induced increase in proliferation and neuronal differentiation of human hippocampal progenitor cells (Anacker et al. 2011b), and may therefore be a molecular mediator around which stress and antidepressants exert their opposing effects on gene transcription, hippocampal neurogenesis, and ultimately, behavior.

7 Neurogenesis-Dependent Effects of Antidepressants

In a groundbreaking study in 2003, Santarelli and colleagues demonstrated that some of the above-described behavioral effects of antidepressants are indeed dependent on neurogenesis. While antidepressants counteracted stress-induced behavioral symptoms in control mice, mice in which neurogenesis was depleted by focal X-ray irradiation of the hippocampus failed to respond to treatment with either fluoxetine or imipramine, demonstrating that neurogenesis is indeed necessary for the behavioral response to pharmacologically different antidepressants (Santarelli et al. 2003). Moreover, the necessity of adult hippocampal neurogenesis for the therapeutic action of antidepressants has also been demonstrated in nonhuman primates, in which no behavioral effects of fluoxetine are observed when neurogenesis is abolished by X-ray irradiation, extending the necessity of hippocampal neurogenesis for mediating the behavioral effects of antidepressants also into higher mammals (Perera et al. 2011).

Despite this aforementioned evidence, conflicting data exist, showing that only some behavioral effects of antidepressants are dependent on neurogenesis. For example, even when neurogenesis is ablated, fluoxetine reduces the immobility in the forced swim test, a test in which mice show freezing behavior when placed in a water basin, and in the novelty-induced hypophagia test, a test in which mice consume less sweet milk when placed in a foreign cage (Bessa et al. 2009; Holick et al. 2008). These findings therefore suggest that some antidepressant effects on behavior may indeed be completely independent of antidepressant-induced neurogenesis. However, neurogenesis-deficient mice consistently fail to improve upon antidepressant treatment in the novelty-suppressed feeding test, suggesting that antidepressants may have both neurogenesis-dependent and -*in*dependent effects on mood and behavior (David et al. 2009).

Interestingly, pharmacological compounds that target the HPA axis, such as CRH antagonists, increase neurogenesis and ameliorate depressive behavior, but the behavioral effect of these compounds is still observed when neurogenesis is depleted (Surget et al. 2008), further supporting the notion that normalizing HPA axis hyperactivity may be a promising therapeutic strategy to increase neurogenesis and to overcome depressive symptoms.

A further criticism for a direct behavioral effect of neurogenesis emerged recently, when an elegant study showed that enhancing neurogenesis by genetic ablation of the proapoptotic gene, *Bax*, does not recapitulate any antidepressant-like behavioral response (Sahay et al. 2011). However, the study investigated the effects of enhancing neurogenesis under baseline conditions, when the animal is not challenged by chronic stress. In the light of some of the above-described studies, which have shown that hippocampal neurogenesis may be particularly important for eliciting behavioral effects only under conditions of stress (Anacker et al. 2011b; Snyder et al. 2011; Surget et al. 2011), it will be crucial to explore "stress" as the critical link between neurogenesis and antidepressant action, and to examine whether increasing neurogenesis may indeed confer antidepressant effects by counteracting stress-induced behavioral abnormalities. Although the study by Sahay and colleagues did not address such neurogenesis-stress interactions, it did reveal an important new function for neurogenesis in the adult hippocampus: Mice in which neurogenesis was increased were more capable of distinguishing a fear-associated situation from a similar situation that does not actually pose a potential threat, a phenomenon called "pattern separation" (Sahay et al. 2011). Such neurogenesis-dependent regulation of pattern separation may be a particularly relevant element of the aforementioned vicious cycle of HPA axis hyperactivity and neurogenesis reduction in depression: ambiguous environmental cues may be evaluated as threatening or as generally negative when neurogenesis is reduced, leading to stress responses and HPA axis activation even when no actual threat is present. This exaggerated stress reactivity may thus increase glucocorticoid levels and further impair hippocampal neurogenesis, ultimately leading to sustained anxiety and depression-like behavior. This may indeed contribute to the development of the above-hypothesized "vicious cycle" of reduced neurogenesis and impaired neuroendocrine function, and may represent yet another aspect of how changes in neurogenesis may cause long-term disturbances in stress reactivity and mood (Anacker and Pariante 2012a, b; Sahay et al. 2011). Accordingly, increasing neurogenesis by antidepressant treatment may improve the cognitive ability to separate such ambiguous cues and thereby help to overcome prolonged heightened sensitivity to developing stressful responses and associated psychological disturbances.

8 Adult Hippocampal Neurogenesis in Depression: Evidence from Human Studies

While rodent studies and cellular models are powerful, widely used tools to investigate the molecular regulators and behavioral implications of hippocampal neurogenesis upon stress, less evidence exists for changes in neurogenesis in depressed patients. The first study that had investigated neurogenesis in hippocampal postmortem brain tissue of depressed patients had found no differences in the number of proliferating neural progenitor cells in the anterior hippocampus of medicated depressed patients (corresponding to the ventral hippocampus in rodents) (Reif et al. 2006). A second study investigated both, the number of proliferating progenitor cells and the total number of stem cells in postmortem brain tissue of depressed patients with and without antidepressant treatment (Boldrini et al. 2009). The strength of this latter study lay in the availability of toxicological data from blood and urine samples, which allowed to screen whether all medicated patients did indeed take their prescribed medication, and permitted exclusion of subjects with substance abuse. Although the study found that depressed patients have around 50 % less proliferating hippocampal stem cells than healthy controls, this effect did not reach significance in their sample of seven controls and five unmedicated depressed patients. However, depressed patients who had received antidepressant treatment showed a strong increase in cell proliferation and in the total number of neural stem cells in the hippocampus when compared to depressed patients without antidepressant treatment as well as unmedicated healthy controls. Interestingly, these antidepressant-induced changes in neurogenesis were more pronounced in the anterior part of the dentate gyrus, which, as described before, is particularly involved in mood and neuroendocrine regulation (Boldrini et al. 2009, 2012). Interestingly, in contrast to the aforementioned studies that investigated changes in neurogenesis in patients ranging from 17 to 62 years of age, elderly depressed patients with an average age of 68 years and known disturbances in HPA axis function, exhibit a significant reduction in total hippocampal stem cell number but no regulation by previous antidepressant treatment (Lucassen et al. 2010).

Taken together, although conflicting data exists, some studies have pointed toward changes in adult hippocampal neurogenesis in the brain of depressed and antidepressant-treated patients. In order to clearly address to what extent adult hippocampal neurogenesis may be involved in depression pathophysiology and antidepressant treatment response, future studies, using larger sample sizes and controlling for patients' medical history, pharmacological treatment, and age, may help to disentangle some of the neurobiological pathways that may lead to disparate results in post-mortem brain studies.

9 Neurogenesis and Mood Regulation: Evolutionary Adaptation Gone Awry?

When discussing the implications of neurogenesis for behavior in situations of stress and in depression, the question arises why such a neurobiological pathway to mental illness exists at all in any organism, and why not every individual has biologically evolved to be resistant to the seemingly detrimental effects of stress on behavior. Glasper and colleagues previously discussed this question and suggested that, from an evolutionary point of view, aversive and possibly life-threatening situations may reduce hippocampal neurogenesis in order to prepare an individual to cope with similar situations effectively in the future. Although neurogenesis-dependent induction of anxiety and fear may initially appear disadvantageous or even harmful to an individual, these behavioral alterations can also be considered adaptive, as they serve the purpose to maximize chances for survival in a threatening environment in which withdrawal and social avoidance are necessary to reduce exposure to predators and to live cautiously with limited resources in a confined but safe environment. Conversely, an individual living in an enriched environment with minimal exposure to predators and other life-threatening situations, does not require such a behavioral strategy in order to survive, but can instead make use of existing opportunities in its safe environment, thereby maximizing health and mating success through increased food intake and social- and sexual interactions (Glasper et al. 2012). Changes in hippocampal neurogenesis and behavior may therefore serve the purpose to maximize the individual's contribution to the gene pool by making best use of all available resources in order to extend the lifespan of an individual and its progeny.

However, the neurobiological mechanisms mediating these environment-behavior interactions in rodents may have been maintained throughout evolution and may still exist in humans. It is therefore important to regard a reduction in neurogenesis not simply as a "malfunction" of the brain, but instead, as an adaptive mechanism, which, from an evolutionary point of view, may have the specific purpose to modulate behavior according to the environment. Considering how dramatically environmental "threats" have changed in modern society, the same adaptive mechanisms that served to maximize survival, may nowadays induce behaviors that do not reflect the appropriate behavioral response to the environmental situation, thus leading to pathological effects, and ultimately, mental illness.

10 Conclusion and Outlook

It has become evident over recent years that adult hippocampal neurogenesis is indeed a crucial mechanism in the behavioral neurobiology of stress-related disorders, particularly depression and anxiety. In William James' words, neurogenesis in the adult hippocampus appears to form part of a neurobiological network that is "weak enough to yield to an influence," and to convert environmental signals into behavioral responses, while at the same time being tightly controlled on the cellular and molecular level, ensuring that this structure is "strong enough not to yield all at once." This tight regulation provides a neurobiological safeguard for the behavioral response to detrimental environmental signals and specifically buffers stress effects on the brain. A considerable amount of animal research on the implications of neurogenesis for mood and behavior has shown that neurogenesis is important for both, the pathogenesis of depressive- and anxiety-like states, as well as the behavioral response to antidepressant treatment. These behavioral effects in rodents indeed appear to be important also for human patients, as postmortem brain studies in depressed patients and in vitro studies on human hippocampal stem cells have shown similar changes in neurogenesis as in chronically stressed rodents. The challenge for future research will lie in identifying relevant molecular mediators that can be targeted in order to reverse reductions in neurogenesis or to prevent the occurrence of neurogenesis abnormalities under conditions of stress. Finding such mechanisms may have great potential to lead to improved antidepressant treatment therapies that are aimed at normalizing HPA axis abnormalities, and ultimately, at achieving long-term remission from depression.

References

Altman J, Das GD (1965) Autoradiographic and histological evidence of postnatal hippocampal neurogenesis in rats. J Comp Neurol 124:319–335

Anacker C, Cattaneo A, Luoni A, Musaelyan K, Zunszain PA, Milanesi E, Rybka J, Berry A, Cirulli F, Thuret S, Price J, Riva MA, Gennarelli M, Pariante CM (2013a) Glucocorticoid-related molecular signaling pathways regulating hippocampal neurogenesis. Neuropsychopharmacology 38:872–883

Anacker C, Cattaneo A, Musaelyan K, Zunszain PA, Horowitz M, Molteni R, Luoni A, Calabrese F, Tansey K, Gennarelli M, Thuret S, Price J, Uher R, Riva MA, Pariante CM (2013b) Role for the kinase SGK1 in stress, depression, and glucocorticoid effects on hippocampal neurogenesis. Proc Natl Acad Sci USA 110:8708–8713

Anacker C, Pariante CM (2012a) Can adult neurogenesis buffer stress responses and depressive behaviour? Mol Psychiatry 17:9–10

Anacker C, Pariante CM (2012b) New models to investigate complex glucocorticoid receptor functions. Front Behav Neurosci 6:90

Anacker C, Zunszain PA, Carvalho LA, Pariante CM (2011a) The glucocorticoid receptor: pivot of depression and of antidepressant treatment? Psychoneuroendocrinology 36:415–425

Anacker C, Zunszain PA, Cattaneo A, Carvalho LA, Garabedian MJ, Thuret S, Price J, Pariante CM (2011b) Antidepressants increase human hippocampal neurogenesis by activating the glucocorticoid receptor. Mol Psychiatry 16:738–750

Appelhof BC, Huyser J, Verweij M, Brouwer JP, van Dyck R, Fliers E, Hoogendijk WJ, Tijssen JG, Wiersinga WM, Schene AH (2006) Glucocorticoids and relapse of major depression (dexamethasone/corticotropin-releasing hormone test in relation to relapse of major depression). Biol Psychiatry 59:696–701

Banasr M, Soumier A, Hery M, Mocaer E, Daszuta A (2006) Agomelatine, a new antidepressant, induces regional changes in hippocampal neurogenesis. Biol Psychiatry 59:1087–1096

Bessa JM, Ferreira D, Melo I, Marques F, Cerqueira JJ, Palha JA, Almeida OF, Sousa N (2009) The mood-improving actions of antidepressants do not depend on neurogenesis but are associated with neuronal remodeling. Mol Psychiatry 14:739, 764–773

Boldrini M, Hen R, Underwood MD, Rosoklija GB, Dwork AJ, Mann JJ, Arango V (2012) Hippocampal angiogenesis and progenitor cell proliferation are increased with antidepressant use in major depression. Biol Psychiatry 72:562–571

Boldrini M, Underwood MD, Hen R, Rosoklija GB, Dwork AJ, John Mann J, Arango V (2009) Antidepressants increase neural progenitor cells in the human hippocampus. Neuropsychopharmacology 34:2376–2389

Boyle MP, Brewer JA, Funatsu M, Wozniak DF, Tsien JZ, Izumi Y, Muglia LJ (2005) Acquired deficit of forebrain glucocorticoid receptor produces depression-like changes in adrenal axis regulation and behavior. Proc Natl Acad Sci USA 102:473–478

Boyle MP, Kolber BJ, Vogt SK, Wozniak DF, Muglia LJ (2006) Forebrain glucocorticoid receptors modulate anxiety-associated locomotor activation and adrenal responsiveness. J Neurosci 26:1971–1978

Brown J, Cooper-Kuhn CM, Kempermann G, Van Praag H, Winkler J, Gage FH, Kuhn HG (2003) Enriched environment and physical activity stimulate hippocampal but not olfactory bulb neurogenesis. Eur J Neurosci 17:2042–2046

Caldji C, Diorio J, Meaney MJ (2000) Variations in maternal care in infancy regulate the development of stress reactivity. Biol Psychiatry 48:1164–1174

Cameron HA, McEwen BS, Gould E (1995) Regulation of adult neurogenesis by excitatory input and NMDA receptor activation in the dentate gyrus. J Neurosci 15:4687–4692

Coe CL, Kramer M, Czeh B, Gould E, Reeves AJ, Kirschbaum C, Fuchs E (2003) Prenatal stress diminishes neurogenesis in the dentate gyrus of juvenile rhesus monkeys. Biol Psychiatry 54:1025–1034

Dagyte G, Trentani A, Postema F, Luiten PG, Den Boer JA, Gabriel C, Mocaër E, Meerlo P, Van der Zee EA (2010) The novel antidepressant agomelatine normalizes hippocampal neuronal activity and promotes neurogenesis in chronically stressed rats. CNS Neurosci Ther 16:195–207

David DJ, Samuels BA, Rainer Q, Wang JW, Marsteller D, Mendez I, Drew M, Craig DA, Guiard BP, Guilloux JP, Artymyshyn RP, Gardier AM, Gerald C, Antonijevic IA, Leonardo ED, Hen R (2009) Neurogenesis-dependent and -independent effects of fluoxetine in an animal model of anxiety/depression. Neuron 62:479–493

Diamond DM, Bennett MC, Fleshner M, Rose GM (1992) Inverted-U relationship between the level of peripheral corticosterone and the magnitude of hippocampal primed burst potentiation. Hippocampus 2:421–430

Duman RS (2004) Role of neurotrophic factors in the etiology and treatment of mood disorders. Neuromolecular Med 5:11–25

Egeland M, Warner-Schmidt J, Greengard P, Svenningsson P (2010) Neurogenic effects of fluoxetine are attenuated in p11 (S100A10) knockout mice. Biol Psychiatry 67:1048–1056

Eriksson PS, Perfilieva E, Bjork-Eriksson T, Alborn AM, Nordborg C, Peterson DA, Gage FH (1998) Neurogenesis in the adult human hippocampus. Nat Med 4:1313–1317

Fitzsimons CP, van Hooijdonk LW, Schouten M, Zalachoras I, Brinks V, Zheng T, Schouten TG, Saaltink DJ, Dijkmans T, Steindler DA, Verhaagen J, Verbeek FJ, Lucassen PJ, de Kloet ER, Meijer OC, Karst H, Joels M, Oitzl MS, Vreugdenhil E (2013) Knockdown of the glucocorticoid receptor alters functional integration of newborn neurons in the adult hippocampus and impairs fear-motivated behavior. Mol Psychiatry 18:993–1005

Gass P, Kretz O, Wolfer DP, Berger S, Tronche F, Reichardt HM, Kellendonk C, Lipp HP, Schmid W, Schutz G (2000) Genetic disruption of mineralocorticoid receptor leads to impaired neurogenesis and granule cell degeneration in the hippocampus of adult mice. EMBO Rep 1:447–451

Glasper ER, Schoenfeld TJ, Gould E (2012) Adult neurogenesis: optimizing hippocampal function to suit the environment. Behav Brain Res 14; 227(2):380–383

Goritz C, Frisen J (2012) Neural stem cells and neurogenesis in the adult. Cell Stem Cell 10:657–659

Gould E, McEwen BS, Tanapat P, Galea LA, Fuchs E (1997) Neurogenesis in the dentate gyrus of the adult tree shrew is regulated by psychosocial stress and NMDA receptor activation. J Neurosci 17:2492–2498

Guidotti G, Calabrese F, Anacker C, Racagni G, Pariante CM, Riva MA (2013) Glucocorticoid receptor and FKBP5 expression is altered following exposure to chronic stress: modulation by antidepressant treatment. Neuropsychopharmacology 38:616–627

Henke PG (1990) Hippocampal pathway to the amygdala and stress ulcer development. Brain Res Bull 25:691–695

Hoglinger GU, Rizk P, Muriel MP, Duyckaerts C, Oertel WH, Caille I, Hirsch EC (2004) Dopamine depletion impairs precursor cell proliferation in Parkinson disease. Nat Neurosci 7:726–735

Holick KA, Lee DC, Hen R, Dulawa SC (2008) Behavioral effects of chronic fluoxetine in BALB/cJ mice do not require adult hippocampal neurogenesis or the serotonin 1A receptor. Neuropsychopharmacology 33:406–417

Holsboer F (2000) The corticosteroid receptor hypothesis of depression. Neuropsychopharmacology 23:477–501

Holsboer F, Von Bardeleben U, Gerken A, Stalla GK, Muller OA (1984) Blunted corticotropin and normal cortisol response to human corticotropin-releasing factor in depression. N Engl J Med 311:1127

Hu P, Oomen C, van Dam AM, Wester J, Zhou JN, Joëls M, Lucassen PJ (2012) A single-day treatment with mifepristone is sufficient to normalize chronic glucocorticoid induced suppression of hippocampal cell proliferation. PLoS ONE 7:e46224

Huang GJ, Herbert J (2006) Stimulation of neurogenesis in the hippocampus of the adult rat by fluoxetine requires rhythmic change in corticosterone. Biol Psychiatry 59:619–624

Jacobson L, Sapolsky R (1991) The role of the hippocampus in feedback regulation of the hypothalamic-pituitary-adrenocortical axis. Endocr Rev 12:118–134

Juruena MF, Cleare AJ, Papadopoulos AS, Poon L, Lightman S, Pariante CM (2006) Different responses to dexamethasone and prednisolone in the same depressed patients. Psychopharmacology 189:225–235

Kaplan MS, Hinds JW (1977) Neurogenesis in the adult rat: electron microscopic analysis of light radioautographs. Science 197:1092–1094

Kappeler L, Meaney MJ (2010) Enriching stress research. Cell 142:15–17

Kheirbek MA, Drew LJ, Burghardt NS, Costantini DO, Tannenholz L, Ahmari SE, Zeng H, Fenton AA, Hen R (2013) Differential control of learning and anxiety along the dorsoventral axis of the dentate gyrus. Neuron 77:955–968

Koehl M, van der Veen R, Gonzales D, Piazza PV, Abrous DN (2012) Interplay of maternal care and genetic influences in programming adult hippocampal neurogenesis. Biol Psychiatry 72:282–289

Kronenberg G, Kirste I, Inta D, Chourbaji S, Heuser I, Endres M, Gass P (2009) Reduced hippocampal neurogenesis in the GR($\pm$) genetic mouse model of depression. Eur Arch Psychiatry Clin Neurosci 259(8):499–504

Lehmann ML, Brachman RA, Martinowich K, Schloesser RJ, Herkenham M (2013) Glucocorticoids orchestrate divergent effects on mood through adult neurogenesis. J Neurosci 33:2961–2972

Lemaire V, Koehl M, Le Moal M, Abrous DN (2000) Prenatal stress produces learning deficits associated with an inhibition of neurogenesis in the hippocampus. Proc Natl Acad Sci U S A 97:11032–11037

Leuner B, Caponiti JM, Gould E (2012) Oxytocin stimulates adult neurogenesis even under conditions of stress and elevated glucocorticoids. Hippocampus 22:861–868

Leuner B, Glasper ER, Gould E (2010) Sexual experience promotes adult neurogenesis in the hippocampus despite an initial elevation in stress hormones. PLoS ONE 5:e11597

Liu D, Diorio J, Day JC, Francis DD, Meaney MJ (2000) Maternal care, hippocampal synaptogenesis and cognitive development in rats. Nat Neurosci 3:799–806

Lucassen PJ, Stumpel MW, Wang Q, Aronica E (2010) Decreased numbers of progenitor cells but no response to antidepressant drugs in the hippocampus of elderly depressed patients. Neuropharmacology 58:940–949

Malberg JE, Eisch AJ, Nestler EJ, Duman RS (2000) Chronic antidepressant treatment increases neurogenesis in adult rat hippocampus. J Neurosci 20:9104–9110

Mateus-Pinheiro A, Pinto L, Bessa JM, Morais M, Alves ND, Monteiro S, Patrício P, Almeida OF, Sousa N (2013) Sustained remission from depressive-like behavior depends on hippocampal neurogenesis. Transl Psychiatry 3:e210

Mayer JL, Klumpers L, Maslam S, de Kloet ER, Joels M, Lucassen PJ (2006) Brief treatment with the glucocorticoid receptor antagonist mifepristone normalises the corticosterone-induced reduction of adult hippocampal neurogenesis. J Neuroendocrinol 18:629–631

Meaney MJ, O'Donnell D, Rowe W, Tannenbaum B, Steverman A, Walker M, Nair NP, Lupien S (1995) Individual differences in hypothalamic-pituitary-adrenal activity in later life and hippocampal aging. Exp Gerontol 30:229–251

Miller AH, Vogt GJ, Pearce BD (2002) The phosphodiesterase type 4 inhibitor, rolipram, enhances glucocorticoid receptor function. Neuropsychopharmacology 27:939–948

Mirescu C, Peters JD, Gould E (2004) Early life experience alters response of adult neurogenesis to stress. Nat Neurosci 7:841–846

Murray F, Smith DW, Hutson PH (2008) Chronic low dose corticosterone exposure decreased hippocampal cell proliferation, volume and induced anxiety and depression like behaviours in mice. Eur J Pharmacol 583:115–127

Nemeroff CB, Krishnan KR, Reed D, Leder R, Beam C, Dunnick NR (1992) Adrenal gland enlargement in major depression. A computed tomographic study: Arch Gen Psychiatry 49:384–387

Oomen CA, Mayer JL, de Kloet ER, Joels M, Lucassen PJ (2007) Brief treatment with the glucocorticoid receptor antagonist mifepristone normalizes the reduction in neurogenesis after chronic stress. Eur J Neurosci 26:3395–3401

Pariante CM (2006) The glucocorticoid receptor: part of the solution or part of the problem? J Psychopharmacol 20:79–84

Pariante CM, Lightman SL (2008) The HPA axis in major depression: classical theories and new developments. Trends Neurosci 31:464–468

Pariante CM, Makoff A, Lovestone S, Feroli S, Heyden A, Miller AH, Kerwin RW (2001) Antidepressants enhance glucocorticoid receptor function in vitro by modulating the membrane steroid transporters. Br J Pharmacol 134:1335–1343

Pariante CM, Miller AH (2001) Glucocorticoid receptors in major depression: relevance to pathophysiology and treatment. Biol Psychiatry 49:391–404

Pariante CM, Pearce BD, Pisell TL, Owens MJ, Miller AH (1997) Steroid-independent translocation of the glucocorticoid receptor by the antidepressant desipramine. Mol Pharmacol 52:571–581

Pascual-Leone A, Amedi A, Fregni F, Merabet LB (2005) The plastic human brain cortex. Annu Rev Neurosci 28:377–401

Perera TD, Coplan JD, Lisanby SH, Lipira CM, Arif M, Carpio C, Spitzer G, Santarelli L, Scharf B, Hen R, Rosoklija G, Sackeim HA, Dwork AJ (2007) Antidepressant-induced neurogenesis in the hippocampus of adult nonhuman primates. J Neurosci 27:4894–4901

Perera TD, Dwork AJ, Keegan KA, Thirumangalakudi L, Lipira CM, Joyce N, Lange C, Higley JD, Rosoklija G, Hen R, Sackeim HA, Coplan JD (2011) Necessity of hippocampal neurogenesis for the therapeutic action of antidepressants in adult nonhuman primates. PLoS ONE 6:e17600

Rakic P (1985) Limits of neurogenesis in primates. Science 227:1054–1056

Ramon y Cajal S (1928) Degeneration and regeneration of the nervous system. Oxford University Press, London

Reif A, Fritzen S, Finger M, Strobel A, Lauer M, Schmitt A, Lesch KP (2006) Neural stem cell proliferation is decreased in schizophrenia, but not in depression. Mol Psychiatry 11:514–522

Revest JM, Dupret D, Koehl M, Funk-Reiter C, Grosjean N, Piazza PV, Abrous DN (2009) Adult hippocampal neurogenesis is involved in anxiety-related behaviors. Mol Psychiatry 14:959–967

Sahay A, Scobie KN, Hill AS, O'Carroll CM, Kheirbek MA, Burghardt NS, Fenton AA, Dranovsky A, Hen R (2011) Increasing adult hippocampal neurogenesis is sufficient to improve pattern separation. Nature 472:466–470

Santarelli L, Saxe M, Gross C, Surget A, Battaglia F, Dulawa S, Weisstaub N, Lee J, Duman R, Arancio O, Belzung C, Hen R (2003) Requirement of hippocampal neurogenesis for the behavioral effects of antidepressants. Science 301:805–809

Sapolsky RM, Meaney MJ, McEwen BS (1985) The development of the glucocorticoid receptor system in the rat limbic brain: III. Negative-feedback regulation. Brain Res 350:169–173

Schloesser RJ, Manji HK, Martinowich K (2009) Suppression of adult neurogenesis leads to an increased hypothalamo-pituitary-adrenal axis response. NeuroReport 20:553–557

Sheline YI, Wang PW, Gado MH, Csernansky JG, Vannier MW (1996) Hippocampal atrophy in recurrent major depression. Proc Natl Acad Sci USA 93:3908–3913

Snyder JS, Soumier A, Brewer M, Pickel J, Cameron HA (2011) Adult hippocampal neurogenesis buffers stress responses and depressive behaviour. Nature 476:458–461

Spalding KL, Bergmann O, Alkass K, Bernard S, Salehpour M, Huttner HB, Boström E, Westerlund I, Vial C, Buchholz BA, Possnert G, Mash DC, Druid H, Frisén J (2013) Dynamics of hippocampal neurogenesis in adult humans. Cell 153:1219–1227

Surget A, Saxe M, Leman S, Ibarguen-Vargas Y, Chalon S, Griebel G, Hen R, Belzung C (2008) Drug-dependent requirement of hippocampal neurogenesis in a model of depression and of antidepressant reversal. Biol Psychiatry 64:293–301

Surget A, Tanti A, Leonardo ED, Laugeray A, Rainer Q, Touma C, Palme R, Griebel G, Ibarguen-Vargas Y, Hen R, Belzung C (2011) Antidepressants recruit new neurons to improve stress response regulation. Mol Psychiatry 16:1177–1188

Tanapat P, Galea LA, Gould E (1998) Stress inhibits the proliferation of granule cell precursors in the developing dentate gyrus. Int J Dev Neurosci 16:235–239

Tanti A, Rainer Q, Minier F, Surget A, Belzung C (2012) Differential environmental regulation of neurogenesis along the septo-temporal axis of the hippocampus. Neuropharmacology 63:374–384

Tronche F, Kellendonk C, Kretz O, Gass P, Anlag K, Orban PC, Bock R, Klein R, Schutz G (1999) Disruption of the glucocorticoid receptor gene in the nervous system results in reduced anxiety. Nat Genet 23:99–103

van Praag H, Christie BR, Sejnowski TJ, Gage FH (1999) Running enhances neurogenesis, learning, and long-term potentiation in mice. Proc Natl Acad Sci USA 96:13427–13431

Watanabe Y, Gould E, Daniels DC, Cameron H, McEwen BS (1992) Tianeptine attenuates stress-induced morphological changes in the hippocampus. Eur J Pharmacol 222:157–162

Weaver IC, Meaney MJ, Szyf M (2006) Maternal care effects on the hippocampal transcriptome and anxiety-mediated behaviors in the offspring that are reversible in adulthood. Proc Natl Acad Sci USA 103:3480–3485

Willner P (2005) Chronic mild stress (CMS) revisited: consistency and behavioural-neurobiological concordance in the effects of CMS. Neuropsychobiology 52:90–110

Woolley CS, Gould E, McEwen BS (1990) Exposure to excess glucocorticoids alters dendritic morphology of adult hippocampal pyramidal neurons. Brain Res 531:225–231

Young EA, Haskett RF, Murphy-Weinberg V, Watson SJ, Akil H (1991) Loss of glucocorticoid fast feedback in depression. Arch Gen Psychiatry 48:693–699

Zobel AW, Nickel T, Sonntag A, Uhr M, Holsboer F, Ising M (2001) Cortisol response in the combined dexamethasone/CRH test as predictor of relapse in patients with remitted depression: a prospective study. J Psychiatr Res 35:83–94

Zunszain PA, Anacker C, Cattaneo A, Carvalho LA, Pariante CM (2011) Glucocorticoids, cytokines and brain abnormalities in depression. Prog Neuropsychopharmacol Biol Psychiatry 35:722–729

Zunszain PA, Anacker C, Cattaneo A, Choudhury S, Musaelyan K, Myint AM, Thuret S, Price J, Pariante CM (2012) Interleukin-1β: a new regulator of the kynurenine pathway affecting human hippocampal neurogenesis. Neuropsychopharmacology 37:939–949

Impact of Stress on Prefrontal Glutamatergic, Monoaminergic and Cannabinoid Systems

M. Danet Lapiz-Bluhm

Abstract Stress has been shown to have marked and divergent effects on learning and memory which involves specific brain regions, such as spatial and declarative memory involving the hippocampus, memory of emotional arousing experiences and fear involving the amygdala, and executive functions and fear extinction involving the prefrontal cortex or the PFC. Response to stress involves a coordinated activation of a constellation of physiological systems including the activation of the hypothalamic-pituitary-adrenal (HPA) axis and other modulatory neurotransmitters and signaling systems. This paper presents a concise review of the effects of stress and glucocorticoids on the glutamatergic and monoaminergic (including noradrenergic, dopaminergic, and serotonergic systems) neurotransmitter systems as well as endocannabinoid signaling. Because of the breadth of the scope of this topic, the review is limited to the effects of stress on these brain systems on the prefrontal cortex, and where relevant, the hippocampus and the amygdala.

Keywords Stress · Glucocorticoids · Glutamatergic · Noradrenergic · Dopaminergic · Serotonergic · Endocannabinoid

Contents

1 Introduction	46
2 The Effects of Stress on the Glutamatergic System	47
3 The Effects of Stress on the Noradrenergic System	49
4 Effects of Stress on the Dopaminergic System	51

M.D. Lapiz-Bluhm (✉)
Department of Family and Community Health Systems, School of Nursing,
University of Texas Health Science Center at San Antonio, Floyd Curl Drive,
San Antonio, TX 78229, USA
e-mail: lapiz@uthscsa.edu

5	Effects of Stress on the Serotonergic System	54
6	Effects of Stress on the Cannabinoid System	57
7	Conclusion	59
References		60

1 Introduction

Stress is a nonspecific response of the body to any demand placed on it (Selye 1936). A stressor therefore is an event or experience that threatens the ability of the individual to adapt and cope. The brain is the central organ responsible for the adaptation to stress. It perceives and determines what is threatening, and orchestrates the behavioral and physiological response to the stressor (McEwen and Gianaros 2011). The stress response involves a coordinated activation of a constellation of physiological systems: an autonomic response and a neuroendocrine response (Hill and Tasker 2012). The autonomic response involves stimulation of the sympathetic motor and hormonal outputs via descending neural circuits originating in hypothalamic preautonomic control centers. The neuroendocrine stress response is mediated by activation of the hypothalamic-pituitary-adrenal (HPA) axis. This response results in an increase of circulating corticosteroids and corticosteroid coordination of activity in multiple target organ systems (Pecoraro et al. 2006).

The general nature of the connectivity of the HPA axis is well-known and an overview is discussed in another chapter of this book (Baumeister et al. 2014). Exposure to a stressor causes the activation of neural inputs to corticotrophin-releasing hormone (CRH) neurons in the hypothalamic paraventricular nucleus (PVN) to release CRH and vasopressin from axonal terminals into the pituitary portal circulation. Subsequently, CRH and vasopressin stimulate cells of the anterior pituitary to produce and release adrenocorticotrophic hormone (ACTH) into the systemic circulation. Circulating ACTH then stimulates the synthesis and secretion of corticosteroids from the cortex of the adrenal glands. Systemic corticosteroids then elicit both rapid and protracted actions in target tissues and organs, including the brain (Pecoraro et al. 2006; Tasker and Herman 2011).

The HPA axis is under negative feedback control by circulating glucocorticoids (Hill and Tasker 2012). This glucocorticoid feedback regulation of the HPA axis can occur directly at the level of the hypothalamus (Evanson et al. 2010) and pituitary (Russell et al. 2010), as well as at upstream limbic structures, such as the hippocampus (Sapolsky et al. 1984; Furay et al. 2008), paraventricular thalamus (Jaferi et al. 2003; Jaferi and Bhatnagar 2006), and prefrontal cortex (Hill et al. 2011; Radley and Sawchenko 2011). Outputs from the prefrontal cortex (PFC) and hippocampus/subiculum comprise excitatory projections from principal neurons that transit to the paraventricular nucleus, and reverse their signal via inhibitory relays in the bed nucleus of the stria terminalis (BNST) and peri-paraventricular hypothalamic regions (Radley and Sawchenko 2011; Ulrich-Lai and Herman 2009). The direct negative feedback actions of glucocorticoids in the PVN and pituitary are inhibitory (Evanson et al. 2010). Interestingly, the involvement of higher limbic structures in the negative glucocorticoid feedback control of the HPA axis appears to be specific response to psychological stressors, and not physiological stressors (Furay et al. 2008).

Stress has been shown to have marked diverse effects on learning and memory which involves specific brain regions (Diamond et al. 2007; Lupien 2009), such as spatial and declarative memory involving the hippocampus, memory of emotional arousing experiences and fear involving the amygdala, and executive functions and fear extinction involving the prefrontal cortex or the PFC. The prefrontal cortex is essential for behavioral adaptation. It is responsible for the inhibition of inappropriate actions and flexible regulation of behavior that enables a proper response to the changes in the environment (Milad et al. 2006; Milad and Quirk 2002).

This chapter aims to address the effects of stress and glucocorticoids on the glutamatergic, monoaminergic (i.e., noradrenergic, dopaminergic, and serotonergic), and cannabinoid systems, with emphasis on the prefrontal cortex. In some instances, effects on other regions such as the hippocampus and amygdala are also mentioned.

2 The Effects of Stress on the Glutamatergic System

Glutamate is the major excitatory neurotransmitter in the brain. A detailed review on the glutamatergic neurotransmission in the nervous system is available elsewhere (Niciu et al. 2013) and will not be discussed here. Glutamate is a key intermediary metabolite in the detoxification of ammonia and a building block in the synthesis of peptides and proteins. Glutamate is present at extremely high concentrations within the cells of the central nervous system. Tight regulatory processes are in place to limit extracellular levels and modulate receptor activity that ensure optimal neurotransmission and prevent against potential excitotoxicity (Niciu et al. 2013). De novo neuronal glutamate is synthesized from glucose via the Krebs cycle and the transamination of a-oxoglutarate (Erecinska and Silver 1990). It can also be recycled through the glutamate-glutamine cycle. Exocytotoxic vesicular release of glutamate underlies the vast majority of excitatory neurotransmission in the brain. This is a strictly regulated process in which the synaptic vesicles that store glutamate merge and then fuse with the presynaptic membrane in response to stimulation. In glutamatergic synapses, presynaptic terminals are normally associated with specialized postsynaptic dendritic spines. Glutamatergic synapses serve as excitatory relay stations between presynaptic nerve terminals and postsynaptic dendritic spines (axo-dendritic synapses) or adjacent nerve endings (axo-axonal synapses) (Niciu et al. 2013).

The core presynaptic machinery for glutamate release is SNARE (soluble N-ethylmaleimide)-sensitive fusion protein attachment protein receptor complex (Lang and Jahn 2008; Sudhof and Rothman 2009). The SNARE complex is formed by the interaction of two synaptic membrane proteins (syntaxin-1 or syntaxin-2 and SNAP-25) and a vesicular protein (synaptobrevin-1 or synaptobrevin-2). It is thought to mediate the fusion of synaptic vesicles with the presynaptic membrane (Sudhof and Rothman 2009).

Glutamate exerts its action through activation of ionotropic and metabotropic glutamate receptors. Three classes of ionotropic glutamate receptors have been identified and named based on their agonist selectivity: N-methyl-D-aspartate (NMDA) and α-amino-3-hydroxy-5-methyl-4-isoxazole proprionic acid (AMPA) and kainate receptors. Ionotropic glutamate receptors form tetrameric complexes of individual/heteromeric subunits. Metabotropic glutamate receptors exert their effects via the recruitment and activation of intracellular trimeric G proteins and downstream signal transduction pathways. There are currently eight metabotropic glutamate receptors identified: mGluR1–8 receptors (Kim et al. 2008).

Glutamate regulates synaptic transmission and plasticity by activating ionotropic glutamate (AMPA and NMDA) and metabotropic glutamate receptors (mGluR1–8). The number and stability of these receptors at the synaptic membrane determine excitatory synaptic efficacy of these receptors. Several mechanisms may control the surface expression of NMDA receptors (NMDARs) and AMPA receptors (AMPARs). As reviewed by Popoli et al. (2011), these mechanisms include PDZ (PDZ-95/Discs-large/ZO-1) domain-mediated interactions between channel subunits and synaptic scaffolding proteins, clathrin-dependent endocytosis regulated by phosphorylation and motor protein-based transport along microtubule or actin cytoskeletons. The Rab family small GTPases, which function as key regulators for all stages of membrane traffic, is involved in the internalization, recycling, and spine delivery of NMDARs and AMPARs. The synthesis and degradation of postsynaptic glutamate receptors are dynamically regulated. Glutamate is cleared from the extracellular space via high-affinity excitatory amino acid transporters (EAATs), which are located on neighboring glial cells (EAAT1–2) and on neurons (EAAT3–5) (O'Shea 2002). In glial cells, glutamate is converted into glutamine by glutamine synthetase. Glutamine is then transported back into the glutamatergic neuron, where it is hydrolyzed into glutamate by glutaminase (Erecinska and Silver 1990). Uptake by EAATs is the primary mechanism through which the action of extracellular glutamate is terminated.

Depending on the type, intensity, and duration of the stressor, stress can have either plasticity-enhancing effects that are associated with improved cognition and function or noxious effects that are associated with impaired function. Studies have elucidated how stress-induced changes in various aspects of glutamate neurotransmission are causally linked to each other and to the glucocorticoid responses to stress. A review (Popoli et al. 2011) provides an in-depth investigation of the nature of the response of the glutamatergic system in response to acute and chronic stress.

Acute stress have been shown to have the general effect of increasing glutamatergic neurotransmission in the PFC and other regions associated with memory, learning, and affect, by inducing both genomic and nongenomic changes at various sites within the glutamatergic synapse (Popoli et al. 2011). Mineralocorticoid or glucocorticoid receptor-mediated effects increase the presynaptic release of glutamate. At the postsynaptic site, acute stress seems to increase the surface expression and density of ionotropic glutamate receptors, resulting in synaptic potentiation, with the mechanism and timing of these effects varying between brain regions.

Acute stress affects glutamate clearance and metabolism through an increased expression of EAAT2 and possibly other glutamate transporters, matching the increased synaptic release of glutamate following acute stress exposure. These changes could possibly contribute to the adaptive stress response on cognitive functions, where moderate acute stress facilitates classical conditioning (Shors et al. 1992), associative learning (Beylin and Shors 2003) and working memory (Yuen et al. 2009).

On the other hand, chronic stress exposure seems to have different effects on the glutamate synapse. Chronic stress has been shown to cause prolonged periods of stimulated glutamate release following acute stress exposure, at least in the hippocampus. This elevated synaptic glutamate activity has been associated with changes in the surface expression of AMPA receptor and NMDA receptor subunits and decreased transmission efficiency and potentially impaired synaptic plasticity. Rodent studies suggest that the PFC may be specifically sensitive to the chronic stress-induced effects on postsynaptic receptor function. Chronic stress has also been shown to have effects on glial cell morphology, metabolism, and function in the PFC and possibly also the hippocampus. These long-lasting chronic stress-induced changes in glutamatergic transmission may be linked to the impairments in spatial and contextual memory performance and attentional control (Liston et al. 2006; McEwen 1999) and the impaired synaptic plasticity in the hippocampus—PFC connection that have been observed in rats after chronic stress (Cerqueira et al. 2007). The decreased ability to clear extracellular glutamate as a result of impaired glial cell uptake and metabolism, combined with stress-induced changes in glutamate release and glutamate receptor function, could provide a pathophysiological mechanism leading to many of the structural changes observed in brain regions of individuals with stress-associated psychiatric disorders, such as mood and anxiety disorders (Popoli et al. 2011).

3 The Effects of Stress on the Noradrenergic System

The noradrenergic (NAergic) system in the brain is considered to play an important role in attention, sleep and wakefulness, learning and memory, emotion, reproduction, and central responses to stress (Berridge and Waterhouse 2003; Sara 2009). The locus coeruleus (LC), a pontine nucleus located near the pontomesencephalic junction, is the largest group of NAergic neurons in the central nervous system (Samuels and Szabadi 2008). There are seven distinctive clusters of NAergic cell bodies (A1–A7), which projects extensively to widespread areas of the brain and spinal cord. Through its action upon corticotrophin-releasing factor neurons in the paraventricular nucleus (PVN) of the hypothalamus (Pacak and Palkovits 2001; Itoi et al. 2004), the NAergic system is able to influence the hypothalamic-pituitary-adrenal stress axis. The NAergic afferents to the paraventricular nucleus originate mainly from the medullary NAergic nuclei and reach the PVN via the ventral NAergic bundle.

The effects of noradrenaline are mediated via two main receptor categories: α- and β-adrenoceptors (Pertovaara 2013; Ruffolo and Hieble 1994). α-Adrenoceptors are classified into subtypes α_{1A}, α_{1B}, α_{1D}, α_{2A}, α_{2B}, and α_{2C}. β-adrenoceptors are classified into subtypes β_1, β_2, and β_3. In general, guanine nucleotide-binding regulatory proteins (G proteins) mediate the actions of adrenoceptors. α_2-Adrenoceptors decrease intracellular adenylcyclase activity through G_i or directly modify activity of ion channels, such as the Na^+/H^+ antiport, Ca^{2+} channels, or K^+ channels (Summers and McMartin 1993). β-Adrenoceptors increase adenylcyclase activity through G_s. α_1-Adrenoceptors are coupled to phospholipase C through G_q or they are coupled directly to Ca^{2+} influx (Summers and McMartin 1993). Adrenoceptors located on the catecholaminergic neurons are considered autoreceptors. α_2-Adrenergic autoreceptors located in the somatodendritic area inhibit impulse discharge of adrenergic neurons and those on axon terminals inhibit the release of the adrenergic neurotransmitter. Adrenoceptors located on nonadrenergic target cells are heteroreceptors that have varying effects depending on the target cell and the subtype of the adrenoceptor.

Research has documented the relationship between stress and locus coeruleus activation. Earlier studies have shown that exposure to stressful stimuli was associated with robust activation of the locus coeruleus in cats (Rasmussen et al. 1986a, b). Exposure to stressors was also associated with markedly increased tyrosine hydroxylase transcripts (Chang et al. 2000) and expression of c-fos, an immediate early gene product (Pirnik et al. 2004) in the locus coeruleus. Interestingly, the stress-induced tyrosine hydroxylase gene expression was influenced by the levels of circulating glucocorticoid (Makino et al. 2002).

Further research is needed to clarify the nature of the neural substrates involved in the stress-induced locus coeruleus activation. Corticotrophin-releasing factor (CRF)-containing neuronal system may participate in this activation (Itoi and Sugimoto 2010). The locus coeruleus receives CRF-immunoreactive afferents. Direct application of CRF to the locus coeruleus has been shown to activate the firing of these neurons (Jedema and Grace 2004). Handling stress-induced noradrenaline release in the prefrontal cortex was inhibited by a CRF receptor antagonist CP-154,526, and an NMDA receptor antagonist CPP, whereas it was potentiated by idazoxan, an alpha-2-adrenergic antagonist (Kawahara et al. 2000). These results suggest that CRF and glutamate may mediate the handling stress-induced prefrontal activation and the alpha-2-adrenergic receptors may inhibit it.

The role of the locus coeruleus in fear and anxiety was first suggested in the 1970s following studies which showed that electrical stimulation of the locus coeruleus resulted in particular behaviors that were observed in fearful or threatening situations in the wild (Redmond et al. 1976). Subsequent studies in rats involved the examination of whether ablation of the locus coeruleus elicits behaviors comparable with those seen in monkeys. However, chemical ablation of the locus coeruleus in rats using the neurotoxin 6-hydroxydipamine (6-OHDA), resulting in depletion of noradrenaline, failed to show any signs of impairment in learning and performance of fear-motivated tasks (Mason and Fibiger 1979). The noradrenaline depleted rats were slower in habituating to novelty. These rats were also more reluctant to leave a familiar place and took longer to consume the food pellets in an unfamiliar place. These data suggest increase in fear following the lesion. This is opposite to the predictions of the fear and anxiety hypothesis derived from the monkey studies (Itoi and Sugimoto 2010). A similar increase in 'neophobia' (i.e., fear in response to novelty) was also observed in subsequent studies from other laboratories (Harro et al. 1995; Lapiz et al. 2001). The discrepancy between the result from monkey and rat studies is yet to be explained, although this could be attributed to species differences.

Studies have supported for the functional involvement of the locus coeruleus in the regulation of the HPA stress axis. 6-OHDA lesion of the locus coeruleus attenuated the plasma ACTH and corticosterone responses induced by acute-restraint stress, suggesting the partial involvement of the locus coeruleus in HPA regulation (Ziegler et al. 1999). On the other hand, inhibition of the locus coeruleus by local infusion of muscimol, a GABAergic agonist, reduced c-fos expression induced by foot shock in the LC, paraventricular nucleus, amygdaloid nuclei, and cingulate cortex (Passerin et al. 2000). These studies support for the role of the LC in positively regulating the HPA stress axis.

Radley et al. (2008) reported that ablation of NAergic inputs to this medial prefrontal cortex resulted in the attenuation of stress-induced c-fos and CRF mRNA expression in the paraventricular nucleus, while the stress-induced c-fos in the medial prefrontal cortex was enhanced. These authors concluded that the noradrenergic inputs originating from the LC to the medial prefrontal cortex maybe inhibitory in nature and thus disinhibitory to the putative gamma-aminobutyric acid or GABAergic inputs to the paraventricular nucleus, leading to the activation of the HPA output.

4 Effects of Stress on the Dopaminergic System

The brain dopaminergic (DAergic) system consists of projections originating from brain areas that synthesize dopamine (DA) for distribution to four axonal pathways, namely, (1) nigrostriatal; (2) mesolimbic; (3) mesocortical; and (4) tuberoinfundibular pathways. These pathways have distinct connections and functions, as per review by Vallone et al. (2000). Projections constituting the nigrostriatal pathway arise from dopamine-synthesizing neurons of the midbrain nucleus, the substantia nigra compacta (SNc) which innervates the dorsal striatum (caudate-putamen). The nigrostriatal pathway is involved in the control of movement and its degeneration causes Parkinson's disease. The mesocortical pathway arises from the ventral tegmental area (VTA) and innervates different regions of the frontal cortex. This pathway seems involved in some aspects of learning and memory (Taghzouti et al. 1991). The mesolimbic pathway originates from the midbrain ventral tegmental area and innervates the ventral striatum (nucleus accumbens), the olfactory tubercle, and parts of the limbic system. It has been implicated in influencing motivated behavior. The tuberoinfundibular pathway arises from cells of the periventricular and arcuate nuclei of the hypothalamus, and is involved in the control of milk production from mammary glands (Doppler 1994).

Dopamine exerts its action by binding to specific membrane receptors. These receptors belong to the family of seven transmembrane domain G protein-coupled receptors. Five distinct dopamine receptors have been isolated, characterized, and subdivided into two subfamilies on the basis of their biochemical and pharmacological properties: D1- and D2-like. The D1-like subfamily comprises D1 and D receptors. The D2-like includes D2-, D3-, and D4- receptors.

The mesocorticolimbic DA system regulates mood, emotional responses, and incentive-based behavior (Doyon et al. 2013; Grace et al. 2007; Schultz 2007). Neurons in the ventral tegmental area (VTA) are the primary source of the mesocorticolimbic DA system. Those neurons project to many cortical and forebrain limbic structures, including the nucleus accumbens, ventral pallidum, amygdala, and the medial prefrontal cortex. The mesocorticolimbic DA system are affected by stress and glucocorticoids released in response to stress. The following section will describe the effects of acute and chronic stress on the dopaminergic system.

Acute stress has been shown to trigger dopamine release in the brain. In humans, dopamine release occurs in the prefrontal cortex in response to psychological stress (Nagano-Saito et al. 2013), in parallel with decreased working memory-related and reward-related prefrontal activation (Ossewaarde et al. 2011; Qin et al. 2009) and impaired working memory performance in males (Schoofs et al. 2013). This suggests that increased prefrontal dopamine secretion may impair, rather than facilitate, prefrontal function during stress. Dopamine release can also occur in the striatum in response to acute psychosocial stress and after stress in the presence of amphetamine (Burghardt et al. 2012; Pruessner et al. 2004; Wand et al. 2007). This stress-induced dopamine increase is positively correlated with the magnitude of salivary cortisol response (Pruessner et al. 2004). It should be noted that the immediate stress-induced dopamine secretion may be influenced not only by glucocorticoids, but also by inputs from brain circuitry evaluating stressors (Joels and Baram 2009; Ulrich-Lai and Herman 2009).

Chronic stress has long-lasting, deleterious effects on dopamine function and dopamine-related behaviors. Previous studies examined the effects of chronic stress using experimental paradigms in rodents such as social defeat (repeated exposure to a dominant aggressor), isolation, restraint, or exposure to aversive odors or environments (Sinclair et al. 2014). Chronic social defeat in adolescent rats induced long-term dopamine-related changes in the PFC in adulthood, such as decreased basal dopamine levels, decreased DA receptor 2 expression, increased DAT binding, increased monoamine oxidase A (MAOA) gene expression, and increased monoamine oxidase A promoter histone acetylation (Marquez et al. 2013; Watt et al. 2009). These rats also exhibited abnormal behavior, such as increased aggression and anxiety-like behaviors (Marquez et al. 2013).

Chronic social defeat in adolescent rats also altered responses to amphetamine, resulting in increased locomotion, decreased corticosterone secretion, decreased medial PFC dopamine levels, increased nucleus accumbens core dopamine levels, and impaired DA receptor 2 downregulation in the nucleus accumbens core (Burke et al. 2010, 2011, 2013). Hence, chronic stress may detrimentally impact the developmental trajectory of dopaminergic circuits, leading to long-term molecular and behavioral maladaptation.

The mechanisms underlying the effects of chronic stress on dopaminergic neurotransmission are not fully understood. However, research suggests that stress may induce changes in neuronal morphology in the PFC, potentially impacting available target sites for incoming dopamine afferents. Interestingly, these effects may be modulated also by sex hormones. For example, 7 days of chronic restraint stress (between PND51 and 58 in males and between PND55 and 62 in females) had gender-specific effects on apical dendritic length of pyramidal neurons in layers II–III in the medial PFC (Garrett and Wellman 2009). Males displayed decreased apical dendritic length after stress, while females displayed increased apical dendritic length which can be ameliorated by ovariectomy and restored by ovariectomy combined with estrogen replacement (Garrett and Wellman 2009).

The effects of chronic stress on dopamine neurotransmission may also arise from control of dopamine-related gene transcription by glucocorticoid receptor in key brain regions. For example, monoamine oxidase A is a glucocorticoid receptor target gene whose expression is rapidly increased by glucocorticoid administration in the adolescent rat hippocampus (Morsink et al. 2006) and is persistently increased, following adolescent chronic stress, in the adult rat PFC (Marquez et al. 2013).

Glucocorticoid receptors within dopamine-responsive neurons in dopamine neuron projection areas mediate the effects of stress on dopamine signaling and cognition in young adult rats (Sinclair et al. 2014). Administration of the glucocorticoid receptor (GR) antagonist RU38486 into the prefrontal cortex, but not the VTA, has been shown to attenuate acute stress-induced dopamine efflux in the prefrontal cortex, which is associated with impairment in working memory impairment in rats (Butts et al. 2011). Similar findings were found in mice, where selective ablation of glucocorticoid receptors in dopamine-responsive neurons of the striatum, nucleus accumbens, and cortex (layers V and VI only) diminished the stress-induced increase in dopamine release in the nucleus accumbens (Barik et al. 2013). Further, this manipulation also abolished the effects of chronic stress on social behaviors and eliminated stress-induced increases in dopamine neuron firing in the ventral tegmental area (Barik et al. 2013). Interestingly, these effects were not shown when the glucocorticoid receptors in dopamine neurons of the substantia nigra and ventral tegmental area were ablated (Barik et al. 2013).

These data support the important role of the glutamatergic system in the body's response to stress, both acute and chronic stressors.

5 Effects of Stress on the Serotonergic System

The brain serotonergic systems control diverse physiologic and behavioral functions including motor control, appetite, sleep-wake cycles, as well as emotional behavior and emotional states (Hale et al. 2012). Majority of the serotonin (5-HT) producing forebrain-projecting neurons are contained in the subregions of the dorsal raphe nucleus and median raphe nucleus. These nuclei have been shown to differentially respond to stress-related stimuli. Comprehensive descriptive reviews of the midbrain and pontine serotonergic systems are available elsewhere (Hale et al. 2012; Lowry et al. 2008) and therefore will not be discussed here.

5-HT exerts its biological activity through interaction with different receptors, currently classified into 7 groups on the basis of their structure, transduction mechanism, and pharmacological profile (Varnas et al. 2004; Stasi et al. 2014): 5-HT1–7. $5\text{-HT}_{1,2,4,5,6,7}$ receptors are coupled to G proteins, while the 5-HT_3 receptor is coupled to an ion channel. The class of 5-HT1 receptors (Pazos et al. 1987a, b) is heterogeneous and includes several subtypes, such as 5-HT_{1A}, 5-HT_{1B}, 5-HT_{1C}, 5-HT_{1E}, 5-HT_{1F}, and 5-HT_1-like. All 5-HT_1 receptor subtypes consist of a single peptide of variable length (from 374 to 421 amino acids), and they share at least 60 % homology in their transmembrane domains. 5-HT_{1A} receptors have a wide distribution in several brain regions involved in the modulation of emotions, such as the hippocampus, septum, dorsal raphe nuclei, and amygdala (Pazos et al. 1987a, b), where they act mainly as inhibitory somatodendritic autoreceptors. However, at limbic level, particularly in the hippocampus, 5-HT1A receptors are located postsynaptically, and here, their activation results in hyperpolarization of somatodendritic neuronal membrane. These receptors have been found also in the neocortex and the gelatinous substance of the spinal cord, which are involved in the regulation of proprioceptive and integrative functions (Pazos et al. 1987a, b).

5-HT_{1B} receptors are predominantly distributed in the striatum of basal ganglia and the prefrontal cortex, where they act as autoreceptors. The 5-HT_{1C} subtype is similar in structure and transduction mechanism to receptors of the 5-HT2 family, and for this reason, it has been renamed 5-HT_{2C}. 5-HT_{1D} receptors display a high degree of homology with 5-HT_{1B} receptors, but they are expressed with lower density. They inhibit neurotransmitter release (Tepper et al. 2002). The highest 5-HT_{1D} receptor densities are found in the raphe nuclei.

The highest densities of 5-HT_{1E} receptor sites have been found in the caudate and putamen (Lowther et al. 1992). 5-HT_{1F} receptors have been identified in the CNS, particularly in the neocortex, where they might contribute to the integration of information associated with limbic functions. 5-HT_1-like receptors are located in the CNS and intracranial vessels, where they inhibit noradrenaline release from sympathetic nerves and vascular smooth muscle cell contraction. 5-HT$_{1P}$ receptors are expressed in the gastrointestinal or GI tract.

There are three known receptor subtypes of 5-HT2 receptor family: 5-HT$_{2A}$, 5-HT$_{2B,}$ and 5-HT$_{2C}$. These receptor subtypes are expressed predominantly in peripheral tissues, such as the stomach, intestine, heart, and kidney. In the CNS, they are found in the cerebellum, lateral septum, hypothalamus, and middle part of the amygdala. 5-HT$_{2A}$ and 5-HT$_{2C}$ receptors are known to mediate the neurochemical and behavioral effects of psychostimulants (Bubar and Cunningham 2006). 5-HT$_{2C}$ receptors are predominantly expressed in epithelial cells of the choroid plexus, cerebral cortex, hippocampus, amygdala, some components of basal ganglia, substantia nigra, substantia innominata, and ventromedial hypothalamus (Pasqualetti et al. 1999).

5-HT$_3$ receptors belong to the ion-channel-linked receptor superfamily, which includes nicotinic, cholinergic, and gamma-aminobutyric acid (GABA) A receptors. They are located in the hippocampus, dorsal motor nucleus of the solitary tract and area postrema. At the CNS level, they are involved in the regulation of emetic responses to various stimuli, including anticancer chemotherapy. The activation of 5-HT$_3$ receptors elicits central effects comparable to those observed after administration of antipsychotic and anxiolytic drugs, due to their ability to modulate the release of other neurotransmitters, such as dopamine, GABA, substance P and acetylcholine (Thompson and Lummis 2007).

5-HT$_4$ receptors are localized in the central nervous system, where it has been suggested they play a role in enhancing memory (Marchetti et al. 2000). 5-HT$_{5A}$ receptors are distributed predominantly in the cortex, hippocampus, hypothalamus, amygdala, and cerebellum. Depending on their localization, 5-HT$_{5A}$ receptors are involved in the regulation of several functions, such as the control of affective states, sensory perception, learning, memory, and neuroendocrine functions (Oliver et al. 2000). 5-HT$_{5B}$ receptors are expressed in mice but not in humans.

5-HT$_6$ receptors are located in the striatum, amygdala, nucleus accumbens, olfactory tubercle and cortex (Marazziti et al. 2013). Many antipsychotic drugs (clozapine, olanzapine, and quetiapine) and antidepressants (clomipramine, amitriptyline, and nortriptyline) act as high-affinity antagonists of 5-HT$_6$ receptors. 5-HT$_7$ receptors are distributed in the limbic system and thalamocortical regions, where they are involved in the modulation of affective states. They are also expressed in smooth muscle cells of peripheral vessels and intestine, where they mediate muscle relaxation (Thomas and Hagan 2004).

Studies suggested that 5-HT$_{1A}$ autoreceptors are an important determinant of basal and stress-induced activity of different subpopulations of serotonergic neurons. 5-HT$_{1A}$ receptor autoinhibition is an important determinant of regional differences in serotonergic neurotransmission. Acute subcutaneous administration of fluoxetine (30 mg/kg) in rats increases extracellular serotonin concentrations in the medial prefrontal cortex, but not the dorsal lateral prefrontal cortex (Beyer and Cremers 2008). However, following pretreatment with the 5-HT$_{1A}$ antagonist, WAY-100635, fluoxetine administration induces a significant, two-fold increase in extracellular serotonin concentrations within the dorsal lateral prefrontal cortex. Further, it potentiated fluoxetine-induced increases in extracellular serotonin concentrations in the medial prefrontal cortex.

The synaptic excitatory and inhibitory inputs to subpopulations of serotonergic neurons, together with 5-HT$_{1A}$ autoreceptor activity, are clearly important determinants of regional differences in basal and stress-induced serotonergic neurotransmission. Other factors are also important, including (1) regional differences in the density of serotonergic nerve terminals, (2) regional differences in tryptophan hydroxylase expression or its activity, (3) regional differences in the density of the presynaptic serotonin transporter at nerve terminals, (4) regional differences in postsynaptic serotonin transporters, such as the corticosterone-sensitive organic cation transporter 3 (OCT3) (Gasser et al. 2006, 2009), and (5) regional differences in the rate of serotonin metabolism (Guptan et al. 1997).

Serotonergic neurons fire at a regular frequency of 3 spikes/second during active waking (Jacobs and Fornal 1999; Sakai and Crochet 2001; Asan et al. 2013). In anesthetized rats, spiking frequencies of 10–14 spikes/10 s were noted in serotonergic dorsal raphe neurons with male rats displaying significantly higher frequencies than freely cycling or ovariectomized female rats (Klink et al. 2002). The regular activity pattern results in a steady serotonergic transmission in target structures including the amygdala during active wakefulness. Further, the steady synaptic concentration of serotonin and a sustained activation of the postsynaptic 5-HT receptors in the target regions may enable active, goal-directed motor and cognitive functions, including response to stress (Klink et al. 2002).

Stress affects the prefrontal serotonergic system. For example, a stressful experience such as restraint in mice induces a time-dependent increase of 5-HT output in the medial prefrontal cortex (Pascucci et al. 2009; Reznikov et al. 2009). The effects of restraint stress were not limited to the prefrontal cortex. It has also been shown to increase GABA output in the basolateral amygdala of the mice (Pascucci et al. 2009; Reznikov et al. 2009). These studies highlighted the major role of medial prefrontal cortex and amygdala in stress-related behaviors (Andolina et al. 2013; Wellman et al. 2007). The medial prefrontal cortex may have a critical role in regulating amygdala-mediated arousal in response to emotionally salient stimuli. Its serotonergic innervation represents a potential molecular mechanism through which the medial prefrontal cortex modulates corticolimbic circuitry. Andolina et al. (2013) proposed that 5-HT transmission in the medial prefrontal cortex may engage the GABAergic transmission in basolateral amygdala in stressful situations in order to determine coping outcomes. Through modulation of GABA in the amygdala prefrontal cortical 5-HT, behavioral responses to stressful experiences are maintained such as sustaining immobility in the forced swimming paradigm that models depressive-like outcomes in rodents (Ebner et al. 2005, 2008; Singewald et al. 2011).

6 Effects of Stress on the Cannabinoid System

First discovered as the biochemical target of delta9-tetrahydrocannabinol (THC), the endocannabinoid system is a lipid signaling system that exerts modulatory actions in both central and peripheral tissues (Pertwee 2008). Two G protein-coupled cannabinoid receptors with distinct patterns of distribution have been characterized: CB1 and CB2 receptors (Matsuda et al. 1990, 1992; Munro et al. 1993). The CB1 receptors are widely expressed throughout the brain and in many peripheral cell types, glands, and organ systems (Herkenham et al. 1991; Bellocchio et al. 2008). The CB2 receptors are traditionally viewed as distributed in the periphery particularly on organs involved in the immune response, such as leukocytes and the spleen (Atwood and Mackie 2010; Patel et al. 2010). However, there is emerging evidence for the expression of CB2 receptors in the brain (Xi et al. 2011; van Sickle et al. 2005).

In the brain, the CB1 receptor is predominantly expressed on axon terminals of a variety of neuronal populations, including glutamatergic, GABAergic, and monoaminergic neurons (Miederer et al. 2013). CB1 receptor signaling suppresses neurotransmitter release into the synapse through suppression of adenylate cyclase activity and calcium influx into the axon terminal. Two endogenous ligands activate the CB1 receptor (also termed as endocannabinoids): *N*-arachidonoyl-ethanolamine (anandamide; AEA) (Devane et al. 1992) and 2-arachidonoyl-glycerol (2-AG) (Sugiura and Waku 2000; Sugiura et al. 2006). These ligands are arachidonate-derived signaling lipids synthesized in the postsynaptic membrane. They are released retrogradely to activate CB1 receptors located in the presynaptic site (Carlson et al. 2002; Wilson and Nicoll 2002). Endocannabinoids are metabolized by specific enzymatic pathways. Fatty acid amide hydrolase (FAAH) is the primary catabolic enzyme of AEA and hydrolyzes AEA into ethanolamine and arachidonic acid (Deutsch et al. 2002; Ueda et al. 2002). 2-AG is primarily metabolized by monoacylglyceride lipase (MAG lipase) to form glycerol and arachidonic acid (Dinh et al. 2002; Ueda et al. 2002). AEA represents a "tonic" signal that gates and regulates transmitter release under steady-state conditions, while 2-AG represents a "phasic" signal that is activated during sustained neuronal depolarization and is involved in many forms of synaptic plasticity (Gorzalka et al. 2008; Hill and Tasker 2012).

The endocannabinoid system is widely distributed throughout the corticolimbic and hypothalamic circuitry that regulates activation of the HPA axis (Hill et al. 2010; Gorzalka et al. 2008). The predominant effects of endocannabinoid signaling are to constrain activation of the HPA axis, although it has been found to also regulate both excitatory and inhibitory transmitter release. Research has identified site-specific roles, and divergent functions of AEA and 2-AG, with respect to HPA axis regulation in the context of basal function, activation in response to stress, and termination during the HPA recovery phase. The roles of the cannabinoid system on stress response have been previously reviewed (Hill and Tasker 2012a;

Hill et al. 2010; Gorzalka et al. 2008; Gorzalka and Hill 2011). A brief synthesis of the literature is presented in the succeeding section.

Under steady-state conditions, the endocannabinoid tone constrains the activation of the HPA axis. However, disruption of this tone causes an increase in HPA axis outflow. For example, acute treatment with a CB1 receptor antagonist to an unstressed rodent increased circulating levels of ACTH and corticosterone in a dose-dependent manner during the *nadir* of the diurnal cycle (Corchero et al. 1999; Wade et al. 2006). This tonic suppression of HPA axis activity may be centrally mediated. The HPA stimulating effects of can be replicated by intracerebroventricular administration of CB1 receptor antagonist (Corchero et al. 1999).

There is evidence that the tonic regulation of the HPA axis by endocannabinoids occurs at an extrahypothalamic site that communicates with the paraventricular nucleus (Hill and Tasker 2012a). For example, there seems to be an endocannabinoid tone within the basolateral amygdala that tonically gates excitation of this structure. The disruption of CB1 receptor function in this region increases its intrinsic excitability. Administration of a CB1 receptor antagonist in the basolateral nucleus of the amygdala increases HPA axis activity in nonstressed animals (Ganon-Elazar and Akirav 2009; Hill et al. 2009), which in turn activates the paraventricular nucleus resulting in an increase in HPA activity. Hence, AEA signaling within the basolateral amygdala may be the "distal gatekeeper" of basal HPA axis activity (Hill and Tasker 2012a).

While AEA signaling within the basolateral amygdala may be the "gatekeeper" of HPA axis activity, natural activation of the HPA axis in response to stress results in a rapid loss of this AEA signal in this region. This modification of AEA signaling following stress exposure facilitates the neuroendocrine response to stress. Specifically, exposure to stress resulted in a reduction in the tissue content of AEA in the amygdala (Rademacher et al. 2008), possibly through a rapid induction of FAAH-mediated AEA hydrolysis (Hill and Tasker 2012a). The magnitude of the decline in AEA content within the amygdala negatively correlated with the extent of HPA axis activation. The larger reductions in amygdala AEA levels in response to stress were associated with greater increases in corticosterone secretion. Furthermore, local administration of a FAAH inhibitor in the basolateral amygdala, but not in the central or medial nuclei, attenuated stress-induced activation of the HPA axis. These data indicated that AEA hydrolysis in the basolateral amygdala in response to stress contributes to activation of the HPA axis (Hill and Tasker 2012a).

There is evidence that the endocannabinoid activity contributes to the feedback inhibition of the HPA axis through a nongenomic glucocorticoid mechanism. Endocannabinoids mediate fast feedback inhibition of the HPA axis by glucocorticoid receptors within the paraventricular nucleus through a mechanism by which glucocorticoids induce endocannabinoid mobilization to suppress excitatory input to corticotrophic neurons. CB1 receptor knock-out mice exhibited a larger peak ACTH and corticosterone response following acute stress, suggesting that a loss of CB1 receptors reduces the fast feedback inhibition and increases the magnitude and duration of the HPA axis response to acute stress (Hill et al. 2011; Barna et al. 2004; Haller et al. 2004; Uriguen et al. 2004; Aso et al. 2008).

Endocannabinoids are also involved in the glucocorticoid feedback inhibition of the HPA axis outside of the paraventricular nucleus, possibly within the prefrontal cortex and hippocampus. Glucocorticoid receptors are abundantly expressed in both these regions (Radley and Sawchenko 2011; Diorio et al. 1993; Herman and Mueller 2006). The prefrontal cortex and hippocampus suppress HPA axis activity through activation of glutamatergic projection neurons to inhibitory relays to the paraventricular nucleus within the bed nucleus of the stria terminalis (Radley and Sawchenko 2011). Glucocorticoids facilitate neuronal activity within the medial prefrontal cortex (Hill et al. 2011; Yuen et al. 2009) and hippocampus (Karst et al. 2005) to suppress HPA axis activity and terminate the stress response. In the prefrontal cortex, a clear role of endocannabinoid signaling in the glucocorticoid-mediated negative feedback inhibition of the HPA axis has been demonstrated. Specifically, exposure to stress was found to increase 2-AG content, but not AEA, within the prefrontal cortex in a glucocorticoid-dependent manner (Hill et al. 2011). This ability of glucocorticoids to increase endocannabinoid content was not rapid and involved genomic actions, as it was blocked by the classical intracellular glucocorticoid receptor antagonist RU-486 (Hill et al. 2011). The ability of endocannabinoids in the prefrontal cortex to contribute to the termination of the stress response appears to be due to modulation of local excitability (Hill and Tasker 2012a). CB1 receptors were found on GABAergic terminals clustered around pyramidal neurons in the prefrontal cortex, and bath application of corticosterone to prefrontal cortical slices resulted in a CB1 receptor-dependent reduction of inhibitory tone in these cells (Hill et al. 2011).

These data demonstrate that endocannabinoids play an important role in glucocorticoid-mediated negative feedback. The endocannabinoid system appears to represent one of the synaptic workhorses of glucocorticoids, bridging postsynaptic effects of glucocorticoids to presynaptic regulation of excitability within a given circuit (Hill et al. 2009; Tasker and Herman 2011).

7 Conclusion

Stress response involves a constellation of physiological responses that promote adaptation with the efficient turning on and off of the response. The activation of the HPA results on the release of glucocorticoids which subsequently influences other brain neurotransmitter and signaling systems including, but not limited, to the prefrontal cortex, amygdala, and hypothalamus. The interactions of these systems with the HPA axis are both complex, and in some instances, reciprocal. They play an integral in the changing expression of the HPA response to a continually changing and often challenging environment. The plasticity of these systems may play an adaptive or maladaptive role in the induction of stress-related pathologies. Further research is needed to fully understand the interactions of these brain neurotransmitter systems and the HPA axis.

Acknowledgments Dr. Lapiz-Bluhm receives funding from the Robert Wood Johnson Nurse Faculty Program. She sincerely thanks Dr. Carrie Jo Braden, Dr. James Michael Bluhm and Dr. Charles Marsden for taking the time to read the manuscript.

References

Andolina D, Maran D, Valzania A, Conversi D, Puglisi-Allegra S (2013) Prefrontal/amygdalar system determines stress coping behavior through 5-HT/GABA connection. Neuropsychopharmacol: Off Publ Am Coll Neuropsychopharmacol 38(10):2057–2067

Asan E, Steinke M, Lesch KP (2013) Serotonergic innervation of the amygdala: targets, receptors, and implications for stress and anxiety. Histochem Cell Biol 139(6):785–813

Aso E, Ozaita A, Valdizan EM, Ledent C, Pazos A, Maldonado R, Valverde O (2008) BDNF impairment in the hippocampus is related to enhanced despair behavior in CB1 knockout mice. J Neurochem 105(2):565–572

Atwood BK, Mackie K (2010) CB2: a cannabinoid receptor with an identity crisis. Br J Pharmacol 160(3):467–479

Barik J, Marti F, Morel C, Fernandez SP, Lanteri C, Godeheu G, Tassin JP, Mombereau C, Faure P, Tronche F (2013) Chronic stress triggers social aversion via glucocorticoid receptor in dopaminoceptive neurons. Science (New York, N.Y.) 339(6117):332–335

Barna I, Zelena D, Arszovszki AC, Ledent C (2004) The role of endogenous cannabinoids in the hypothalamo-pituitary-adrenal axis regulation: in vivo and in vitro studies in CB1 receptor knockout mice. Life Sci 75(24):2959–2970

Baumeister D, Lightman SL, Pariante CM (2014) The interface of stress and the HPA axis in behavioural phenotypes of mental illness. In: Current topics in behavioral neurosciences. doi:10.1007/7854_2014_304

Bellocchio L, Cervino C, Pasquali R, Pagotto U (2008) The endocannabinoid system and energy metabolism. J Neuroendocrinol 20(6):850–857

Berridge CW, Waterhouse BD (2003) The locus coeruleus-noradrenergic system: modulation of behavioral state and state-dependent cognitive processes. Brain Res Brain Res Rev 42(1):33–84

Beyer CE, Cremers TI (2008) Do selective serotonin reuptake inhibitors acutely increase frontal cortex levels of serotonin? Eur J Pharmacol 580(3):350–354

Beylin AV, Shors TJ (2003) Glucocorticoids are necessary for enhancing the acquisition of associative memories after acute stressful experience. Horm Behav 43(1):124–131

Bubar MJ, Cunningham KA (2006) Serotonin 5-HT2A and 5-HT2C receptors as potential targets for modulation of psychostimulant use and dependence. Curr Top Med Chem 6(18):1971–1985

Burghardt PR, Love TM, Stohler CS, Hodgkinson C, Shen PH, Enoch MA, Goldman D, Zubieta JK (2012) Leptin regulates dopamine responses to sustained stress in humans. J Neurosci: Off J Soc Neurosci 32(44):15369–15376

Burke AR, Forster GL, Novick AM, Roberts CL, Watt MJ (2013) Effects of adolescent social defeat on adult amphetamine-induced locomotion and corticoaccumbal dopamine release in male rats. Neuropharmacology 67:359–369

Burke AR, Renner KJ, Forster GL, Watt MJ (2010) Adolescent social defeat alters neural, endocrine and behavioral responses to amphetamine in adult male rats. Brain Res 1352:147–156

Burke AR, Watt MJ, Forster GL (2011) Adolescent social defeat increases adult amphetamine conditioned place preference and alters D2 dopamine receptor expression. Neuroscience 197:269–279

Butts KA, Weinberg J, Young AH, Phillips AG (2011) Glucocorticoid receptors in the prefrontal cortex regulate stress-evoked dopamine efflux and aspects of executive function. Proc Natl Acad Sci U S A 108(45):18459–18464

Carlson G, Wang Y, Alger BE (2002) Endocannabinoids facilitate the induction of LTP in the hippocampus. Nat Neurosci 5(8):723–724

Cerqueira JJ, Mailliet F, Almeida OF, Jay TM, Sousa N (2007) The prefrontal cortex as a key target of the maladaptive response to stress. J Neurosci: Off J Soc Neurosci 27(11):2781–2787

Chang MS, Sved AF, Zigmond MJ, Austin MC (2000) Increased transcription of the tyrosine hydroxylase gene in individual locus coeruleus neurons following footshock stress. Neuroscience 101(1):131–139

Corchero J, Romero J, Berrendero F, Fernandez-Ruiz J, Ramos JA, Fuentes JA, Manzanares J (1999) Time-dependent differences of repeated administration with Delta9-tetrahydrocannabinol in proenkephalin and cannabinoid receptor gene expression and G-protein activation by mu-opioid and CB1-cannabinoid receptors in the caudate-putamen. Brain Res Mol Brain Res 67(1):148–157

Deutsch DG, Ueda N, Yamamoto S (2002) The fatty acid amide hydrolase (FAAH). Prostaglandins Leukot Essent Fatty Acids 66(2–3):201–210

Devane WA, Hanus L, Breuer A, Pertwee RG, Stevenson LA, Griffin G, Gibson D, Mandelbaum A, Etinger A, Mechoulam R (1992) Isolation and structure of a brain constituent that binds to the cannabinoid receptor. Science (New York, N.Y.) 258(5090):1946–1949

Diamond DM, Campbell AM, Park CR, Halonen J, Zoladz PR (2007) The temporal dynamics model of emotional memory processing: a synthesis on the neurobiological basis of stress-induced amnesia, flashbulb and traumatic memories, and the Yerkes-Dodson law. Neural Plast 2007:60803

Dinh TP, Freund TF, Piomelli D (2002) A role for monoglyceride lipase in 2-arachidonoylglycerol inactivation. Chem Phys Lipids 121(1–2):149–158

Diorio D, Viau V, Meaney MJ (1993) The role of the medial prefrontal cortex (cingulate gyrus) in the regulation of hypothalamic-pituitary-adrenal responses to stress. J Neurosci: Off J Soc Neurosci 13(9):3839–3847

Doppler W (1994) Regulation of gene expression by prolactin. Rev Physiol Biochem Pharmacol 124:93–130

Doyon WM, Thomas AM, Ostroumov A, Dong Y, Dani JA (2013) Potential substrates for nicotine and alcohol interactions: a focus on the mesocorticolimbic dopamine system. Biochem Pharmacol 86(8):1181–1193

Ebner K, Bosch OJ, Kromer SA, Singewald N, Neumann ID (2005) Release of oxytocin in the rat central amygdala modulates stress-coping behavior and the release of excitatory amino acids. Neuropsychopharmacol: Off Publ Am Coll Neuropsychopharmacol 30(2):223–230

Ebner K, Singewald GM, Whittle N, Ferraguti F, Singewald N (2008) Neurokinin 1 receptor antagonism promotes active stress coping via enhanced septal 5-HT transmission. Neuropsychopharmacol: Off Publ Am Coll Neuropsychopharmacol 33(8):1929–1941

Erecinska M, Silver IA (1990) Metabolism and role of glutamate in mammalian brain. Prog Neurobiol 35(4):245–296

Evanson NK, Tasker JG, Hill MN, Hillard CJ, Herman JP (2010) Fast feedback inhibition of the HPA axis by glucocorticoids is mediated by endocannabinoid signaling. Endocrinology 151(10):4811–4819

Furay AR, Bruestle AE, Herman JP (2008) The role of the forebrain glucocorticoid receptor in acute and chronic stress. Endocrinology 149(11):5482–5490

Ganon-Elazar E, Akirav I (2009) Cannabinoid receptor activation in the basolateral amygdala blocks the effects of stress on the conditioning and extinction of inhibitory avoidance. J Neurosci: Off J Soc Neurosci 29(36):11078–11088

Garrett JE, Wellman CL (2009) Chronic stress effects on dendritic morphology in medial prefrontal cortex: sex differences and estrogen dependence. Neuroscience 162(1):195–207

Gasser PJ, Lowry CA, Orchinik M (2006) Corticosterone-sensitive monoamine transport in the rat dorsomedial hypothalamus: potential role for organic cation transporter 3 in stress-induced modulation of monoaminergic neurotransmission. J Neurosci: Off J Soc Neurosci 26(34):8758–8766

Gasser PJ, Orchinik M, Raju I, Lowry CA (2009) Distribution of organic cation transporter 3, a corticosterone-sensitive monoamine transporter, in the rat brain. J Comp Neurol 512(4): 529–555

Gorzalka BB, Hill MN (2011) Putative role of endocannabinoid signaling in the etiology of depression and actions of antidepressants. Prog Neuropsychopharmacol Biol Psychiatry 35(7):1575–1585

Gorzalka BB, Hill MN, Hillard CJ (2008) Regulation of endocannabinoid signaling by stress: implications for stress-related affective disorders. Neurosci Biobehav Rev 32(6):1152–1160

Grace AA, Floresco SB, Goto Y, Lodge DJ (2007) Regulation of firing of dopaminergic neurons and control of goal-directed behaviors. Trends Neurosci 30(5):220–227

Guptan P, Dhingra A, Panicker MM (1997) Multiple transcripts encode the 5-HT1F receptor in rodent brain. Neuroreport 8(15):3317–3321

Hale MW, Shekhar A, Lowry CA (2012) Stress-related serotonergic systems: implications for symptomatology of anxiety and affective disorders. Cell Mol Neurobiol 32(5):695–708

Haller J, Varga B, Ledent C, Barna I, Freund TF (2004) Context-dependent effects of CB1 cannabinoid gene disruption on anxiety-like and social behaviour in mice. Eur J Neurosci 19(7):1906–1912

Harro J, Oreland L, Vasar E, Bradwejn J (1995) Impaired exploratory behaviour after DSP-4 treatment in rats: implications for the increased anxiety after noradrenergic denervation. Eur Neuropsychopharmacol: J Eur Coll Neuropsychopharmacol 5(4):447–455

Herkenham M, Lynn AB, Johnson MR, Melvin LS, de Costa BR, Rice KC (1991) Characterization and localization of cannabinoid receptors in rat brain: a quantitative in vitro autoradiographic study. J Neurosci: Off J Soc Neurosci 11(2):563–583

Herman JP, Mueller NK (2006) Role of the ventral subiculum in stress integration. Behav Brain Res 174(2):215–224

Hill MN, Hillard CJ, MCEwen BS (2011) Alterations in corticolimbic dendritic morphology and emotional behavior in cannabinoid CB1 receptor-deficient mice parallel the effects of chronic stress. Cereb cortex (New York, N.Y.: 1991) 21(9):2056–2064

Hill MN, Hunter RG, McEwen BS (2009) Chronic stress differentially regulates cannabinoid CB1 receptor binding in distinct hippocampal subfields. Eur J Pharmacol 614(1–3):66–69

Hill MN, Patel S, Campolongo P, Tasker JG, Wotjak CT, Bains JS (2010) Functional interactions between stress and the endocannabinoid system: from synaptic signaling to behavioral output. J Neurosci: Off J Soc Neurosci 30(45):14980–14986

Hill MN, Tasker JG (2012) Endocannabinoid signaling, glucocorticoid-mediated negative feedback, and regulation of the hypothalamic-pituitary-adrenal axis. Neuroscience 204:5–16

Itoi K, Jiang YQ, Iwasaki Y, Watson SJ (2004) Regulatory mechanisms of corticotropin-releasing hormone and vasopressin gene expression in the hypothalamus. J Neuroendocrinol 16(4): 348–355

Itoi K, Sugimoto N (2010) The brainstem noradrenergic systems in stress, anxiety and depression. J Neuroendocrinol 22(5):355–361

Jacobs BL, Fornal CA (1999) Activity of serotonergic neurons in behaving animals. Neuropsychopharmacol: Off Publ Am Coll Neuropsychopharmacol 21(2 Suppl):9S–15S

Jaferi A, Bhatnagar S (2006) Corticosterone can act at the posterior paraventricular thalamus to inhibit hypothalamic-pituitary-adrenal activity in animals that habituate to repeated stress. Endocrinology 147(10):4917–4930

Jaferi A, Nowak N, Bhatnagar S (2003) Negative feedback functions in chronically stressed rats: role of the posterior paraventricular thalamus. Physiol Behav 78(3):365–373

Jedema HP, Grace AA (2004) Corticotropin-releasing hormone directly activates noradrenergic neurons of the locus ceruleus recorded in vitro. J Neurosci: Off J Soc Neurosci 24(43): 9703–9713

Joels M, Baram TZ (2009) The neuro-symphony of stress. Nat Rev Neurosci 10(6):459–466

Karst H, Berger S, Turiault M, Tronche F, Schutz G, Joels M (2005) Mineralocorticoid receptors are indispensable for nongenomic modulation of hippocampal glutamate transmission by corticosterone. Proc Natl Acad Sci U S A 102(52):19204–19207

Kawahara H, Kawahara Y, Westerink BH (2000) The role of afferents to the locus coeruleus in the handling stress-induced increase in the release of noradrenaline in the medial prefrontal cortex: a dual-probe microdialysis study in the rat brain. Eur J Pharmacol 387(3):279–286

Kim CH, Lee J, Lee JY, Roche KW (2008) Metabotropic glutamate receptors: phosphorylation and receptor signaling. J Neurosci Res 86(1):1–10

Klink R, Robichaud M, Debonnel G (2002) Gender and gonadal status modulation of dorsal raphe nucleus serotonergic neurons. Part II. Regulatory mechanisms. Neuropharmacology 43(7):1129–1138

Lang T, Jahn R (2008) Core proteins of the secretory machinery. Handb Exp Pharmacol 184:107–127

Lapiz MD, Mateo Y, Durkin S, Parker T, Marsden CA (2001) Effects of central noradrenaline depletion by the selective neurotoxin DSP-4 on the behaviour of the isolated rat in the elevated plus maze and water maze. Psychopharmacology 155(3):251–259

Liston C, Miller MM, Goldwater DS, Radley JJ, Rocher AB, Hof PR, Morrison JH, McEwen BS (2006) Stress-induced alterations in prefrontal cortical dendritic morphology predict selective impairments in perceptual attentional set-shifting. J Neurosci: Off J Soc Neurosci 26(30): 7870–7874

Lowry CA, Hale MW, Evans AK, Heerkens J, Staub DR, Gasser PJ, Shekhar A (2008) Serotonergic systems, anxiety, and affective disorder: focus on the dorsomedial part of the dorsal raphe nucleus. Ann N Y Acad Sci 1148:86–94

Lowther S, de Paermentier F, Crompton MR, Horton RW (1992) The distribution of 5-HT1D and 5-HT1E binding sites in human brain. Eur J Pharmacol 222(1):137–142

Lupien SJ (2009) Brains under stress. Can J Psychiatry Revue Can Psychiatr 54(1):4–5

Makino S, Smith MA, Gold PW (2002) Regulatory role of glucocorticoids and glucocorticoid receptor mRNA levels on tyrosine hydroxylase gene expression in the locus coeruleus during repeated immobilization stress. Brain Res 943(2):216–223

Marazziti D, Baroni S, Borsini F, Picchetti M, Vatteroni E, Falaschi V, Catena-Dell'Osso M (2013) Serotonin receptors of type 6 (5-HT6): from neuroscience to clinical pharmacology. Curr Med Chem 20(3):371–377

Marchetti E, Dumuis A, Bockaert J, Soumireu-Mourat B, Roman FS (2000) Differential modulation of the 5-HT(4) receptor agonists and antagonist on rat learning and memory. Neuropharmacology 39(11):2017–2027

Marquez C, Poirier GL, Cordero MI, Larsen MH, Groner A, Marquis J, Magistretti PJ, Trono D, Sandi C (2013) Peripuberty stress leads to abnormal aggression, altered amygdala and orbitofrontal reactivity and increased prefrontal MAOA gene expression. Transl Psychiatry 3:e216

Mason ST, Fibiger HC (1979) Current concepts. I. Anxiety: the locus coeruleus disconnection. Life Sci 25(26):2141–2147

Matsuda LA, Bonner TI, Lolait SJ (1992) Cannabinoid receptors: which cells, where, how, and why? NIDA Res Monogr 126:48–56

Matsuda LA, Lolait SJ, Brownstein MJ, Young AC, Bonner TI (1990) Structure of a cannabinoid receptor and functional expression of the cloned cDNA. Nature 346(6284):561–564

McEwen B (1999) Development of the cerebral cortex: XIII. Stress and brain development: II. J Am Acad Child Adolesc Psychiatry 38(1):101–103

McEwen BS, Gianaros PJ (2011) Stress- and allostasis-induced brain plasticity. Annu Rev Med 62:431–445

Miederer I, Maus S, Zwiener I, Podoprygorina G, Meshcheryakov D, Lutz B, Schreckenberger M (2013) Evaluation of cannabinoid type 1 receptor expression in the rat brain using [(1)(8)F]MK-9470 microPET. Eur J Nucl Med Mol Imag 40(11):1739–1747

Milad MR, Quirk GJ (2002) Neurons in medial prefrontal cortex signal memory for fear extinction. Nature 420(6911):70–74

Milad MR, Rauch SL, Pitman RK, Quirk GJ (2006) Fear extinction in rats: implications for human brain imaging and anxiety disorders. Biol Psychol 73(1):61–71

Morsink MC, Joels M, Sarabdjitsingh RA, Meijer OC, de Kloet ER, Datson NA (2006) The dynamic pattern of glucocorticoid receptor-mediated transcriptional responses in neuronal PC12 cells. J Neurochem 99(4):1282–1298

Munro S, Thomas KL, Abu-Shaar M (1993) Molecular characterization of a peripheral receptor for cannabinoids. Nature 365(6441):61–65

Nagano-Saito A, Dagher A, Booij L, Gravel P, Welfeld K, Casey KF, Leyton M, Benkelfat C (2013) Stress-induced dopamine release in human medial prefrontal cortex–18F-fallypride/PET study in healthy volunteers. Synapse (New York, N.Y.) 67(12):821–830

Niciu MJ, Ionescu DF, Richards EM, Zarate CA Jr (2013) Glutamate and its receptors in the pathophysiology and treatment of major depressive disorder. J Neural Transm. doi:10.1007/s00702-013-1130-x

Oliver KR, Kinsey AM, Wainwright A, Sirinathsinghji DJ (2000) Localization of 5-ht(5A) receptor-like immunoreactivity in the rat brain. Brain Res 867(1–2):131–142

O'Shea RD (2002) Roles and regulation of glutamate transporters in the central nervous system. Clin Exp Pharmacol Physiol 29(11):1018–1023

Ossewaarde L, Qin S, van Marle HJ, van Wingen GA, Fernandez G, Hermans EJ (2011) Stress-induced reduction in reward-related prefrontal cortex function. NeuroImage 55(1):345–352

Pacak K, Palkovits M (2001) Stressor specificity of central neuroendocrine responses: implications for stress-related disorders. Endocr Rev 22(4):502–548

Pascucci T, Andolina D, Mela IL, Conversi D, Latagliata C, Ventura R, Puglisi-Allegra S, Cabib S (2009) 5-Hydroxytryptophan rescues serotonin response to stress in prefrontal cortex of hyperphenylalaninaemic mice. Int J Neuropsychopharmacol/official scientific journal of the Collegium Internationale Neuropsychopharmacologicum (CINP) 12(8):1067–1079

Pasqualetti M, Ori M, Castagna M, Marazziti D, Cassano GB, Nardi I (1999) Distribution and cellular localization of the serotonin type 2C receptor messenger RNA in human brain. Neuroscience 92(2):601–611

Passerin AM, Cano G, Rabin BS, Delano BA, Napier JL, Sved AF (2000) Role of locus coeruleus in foot shock-evoked Fos expression in rat brain. Neuroscience 101(4):1071–1082

Patel KD, Davison JS, Pittman QJ, Sharkey KA (2010) Cannabinoid CB(2) receptors in health and disease. Curr Med Chem 17(14):1393–1410

Pazos A, Probst A, Palacios JM (1987a) Serotonin receptors in the human brain—III. Autoradiographic mapping of serotonin-1 receptors. Neuroscience 21(1):97–122

Pazos A, Probst A, Palacios JM (1987b) Serotonin receptors in the human brain—IV. Autoradiographic mapping of serotonin-2 receptors. Neuroscience 21(1):123–139

Pecoraro N, Dallman MF, Warne JP, Ginsberg AB, Laugero KD, la Fleur SE, Houshyar H, Gomez F, Bhargava A, Akana SF (2006) From Malthus to motive: how the HPA axis engineers the phenotype, yoking needs to wants. Prog Neurobiol 79(5–6):247–340

Pertovaara A (2013) The noradrenergic pain regulation system: a potential target for pain therapy. Eur J Pharmacol 716(1–3):2–7

Pertwee RG (2008) Ligands that target cannabinoid receptors in the brain: from THC to anandamide and beyond. Addict Biol 13(2):147–159

Pirnik Z, Mravec B, Kiss A (2004) Fos protein expression in mouse hypothalamic paraventricular (PVN) and supraoptic (SON) nuclei upon osmotic stimulus: colocalization with vasopressin, oxytocin, and tyrosine hydroxylase. Neurochem Int 45(5):597–607

Popoli M, Yan Z, McEwen BS, Sanacora G (2011) The stressed synapse: the impact of stress and glucocorticoids on glutamate transmission. Nat Rev Neurosci 13(1):22–37

Pruessner JC, Champagne F, Meaney MJ, Dagher A (2004) Dopamine release in response to a psychological stress in humans and its relationship to early life maternal care: a positron emission tomography study using [11C]raclopride. J Neurosci: Off J Soc Neurosci 24(11):2825–2831

Qin S, Hermans EJ, van Marle HJ, Luo J, Fernandez G (2009) Acute psychological stress reduces working memory-related activity in the dorsolateral prefrontal cortex. Biol Psychiatry 66(1):25–32

Rademacher DJ, Meier SE, Shi L, Ho WS, Jarrahian A, Hillard CJ (2008) Effects of acute and repeated restraint stress on endocannabinoid content in the amygdala, ventral striatum, and medial prefrontal cortex in mice. Neuropharmacology 54(1):108–116

Radley JJ, Sawchenko PE (2011) A common substrate for prefrontal and hippocampal inhibition of the neuroendocrine stress response. J Neurosci: Off J Soc Neurosci 31(26):9683–9695

Radley JJ, Williams B, Sawchenko PE (2008) Noradrenergic innervation of the dorsal medial prefrontal cortex modulates hypothalamo-pituitary-adrenal responses to acute emotional stress. J Neurosci: Off J Soc Neurosci 28(22):5806–5816

Rasmussen K, Morilak DA, Jacobs BL (1986a) Single unit activity of locus coeruleus neurons in the freely moving cat. I. During naturalistic behaviors and in response to simple and complex stimuli. Brain Res 371(2):324–334

Rasmussen K, Strecker RE, Jacobs BL (1986b) Single unit response of noradrenergic, serotonergic and dopaminergic neurons in freely moving cats to simple sensory stimuli. Brain Res 369(1–2):336–340

Redmond DE Jr, Huang YH, Snyder DR, Maas JW (1976) Behavioral effects of stimulation of the nucleus locus coeruleus in the stump-tailed monkey Macaca arctoides. Brain Res 116(3):502–510

Reznikov LR, Reagan LP, Fadel JR (2009) Effects of acute and repeated restraint stress on GABA efflux in the rat basolateral and central amygdala. Brain Res 1256:61–68

Ruffolo RR Jr, Hieble JP (1994) Alpha-adrenoceptors. Pharmacol Ther 61(1–2):1–64

Russell GM, Henley DE, Leendertz J, Douthwaite JA, Wood SA, Stevens A, Woltersdorf WW, Peeters BW, Ruigt GS, White A, Veldhuis JD, Lightman SL (2010) Rapid glucocorticoid receptor-mediated inhibition of hypothalamic-pituitary-adrenal ultradian activity in healthy males. J Neurosci: Off J Soc Neurosci 30(17):6106–6115

Sakai K, Crochet S (2001) Differentiation of presumed serotonergic dorsal raphe neurons in relation to behavior and wake-sleep states. Neuroscience 104(4):1141–1155

Samuels ER, Szabadi E (2008) Functional neuroanatomy of the noradrenergic locus coeruleus: its roles in the regulation of arousal and autonomic function part I: principles of functional organisation. Curr Neuropharmacol 6(3):235–253

Sapolsky RM, Krey LC, McEwen BS, Rainbow TC (1984) Do vasopressin-related peptides induce hippocampal corticosterone receptors? Implications for aging. J Neurosci: Off J Soc Neurosci 4(6):1479–1485

Sara SJ (2009) The locus coeruleus and noradrenergic modulation of cognition. Nat Rev Neurosci 10(3):211–223

Schoofs D, Pabst S, Brand M, Wolf OT (2013) Working memory is differentially affected by stress in men and women. Behav Brain Res 241:144–153

Schultz W (2007) Behavioral dopamine signals. Trends Neurosci 30(5):203–210

Selye H (1936) A syndrome produced by diverse nocuous agents. Nature 138:32

Shors TJ, Weiss C, Thompson RF (1992) Stress-induced facilitation of classical conditioning. Science (New York, N.Y.) 257(5069):537–539

Sinclair D, Purves-Tyson TD, Allen KM, Weickert CS (2014) Impacts of stress and sex hormones on dopamine neurotransmission in the adolescent brain. Psychopharmacology 231(8):1581–1599

Singewald GM, Rjabokon A, Singewald N, Ebner K (2011) The modulatory role of the lateral septum on neuroendocrine and behavioral stress responses. Neuropsychopharmacol: Off Publ Am Coll Neuropsychopharmacol 36(4):793–804

Stasi C, Bellini M, Bassotti G, Blandizzi C, Milani S (2014) Serotonin receptors and their role in the pathophysiology and therapy of irritable bowel syndrome. Tech Coloproctol 18(17):613–621

Sudhof TC, Rothman JE (2009) Membrane fusion: grappling with SNARE and SM proteins. Science (New York, N.Y.) 323(5913):474–477

Sugiura T, Kishimoto S, Oka S, Gokoh M (2006) Biochemistry, pharmacology and physiology of 2-arachidonoylglycerol, an endogenous cannabinoid receptor ligand. Prog Lipid Res 45(5):405–446

Sugiura T, Waku K (2000) 2-Arachidonoylglycerol and the cannabinoid receptors. Chem Phys Lipids 108(1–2):89–106

Summers RJ, McMartin LR (1993) Adrenoceptors and their second messenger systems. J Neurochem 60(1):10–23

Taghzouti K, le Moal M, Simon H (1991) Suppression of noradrenergic innervation compensates for behavioral deficits induced by lesion of dopaminergic terminals in the lateral septum. Brain Res 552(1):124–128

Tasker JG, Herman JP (2011) Mechanisms of rapid glucocorticoid feedback inhibition of the hypothalamic-pituitary-adrenal axis. Stress (Amsterdam, Netherlands) 14(4):398–406

Tepper SJ, Rapoport AM, Sheftell FD (2002) Mechanisms of action of the 5-HT1B/1D receptor agonists. Arch Neurol 59(7):1084–1088

Thomas DR, Hagan JJ (2004) 5-HT7 receptors. Curr Drug Targets CNS Neurol Disord 3(1):81–90

Thompson AJ, Lummis SC (2007) The 5-HT3 receptor as a therapeutic target. Expert Opin Ther Targets 11(4):527–540

Ueda N, Katayama K, Goparaju SK, Kurahashi Y, Yamanaka K, Suzuki H, Yamamoto S (2002) Catalytic properties of purified recombinant anandamide amidohydrolase. Adv Exp Med Biol 507:251–256

Ulrich-Lai YM, Herman JP (2009) Neural regulation of endocrine and autonomic stress responses. Nat Rev Neurosci 10(6):397–409

Uriguen L, Perez-Rial S, Ledent C, Palomo T, Manzanares J (2004) Impaired action of anxiolytic drugs in mice deficient in cannabinoid CB1 receptors. Neuropharmacology 46(7):966–973

Vallone D, Picetti R, Borrelli E (2000) Structure and function of dopamine receptors. Neurosci Biobehav Rev 24(1):125–132

van Sickle MD, Duncan M, Kingsley PJ, Mouihate A, Urbani P, Mackie K, Stella N, Makriyannis A, Piomelli D, Davison JS, Marnett LJ, di Marzo V, Pittman QJ, Patel KD, Sharkey KA (2005) Identification and functional characterization of brainstem cannabinoid CB2 receptors. Science (New York, N.Y.) 310(5746):329–332

Varnas K, Halldin C, Hall H (2004) Autoradiographic distribution of serotonin transporters and receptor subtypes in human brain. Hum Brain Mapp 22(3):246–260

Wade MR, Degroot A, Nomikos GG (2006) Cannabinoid CB1 receptor antagonism modulates plasma corticosterone in rodents. Eur J Pharmacol 551(1–3):162–167

Wand GS, Oswald LM, McCaul ME, Wong DF, Johnson E, Zhou Y, Kuwabara H, Kumar A (2007) Association of amphetamine-induced striatal dopamine release and cortisol responses to psychological stress. Neuropsychopharmacol: Off Publ Am Coll Neuropsychopharmacol 32(11):2310–2320

Watt MJ, Burke AR, Renner KJ, Forster GL (2009) Adolescent male rats exposed to social defeat exhibit altered anxiety behavior and limbic monoamines as adults. Behav Neurosci 123(3):564–576

Wellman CL, Izquierdo A, Garrett JE, Martin KP, Carroll J, Millstein R, Lesch KP, Murphy DL, Holmes A (2007) Impaired stress-coping and fear extinction and abnormal corticolimbic morphology in serotonin transporter knock-out mice. J Neurosci: Off J Soc Neurosci 27(3):684–691

Wilson RI, Nicoll RA (2002) Endocannabinoid signaling in the brain. Science (New York, N.Y.) 296(5568):678–682

Xi ZX, Peng XQ, Li X, Song R, Zhang HY, Liu QR, Yang HJ, Bi GH, Li J, Gardner EL (2011) Brain cannabinoid CB(2) receptors modulate cocaine's actions in mice. Nat Neurosci 14(9):1160–1166

Yuen EY, Liu W, Karatsoreos IN, Feng J, McEwen BS, Yan Z (2009) Acute stress enhances glutamatergic transmission in prefrontal cortex and facilitates working memory. Proc Natl Acad Sci U S A 106(33):14075–14079

Ziegler DR, Cass WA, Herman JP (1999) Excitatory influence of the locus coeruleus in hypothalamic-pituitary-adrenocortical axis responses to stress. J Neuroendocrinol 11(5):61–369

Interaction of Stress, Corticotropin-Releasing Factor, Arginine Vasopressin and Behaviour

Eléonore Beurel and Charles B. Nemeroff

Abstract Stress mediates the activation of a variety of systems ranging from inflammatory to behavioral responses. In this review we focus on two neuropeptide systems, corticotropin-releasing factor (CRF) and arginine vasopressin (AVP), and their roles in regulating stress responses. Both peptides have been demonstrated to be involved in anxiogenic and depressive effects, actions mediated in part through their regulation of the hypothalamic-pituitary-adrenal axis and the release of adrenocorticotropic hormone. Because of the depressive effects of CRF and AVP, drugs modifying the stress-associated detrimental actions of CRF and AVP are under development, particularly drugs antagonizing CRF and AVP receptors for therapy in depression.

Keywords Stress · Corticotropin-releasing factor · Arginine vasopressin · Behaviour

Contents

1 Stress and Its Consequences on Behaviour	68
2 HPA Axis	69
2.1 Corticotropin-Releasing Factor System	70
2.2 Arginine Vasopressin System	72
References	74

E. Beurel · C. B. Nemeroff (✉)
Department of Psychiatry and Behavioral Sciences,
Leonard M. Miller School of Medicine, University of Miami,
Miami, FL 33136, USA
e-mail: cnemeroff@med.miami.edu

1 Stress and Its Consequences on Behaviour

Although stress is now considered a common component of life in modern societies (Joels and Baram 2009), its definition remains somewhat vague. Stress is generally considered to involve external challenges to the organism, which can include psychogenic stressors that may be actual or potential adverse situations, as well as physical stressors (e.g. immune challenge, hypovolemia or cold exposure) (Dayas et al. 2001; Pacak and Palkovits 2001). Although these disparate stressors activate different brain circuits, adaptive responses to these stressors often include similar mediators. In the short term, the organism tends to adapt to the stress to maintain homeostasis, for example by eliminating the challenge or by avoidance (McEwen 1998, 2007). Over time, maintaining physiological stability becomes more difficult. It is now well-established that exposure to extraordinary levels of stress, chronic stress or repeated exposures to stress can markedly increase vulnerability to serious mental illness, and cardiovascular disorders (Rosengren et al. 2004).

This subject is a vast one with entire volumes and meeting proceedings dedicated to it. Instead of trying to cover stress neurobiology in any comprehensive manner, we focus on two neuropeptide systems, corticotropin-releasing factor (CRF) and arginine vasopressin (AVP). Nevertheless, it is important to note that two major systems have long been known to play prominent roles in mediating the stress response: the hypothalamic-pituitary-adrenal (HPA) axis (Herman and Cullinan 1997) and the sympatho-adrenal-medullary system. Thus, hypothalamic and extra-hypothalamic CRF is the preeminent example of a stress-related neuropeptide system that promotes withdrawal and attenuates appetitive behaviors, while there is evidence that neuopeptide Y (NPY) exerts the opposite effect. CRF is thought to mediate the acute stress response in cooperation with AVP (Gillies et al. 1982; Jaferi and Bhatnagar 2007; Lightman 2008; Ma et al. 1997; van Gaalen et al. 2002). The latter appears to be contributing to the long term stress response which likely leads to depression (Dinan and Scott 2005). It is important to note in any discussion of stress that different individuals encounter different magnitudes of stress exposures and the perception of stress varies significantly from individual to individual. Two divergent hypotheses have been proposed to explain the variable outcomes of stress in different individuals (Nederhof and Schmidt 2012). The first one states that stress exposure early in life increases the risk of vulnerability to detrimental stress responses later in life (McEwen 1998; Heim et al. 2008). In contrast, the second hypothesis focuses on resilience, suggesting that repeated exposures to adverse situations during development can be beneficial by promoting resilience even if the environment remains aversive (Schmidt 2011). Most studies in laboratory animals have focused on vulnerability rather than resilience (Veenema et al. 2008; Zobel et al. 2000) and have been interpreted from the point of view that the molecular modifications that ensue in response to stress result from changes in vulnerability. This is at least in part due to the difficulty of distinguishing resilient animals from controls (Schmidt et al. 2010;

Stedenfeld et al. 2011). However, resilience mechanisms are now the focus of considerable investigation (Bilbo et al. 2008; Champagne et al. 2008) because they represent an innovative approach to both understanding pathophysiology as well as drug development for a range of stress-related syndromes.

Many behaviors that are assessed in rodents in response to stress have been interpreted to resemble symptoms exhibited by patients with post-traumatic stress disorder (PTSD) or major depressive disorder (MDD). Although emotional and psychological stress are difficult to evaluate in rodents, a variety of stressors have been shown to induce "depressive-like behavior". These behaviors include loss of enjoyment (anhedonia), loss of motivation, sleep disturbances, deficient sociability skills, anxiety, changes in appetitive behavior, or cognitive deficits, which have all been associated with prolonged stress exposure. For example, anhedonia, learned helplessness, and sociability deficiencies in animal models have been induced by a variety of stressors, such as chronic restraint stress, in which rodents are immobilized repeatedly for hours in a tube, the learned helplessness paradigm, where rodents receive inescapable footshock, the chronic social defeat paradigm, where rodents are repeatedly exposed to aggression by dominant animals, or chronic unpredictable stress, where rodents receive different (heterotypic) stressors every day. A number of neurobiological consequences of chronic stress have been observed including dysregulation of the HPA axis, reduced hippocampal neurogenesis and reduction of brain-derived neurotrophic factor (BDNF), which is required for synaptogenesis (Maras and Baram 2012). The composition of the microbiota of the gut is also affected by the HPA axis through the release of stress hormones and the sympatho-adrenal medullary system (Collins et al. 2012; Dinan and Cryan 2012). The microbiota is a major regulator of the immune system and the immune system has now been unequivocally shown to be altered in patients with mood disorders. Indeed, administration of a low dose of the inflammatory stimulant lipopolysaccharide (LPS) is sufficient to induce sickness behavior, which shares many characteristics with human major depressive behavior. We review here the involvement of the HPA and the sympatho-adrenal system in stress related disorders.

2 HPA Axis

Activation of the HPA axis in response to stress results in widespread hormonal, neurochemical and physiological alterations (Herman and Cullinan 1997). Activation of the HPA axis is mediated by the release of neuropeptides, including CRF and vasopressin into the hypothalamo-hypophyseal portal system, which stimulates the release from the anterior pituitary of adrenocorticotropic hormone (ACTH). ACTH in turn promotes the synthesis and secretion of glucocorticoids from the adrenal cortex (Aguilera 1994; Antoni 1986a). Thus, glucocorticoids (cortisol in humans, and corticosterone in most rodents) are released upon activation of the HPA axis. Glucocorticoid receptors or mineralocorticoid receptors, both of which are activated by glucocorticoids, are expressed in several brain regions (e.g. prefrontal cortex, amygdala, hippocampus and other limbic and midbrain structures). They are steroid receptors that function as transcription factors that regulate cell function even after acute stress is terminated. The magnitude, type, and duration of the stress are important in determining the HPA axis response. The HPA axis has been most scrutinized in PTSD and MDD. Thus, elevated plasma glucocorticoid concentrations have been observed in patients with MDD, particularly those with more severe and/or psychotic symptoms; in contrast a small population of MDD patients show reduced levels of glucocorticoids, which seems to be associated with milder symptoms (Stetler and Miller 2011). In PTSD, in contrast, a tendency for lower levels of glucocorticoids has been reported, but these findings are also mixed (Meewisse et al. 2007). These discordant findings are undoubtedly in part due to the confounding effects of child abuse and neglect on HPA axis activity in adulthood (Heim et al. 2008). This concatenation of findings renders difficult a comprehensive understanding of the role of glucocorticoids in the development of stress-related disorders. It is important to note the broad effects on the brain of glucocorticoids, which are released peripherally in response to stress, which contrasts with the local release of neurotransmitters and neuropeptides that provide a more restricted synaptic modulation. Thus, increased cerebrospinal fluid (CSF) levels of the neuropeptide CRF seem to correlate more closely than do glucocorticoid levels with stress-related disorders (Heim et al. 2000, 2008; Yehuda et al. 2005).

2.1 Corticotropin-Releasing Factor System

Corticotropin-Releasing Factor (CRF), a 41 amino acids peptide was discovered in 1981, and since then three related ligands and two receptors have been identified. The canonical role of CRF is to initiate the endocrine response to stress by releasing ACTH from the anterior pituitary. This neuropeptide is released from cell bodies within the hypothalamic paraventricular nucleus (PVN) to activate the HPA axis (Korosi and Baram 2008), but neurons express CRF in several extra-hypothalamic brain regions (amygdala, cerebrocortical areas, septum, hippocampus) where they play a key role in the autonomic, immune and behavioral effects of stress (Chen et al. 2000, 2001, 2004; Korosi and Baram 2008; Sawchenko et al. 1993). CRF is also expressed in the periphery (blood vessels, skin, lung, testes, ovaries or placenta). Its three related ligands, urocortin 1, urocortin 2 (stresscopin-related peptide) and urocortin 3 (stresscopin) are also expressed both in the periphery and in the brain. Although urocortin 1 and urocortin 2 share a hypothalamic distribution with CRF, urocortin 3 seems to have minimal overlapping expression with CRF (Hauger et al. 2003). CRF and urocortin 1 both bind preferentially to CRF-R1 receptors, whereas urocortins 1, 2 and 3 bind to CRF-R2 receptors with a high affinity. CRF-R1 is expressed mainly in the brain (Swanson et al. 1983), while CRF-R2 is expressed mainly in the periphery. CRF-R1 and

CRF-R2 have 70 % amino-acid sequence homology, but diverge greatly in their N-terminal sequences and belong to the class B1 of 7-transmembrane G-protein coupled receptors. CRF receptors also regulate a diverse group of other intracellular signaling pathways that involve intracellular effectors such as cAMP and an array of protein kinases. This allows them to exert unique roles in the integration of homeostatic mechanisms. It is thought that CRF-R1 principally mediates the stress response. The CRF system is also regulated by a CRF-binding protein (CRFBP), which is highly conserved and present in the circulation as a 37 kDa glycoprotein that binds CRF and related peptides, reducing their availability.

Thus, the CRF system has a multitude of physiological functions, all related to the orchestration of the stress response. CRF stimulates ACTH synthesis and release in the pituitary, thus controlling the HPA axis, but also activates the noradrenergic and sympathetic systems. Locally, CRF regulates adrenal steroidogenesis and catecholamine synthesis in the adrenal gland. In addition, CRF acts in limbic areas in modulating alertness and fear, and appetite and libido, all dysregulated in depressive disorders. Direct regional brain-specific injections of CRF in rodents promotes anxiety, reduces slow wave sleep, is associated with psychomotor alterations (less time spent in the center of an open field) (Sutton et al. 1982), increased grooming and anhedonia (Dunn et al. 1987; Heinrichs et al. 1991), enhanced novelty-suppressed feeding (Britton et al. 1982), decreased appetite and libido, and decreased exploratory behavior (Berridge and Dunn 1989). These effects of CRF are not mediated by HPA axis activation. This was confirmed using transgenic mouse models where CRF was either knocked out (Muller et al. 2003; Smith et al. 1998; Timpl et al. 1998) or overexpressed, and by using selective CRF receptor antagonists (Steckler and Holsboer 1999).

The role of the CRF system in depression has been supported by clinical studies showing that depressed patients have higher CSF concentrations of CRF (Nemeroff et al. 1984), and depressed patients who died by suicide exhibit increased expression of CRF mRNA in the hypothalamus and PFC (Austin et al. 2003; Merali et al. 2004; Nemeroff et al. 1988; Raadsheer et al. 1994) as well as a reduction in CRF receptor binding density (Owens et al. 1991) and CRF receptor mRNA expression (Merali et al. 2004). Moreover, CSF concentrations of CRF are reduced by electroconvulsive therapy (Nemeroff et al. 1991) and antidepressant treatments (De Bellis et al. 1993; Heuser et al. 1998; Veith et al. 1993). Early relapse of depression is also associated with elevated concentrations of CSF CRF (Banki et al. 1992). Altogether, these findings as well as the neuroendocrine data clearly suggest an overactive CRF/CRF-R1 system in depressed patients (Merali et al. 2004; Nemeroff et al. 1988).

These findings supported the development of CRF receptor antagonists as a new therapeutic strategy for depression (Grigoriadis 2005). Small molecule inhibitors of CRF-R1 have been developed as potential therapies (Holsboer and Ising 2008), because CRF has a 15 times higher affinity for CRF-R1, than CRF-R2. Some CRF-R1 antagonists have been tested clinically, and although there is some evidence for anti-anxiety and antidepressant effects in a few studies without evidence of adverse effects (Ising et al. 2007; Kunzel et al. 2003; Nickel et al. 2003; Zobel et al. 2000), the results of the randomized controlled studies have been very disappointing (for review Brothers et al. 2012, Koshimizu et al. 2012). Unfortunately none of the studies enriched their studies with patients who hypersecreted CRF.

2.2 Arginine Vasopressin System

(AVP) and oxytocin are cyclic nonapeptides. There are two major AVP systems in the brain: one responsible for AVP-dependent actions on blood pressure and water conservation, comprising the magnocellular neurons in the paraventricular (PVN) and supraoptic nuclei secreting AVP and oxytocin into the peripheral circulation from the posterior pituitary. The second is responsible for the regulation of the HPA axis via the PVN secreting AVP into the hypothalamo-hypophyseal portal circulation (Aguilera and Rabadan-Diehl 2000; Antoni 1993). AVP-expressing neurons in the amygdala also influence memory and behavior (Alescio-Lautier et al. 2000; Caffe et al. 1987), and in the suprachiasmatic nucleus, AVP regulates circadian rhythms (Arima et al. 2002; Kalsbeek et al. 2010; Li et al. 2009). The actions of AVP are mediated through two main G-protein coupled receptors: V1 receptors (V1a and V1b) are coupled to phospholipase C, which increases intracellular Ca^{2+} and protein kinase C activity (Jard et al. 1987), and V2 receptors are coupled to the adenylyl cyclase/protein kinase A pathway to regulate water homeostasis in the kidney (Frank and Landgraf 2008). The mitogen activated protein kinase (MAPK)/extracellular signal-regulated kinase (ERK) and the phosphatidylinositol 3 kinase (PI3 K)/Akt pathways are also regulated by AVP during neuronal development and survival, synaptic plasticity and memory formation (Chen and Aguilera 2010; Chen et al. 2008, 2009; de Wied et al. 1993). In addition to protecting neurons against apoptosis, AVP inhibits the production of the pro-inflammatory cytokines interleukin-1 and tumor necrosis factor-α in astrocytes, therefore providing another mechanism to protect neurons (Zhao and Brinton 2004).

Using a variety of experimental approaches, it has been clearly shown that AVP is anxiogenic (Neumann and Landgraf 2012). These approaches include central or peripheral administration of V1 receptor antagonists, siRNA, knockout mice, and adenoviral overexpression of V1 receptors (Landgraf 2006; Mak et al. 2012; Pitkow et al. 2001; Ring 2005; Ryckmans 2010; Simon et al. 2008). Hyperactivity of the AVP system shifts behavior towards hyper-anxiety and passive coping. Indeed, some of the untoward consequences of early-life stress appear to be mediated by AVP (Murgatroyd et al. 2010; Veenema et al. 2006).

Because of the close association of anxiety and depression, AVP has been suggested to mediate both conditions. CNS AVP circuits also promote depressive behavior in rats, and these effects are blocked by the administration of antidepressants (Keck et al. 2003). In postmortem tissue of depressed patients, an increase in mRNA expression of AVP and V1 receptors was observed, as well as an increase in the number of PVN neurons expressing AVP (Bao and Swaab 2010;

Wang et al. 2008). It is also important to note that AVP augments the effects of CRF on ACTH release from the anterior pituitary, thereby increasing HPA axis activity (Holsboer et al. 1984). Thus, this may contribute to the hypercortisolemia observed in depression. These anxiogenic and fear effects are thought to act upon a specific population of neurons in the central amygdala in rats (Huber et al. 2005). Intranasal injection of AVP modulates neurons in the prefrontal cortex-amygdala regions, which are thought to mediate threat perception, social behavior, anxiety, and fear processing (Zink et al. 2010).

AVP also regulates affiliative behaviors (Winslow et al. 1993), in particular paternal behaviors in voles, such as crouching over and contacting or grooming pups (Wang et al. 1994). AVP is important in a variety of species for partner preference and pair bonding (Donaldson and Young 2008; Lim and Young 2004) and is thought to influence social memory in males (Ferguson et al. 2002; Lim and Young 2004). AVP also promotes inter-male aggression (Caldwell et al. 2008) and maternal aggression (Bosch and Neumann 2010).

Intranasal administration of AVP has been shown in men to facilitate the encoding of facial identification (Guastella et al. 2010), to have sex-specific influences on social communication, in particular regarding aggression (Thompson et al. 2006). As in animals, AVP promotes stress responses in humans, increasing the cortisol response to social stressors (Shalev et al. 2011). However, the mechanism whereby AVP affects human behaviors remains unknown (McCall and Singer 2012).

Therefore, targeting the AVP system may open new therapeutic avenues. For example, there is an antagonist of V1 receptors (SSR149415) that has shown anxiolytic, antidepressant and anti-stress effects (Griebel et al. 2002; Hodgson et al. 2007; Iijima and Chaki 2007; Litvin et al. 2011; Overstreet and Griebel 2005; Shimazaki et al. 2006; Simon et al. 2008; Stemmelin et al. 2005; Urani et al. 2011). Unfortunately, the clinical trials in depression have been unsuccessful (for review Brothers et al. 2012, Koshimizu et al. 2012). However, SSR149415 also binds the oxytocin receptor (OXTR) (selectivity ratio of 3.2 V1b/OXTR) (Antoni 1986b; Chadio and Antoni 1989; Griffante et al. 2005; Samson and Schell 1995), which explains certain of the effects of this antagonist; oxytocin is known to antagonize the effects of AVP in anxiety and depression (Neumann and Landgraf 2012). Other V1b receptor antagonists are currently under study. The subjacent strategy is to promote the oxytocin system, which has been shown to exert opposite actions of AVP in anxiety and depression by modulating different neuronal circuitry. Although in development, no lipophilic oxytocin receptor agonists have yet to be developed.

Acknowledgements The authors apologize to the many investigators whose work could not be cited due to the limited number of references permitted. Research in the authors laboratories were supported by NIH grants MH090236 and MH094759.

Financial Disclosures EB report no financial interests or potential conflicts of interest. CBN reports the following. Research/Grants: National Institutes of Health (NIH). Speakers Bureau: None. Consulting: Xhale, Takeda, SK Pharma, Shire, Roche, Lilly, Allergan, Mitsubishi Tanabe

Pharma. Development America, Taisho Pharmaceutical Inc., Lundbeck. Stockholder: CeNeRx BioPharma, PharmaNeuroBoost, Revaax Pharma, Xhale. Other Financial Interests: CeNeRx BioPharma, PharmaNeuroBoost. Patents: Method and devices for transdermal delivery of lithium (US 6,375,990B1), Method of assessing antidepressant drug therapy via transport inhibition of monoamine neurotransmitters by ex vivo assay (US 7,148,027B2). Scientific Advisory Boards: American Foundation for Suicide Prevention (AFSP), CeNeRx BioPharma (2012), National. Alliance for Research on Schizophrenia and Depression (NARSAD), Xhale, PharmaNeuroBoost. (2012) Anxiety Disorders Association of America (ADAA), Skyland Trail. Board of Directors: AFSP, NovaDel (2011), Skyland Trail, Gratitude America, ADAA. Income sources or equity of $10,000 or more: PharmaNeuroBoost, CeNeRx BioPharma, NovaDel Pharma, Reevax Pharma, American. Psychiatric Publishing, Xhale. Honoraria: Various. Royalties: Various. Expert Witness: Various.

References

Aguilera G (1994) Regulation of pituitary ACTH secretion during chronic stress. Front Neuroendocrinol 15:321–350

Aguilera G, Rabadan-Diehl C (2000) Vasopressinergic regulation of the hypothalamic-pituitary-adrenal axis: implications for stress adaptation. Regul Pept 96:23–29

Alescio-Lautier B, Paban V, Soumireu-Mourat B (2000) Neuromodulation of memory in the hippocampus by vasopressin. Eur J Pharmacol 405:63–72

Antoni FA (1986a) Hypothalamic control of adrenocorticotropin secretion: advances since the discovery of 41-residue corticotropin-releasing factor. Endocr Rev 7:351–378

Antoni FA (1986b) Oxytocin receptors in rat adenohypophysis: evidence from radioligand binding studies. Endocrinology 119:2393–2395

Antoni FA (1993) Vasopressinergic control of pituitary adrenocorticotropin secretion comes of age. Front Neuroendocrinol 14:76–122

Arima H, House SB, Gainer H, Aguilera G (2002) Neuronal activity is required for the circadian rhythm of vasopressin gene transcription in the suprachiasmatic nucleus in vitro. Endocrinology 143:4165–4171

Austin MC, Janosky JE, Murphy HA (2003) Increased corticotropin-releasing hormone immunoreactivity in monoamine-containing pontine nuclei of depressed suicide men. Mol Psychiatry 8:324–332

Banki CM, Karmacsi L, Bissette G, Nemeroff CB (1992) CSF corticotropin-releasing hormone and somatostatin in major depression: response to antidepressant treatment and relapse. Eur neuropsychopharmacol 2:107–113 the journal of the European College of Neuropsychopharmacology

Bao AM, Swaab DF (2010) Corticotropin-releasing hormone and arginine vasopressin in depression focus on the human postmortem hypothalamus. Vitam Horm 82:339–365

Berridge CW, Dunn AJ (1989) CRF and restraint-stress decrease exploratory behavior in hypophysectomized mice. Pharmacol Biochem Behav 34:517–519

Bilbo SD, Yirmiya R, Amat J, Paul ED, Watkins LR, Maier SF (2008) Bacterial infection early in life protects against stressor-induced depressive-like symptoms in adult rats. Psychoneuroendocrinology 33:261–269

Bosch OJ, Neumann ID (2010) Vasopressin released within the central amygdala promotes maternal aggression. Eur J Neurosci 31:883–891

Britton DR, Koob GF, Rivier J, Vale W (1982) Intraventricular corticotropin-releasing factor enhances behavioral effects of novelty. Life Sci 31:363–367

Brothers SP, Wahlestedt C, Nemeroff CB (2012) Modulation of HPA axis function for treatment of mood disorders. In: Z Rankovic, M Bingham, EJ Nestler, R Hargreaves (eds) RSC drug discovery series no. 28, drug discovery for psychiatric disorders, The royal society of chemistry 2012

Caffe AR, van Leeuwen FW, Luiten PG (1987) Vasopressin cells in the medial amygdala of the rat project to the lateral septum and ventral hippocampus. J Comp Neurol 261:237–252

Caldwell HK, Lee HJ, Macbeth AH, Young WS 3rd (2008) Vasopressin: behavioral roles of an "original" neuropeptide. Prog Neurobiol 84:1–24

Chadio SE, Antoni FA (1989) Characterization of oxytocin receptors in rat adenohypophysis using a radioiodinated receptor antagonist peptide. J Endocrinol 122:465–470

Champagne DL, Bagot RC, van Hasselt F, Ramakers G, Meaney MJ, de Kloet ER, Joels M, Krugers H (2008) Maternal care and hippocampal plasticity: evidence for experience-dependent structural plasticity, altered synaptic functioning, and differential responsiveness to glucocorticoids and stress. J Neurosci 28:6037–6045

Chen J, Aguilera G (2010) Vasopressin protects hippocampal neurones in culture against nutrient deprivation or glutamate-induced apoptosis. J Neuroendocrinol 22:1072–1081

Chen J, Liu Y, Soh JW, Aguilera G (2009) Antiapoptotic effects of vasopressin in the neuronal cell line H32 involve protein kinase Calpha and beta. J Neurochem 110:1310–1320

Chen J, Young S, Subburaju S, Sheppard J, Kiss A, Atkinson H, Wood S, Lightman S, Serradeil-Le Gal C, Aguilera G (2008) Vasopressin does not mediate hypersensitivity of the hypothalamic pituitary adrenal axis during chronic stress. Ann N Y Acad Sci 1148:349–359

Chen Y, Bender RA, Frotscher M, Baram TZ (2001) Novel and transient populations of corticotropin-releasing hormone-expressing neurons in developing hippocampus suggest unique functional roles: a quantitative spatiotemporal analysis. J Neurosci 21:7171–7181

Chen Y, Brunson KL, Adelmann G, Bender RA, Frotscher M, Baram TZ (2004) Hippocampal corticotropin releasing hormone: pre- and postsynaptic location and release by stress. Neuroscience 126:533–540

Chen Y, Brunson KL, Muller MB, Cariaga W, Baram TZ (2000) Immunocytochemical distribution of corticotropin-releasing hormone receptor type-1 [CRF(1)]-like immunoreactivity in the mouse brain: light microscopy analysis using an antibody directed against the C-terminus. J Comp Neurol 420:305–323

Collins SM, Surette M, Bercik P (2012) The interplay between the intestinal microbiota and the brain. Nat Rev Microbiol 10:735–742

Dayas CV, Buller KM, Crane JW, Xu Y, Day TA (2001) Stressor categorization: acute physical and psychological stressors elicit distinctive recruitment patterns in the amygdala and in medullary noradrenergic cell groups. Eur J Neurosci 14:1143–1152

De Bellis MD, Gold PW, Geracioti TD Jr, Listwak SJ, Kling MA (1993) Association of fluoxetine treatment with reductions in CSF concentrations of corticotropin-releasing hormone and arginine vasopressin in patients with major depression. Am j psychiatry 150:656–657

de Wied D, Diamant M, Fodor M (1993) Central nervous system effects of the neurohypophyseal hormones and related peptides. Front Neuroendocrinol 14:251–302

Dinan TG, Cryan JF (2012) Regulation of the stress response by the gut microbiota: implications for psychoneuroendocrinology. Psychoneuroendocrinology 37:1369–1378

Dinan TG, Scott LV (2005) Anatomy of melancholia: focus on hypothalamic-pituitary-adrenal axis overactivity and the role of vasopressin. J Anat 207:259–264

Donaldson ZR, Young LJ (2008) Oxytocin, vasopressin, and the neurogenetics of sociality. Science 322:900–904

Dunn AJ, Berridge CW, Lai YI, Yachabach TL (1987) CRF-induced excessive grooming behavior in rats and mice. Peptides 8:841–844

Ferguson JN, Young LJ, Insel TR (2002) The neuroendocrine basis of social recognition. Front Neuroendocrinol 23:200–224

Frank E, Landgraf R (2008) The vasopressin system–from antidiuresis to psychopathology. Eur J Pharmacol 583:226–242

Gillies GE, Linton EA, Lowry PJ (1982) Corticotropin releasing activity of the new CRF is potentiated several times by vasopressin. Nature 299:355–357

Griebel G, Simiand J, Serradeil-Le Gal C, Wagnon J, Pascal M, Scatton B, Maffrand JP, Soubrie P (2002) Anxiolytic- and antidepressant-like effects of the non-peptide vasopressin V1b receptor antagonist, SSR149415, suggest an innovative approach for the treatment of stress-related disorders. Proc Nat Acad Sci USA 99:6370–6375

Griffante C, Green A, Curcuruto O, Haslam CP, Dickinson BA, Arban R (2005) Selectivity of d(Cha4)AVP and SSR149415 at human vasopressin and oxytocin receptors: evidence that SSR149415 is a mixed vasopressin V1b/oxytocin receptor antagonist. Br J Pharmacol 146:744–751

Grigoriadis DE (2005) The corticotropin-releasing factor receptor: a novel target for the treatment of depression and anxiety-related disorders. Expert Opin Ther Targets 9:651–684

Guastella AJ, Kenyon AR, Alvares GA, Carson DS, Hickie IB (2010) Intranasal arginine vasopressin enhances the encoding of happy and angry faces in humans. Biol Psychiatry 67:1220–1222

Hauger RL, Grigoriadis DE, Dallman MF, Plotsky PM, Vale WW, Dautzenberg FM (2003) International union of pharmacology XXXVI. Current status of the nomenclature for receptors for corticotropin-releasing factor and their ligands. Pharmacol Rev 55:21–26

Heim C, Newport DJ, Miller AH, Nemeroff CB (2000) Long-term neuroendocrine effects of childhood maltreatment. JAMA J Am Med Assoc 284:2321

Heim C, Newport DJ, Mletzko T, Miller AH, Nemeroff CB (2008) The link between childhood trauma and depression: insights from HPA axis studies in humans. Psychoneuroendocrinology 33:693–710

Heinrichs SC, Britton KT, Koob GF (1991) Both conditioned taste preference and aversion induced by corticotropin-releasing factor. Pharmacol Biochem Behav 40:717–721

Herman JP, Cullinan WE (1997) Neurocircuitry of stress: central control of the hypothalamo-pituitary-adrenocortical axis. Trends Neurosci 20:78–84

Heuser I, Bissette G, Dettling M, Schweiger U, Gotthardt U, Schmider J, Lammers CH, Nemeroff CB, Holsboer F (1998) Cerebrospinal fluid concentrations of corticotropin-releasing hormone, vasopressin, and somatostatin in depressed patients and healthy controls: response to amitriptyline treatment. Depress Anxiety 8:71–79

Hodgson RA, Higgins GA, Guthrie DH, Lu SX, Pond AJ, Mullins DE, Guzzi MF, Parker EM, Varty GB (2007) Comparison of the V1b antagonist, SSR149415, and the CRF1 antagonist, CP-154,526, in rodent models of anxiety and depression. Pharmacol Biochem Behav 86:431–440

Holsboer F, Gerken A, Steiger A, Benkert O, Müller OA, Stalla GK (1984) Corticotropin-releasing-factor induced pituitary-adrenal response in depression. Lancet 1:55

Holsboer F, Ising M (2008) Central CRH system in depression and anxiety–evidence from clinical studies with CRH1 receptor antagonists. Eur J Pharmacol 583:350–357

Huber D, Veinante P, Stoop R (2005) Vasopressin and oxytocin excite distinct neuronal populations in the central amygdala. Science 308:245–248

Iijima M, Chaki S (2007) An arginine vasopressin V1b antagonist, SSR149415 elicits antidepressant-like effects in an olfactory bulbectomy model. Prog Neuropsychopharmacol Biol Psychiatry 31:622–627

Ising M, Zimmermann US, Kunzel HE, Uhr M, Foster AC, Learned-Coughlin SM, Holsboer F, Grigoriadis DE (2007) High-affinity CRF1 receptor antagonist NBI-34041: preclinical and clinical data suggest safety and efficacy in attenuating elevated stress response. Neuropsychopharmacology 32:1941–1949

Jaferi A, Bhatnagar S (2007) Corticotropin-releasing hormone receptors in the medial prefrontal cortex regulate hypothalamic-pituitary-adrenal activity and anxiety-related behavior regardless of prior stress experience. Brain Res 1186:212–223

Jard S, Barberis C, Audigier S, Tribollet E (1987) Neurohypophyseal hormone receptor systems in brain and periphery. Prog Brain Res 72:173–187

Joels M, Baram TZ (2009) The neuro-symphony of stress. Nat Rev Neurosci 10:459–466

Kalsbeek A, Fliers E, Hofman MA, Swaab DF, Buijs RM (2010) Vasopressin and the output of the hypothalamic biological clock. J Neuroendocrinol 22:362–372

Keck ME, Welt T, Muller MB, Uhr M, Ohl F, Wigger A, Toschi N, Holsboer F, Landgraf R (2003) Reduction of hypothalamic vasopressinergic hyperdrive contributes to clinically relevant behavioral and neuroendocrine effects of chronic paroxetine treatment in a psychopathological rat model. Neuropsychopharmacology 28:235–243

Korosi A, Baram TZ (2008) The central corticotropin releasing factor system during development and adulthood. Eur J Pharmacol 583:204–214

Koshimizu TA, Nakamura K, Egashira N, Hiroyama M, Nonoguchi H, Tanoue A (2012) Vasopressin V1a and V1b receptors: from molecules to physiological systems. Physiol Rev 92:1813–1864

Kunzel HE, Zobel AW, Nickel T, Ackl N, Uhr M, Sonntag A, Ising M, Holsboer F (2003) Treatment of depression with the CRH-1-receptor antagonist R121919: endocrine changes and side effects. J Psychiatr Res 37:525–533

Landgraf R (2006) The involvement of the vasopressin system in stress-related disorders. CNS Neurol Disord Drug Targets 5:167–179

Li JD, Burton KJ, Zhang C, Hu SB, Zhou QY (2009) Vasopressin receptor V1a regulates circadian rhythms of locomotor activity and expression of clock-controlled genes in the suprachiasmatic nuclei. Am J Physiol Regul Integr Comp Physiol 296:R824–R830

Lightman SL (2008) The neuroendocrinology of stress: a never ending story. J Neuroendocrinol 20:880–884

Lim MM, Young LJ (2004) Vasopressin-dependent neural circuits underlying pair bond formation in the monogamous prairie vole. Neuroscience 125:35–45

Litvin Y, Murakami G, Pfaff DW (2011) Effects of chronic social defeat on behavioral and neural correlates of sociality: vasopressin, oxytocin and the vasopressinergic V1b receptor. Physiol Behav 103:393–403

Ma XM, Levy A, Lightman SL (1997) Emergence of an isolated arginine vasopressin (AVP) response to stress after repeated restraint: a study of both AVP and corticotropin-releasing hormone messenger ribonucleic acid (RNA) and heteronuclear RNA. Endocrinology 138:4351–4357

Mak P, Broussard C, Vacy K, Broadbear JH (2012) Modulation of anxiety behavior in the elevated plus maze using peptidic oxytocin and vasopressin receptor ligands in the rat. J psychopharmacol 26:532–542

Maras PM, Baram TZ (2012) Sculpting the hippocampus from within: stress, spines, and CRH. Trends Neurosci 35:315–324

McCall C, Singer T (2012) The animal and human neuroendocrinology of social cognition, motivation and behavior. Nat Neurosci 15:681–688

McEwen BS (1998) Stress, adaptation, and disease: allostasis and allostatic load. Ann N Y Acad Sci 840:33–44

McEwen BS (2007) Physiology and neurobiology of stress and adaptation: central role of the brain. Physiol Rev 87:873–904

Meewisse ML, Reitsma JB, de Vries GJ, Gersons BP, Olff M (2007) Cortisol and post-traumatic stress disorder in adults: systematic review and meta-analysis. Br j psychiatry 191:387–392

Merali Z, Du L, Hrdina P, Palkovits M, Faludi G, Poulter MO, Anisman H (2004) Dysregulation in the suicide brain: mRNA expression of corticotropin-releasing hormone receptors and GABA(A) receptor subunits in frontal cortical brain region. J Neurosci 24:1478–1485

Muller MB, Zimmermann S, Sillaber I, Hagemeyer TP, Deussing JM, Timpl P, Kormann MS, Droste SK, Kuhn R, Reul JM et al (2003) Limbic corticotropin-releasing hormone receptor 1 mediates anxiety-related behavior and hormonal adaptation to stress. Nat Neurosci 6:1100–1107

Murgatroyd C, Wu Y, Bockmuhl Y, Spengler D (2010) Genes learn from stress: how infantile trauma programs us for depression. Epigenetics 5(3)

Nederhof E, Schmidt MV (2012) Mismatch or cumulative stress: toward an integrated hypothesis of programming effects. Physiol Behav 106:691–700

Nemeroff CB, Bissette G, Akil H, Fink M (1991) Neuropeptide concentrations in the cerebrospinal fluid of depressed patients treated with electroconvulsive therapy: corticotrophin-releasing factor, beta-endorphin and somatostatin. Br J Psychiatry 158:59–63

Nemeroff CB, Owens MJ, Bissette G, Andorn AC, Stanley M (1988) Reduced corticotropin releasing factor binding sites in the frontal cortex of suicide victims. Arch Gen Psychiatry 45:577–579

Nemeroff CB, Widerlov E, Bissette G, Walleus H, Karlsson I, Eklund K, Kilts CD, Loosen PT, Vale W (1984) Elevated concentrations of CSF corticotropin-releasing factor-like immunoreactivity in depressed patients. Science 226:1342–1344

Neumann ID, Landgraf R (2012) Balance of brain oxytocin and vasopressin: implications for anxiety, depression, and social behaviors. Trends Neurosci 35:649–659

Nickel T, Sonntag A, Schill J, Zobel AW, Ackl N, Brunnauer A, Murck H, Ising M, Yassouridis A, Steiger A et al (2003) Clinical and neurobiological effects of tianeptine and paroxetine in major depression. J Clin Psychopharmacol 23:155–168

Overstreet DH, Griebel G (2005) Antidepressant-like effects of the vasopressin V1b receptor antagonist SSR149415 in the flinders sensitive line rat. Pharmacol Biochem Behav 82:223–227

Owens MJ, Overstreet DH, Knight DL, Rezvani AH, Ritchie JC, Bissette G, Janowsky DS, Nemeroff CB (1991) Alterations in the hypothalamic-pituitary-adrenal axis in a proposed animal model of depression with genetic muscarinic supersensitivity. Neuropsychopharmacol 4:87–93

Pacak K, Palkovits M (2001) Stressor specificity of central neuroendocrine responses: implications for stress-related disorders. Endocr Rev 22:502–548

Pitkow LJ, Sharer CA, Ren X, Insel TR, Terwilliger EF, Young LJ (2001) Facilitation of affiliation and pair-bond formation by vasopressin receptor gene transfer into the ventral forebrain of a monogamous vole. J Neurosci 21:7392–7396

Raadsheer FC, Hoogendijk WJ, Stam FC, Tilders FJ, Swaab DF (1994) Increased numbers of corticotropin-releasing hormone expressing neurons in the hypothalamic paraventricular nucleus of depressed patients. Neuroendocrinology 60:436–444

Ring RH (2005) The central vasopressinergic system: examining the opportunities for psychiatric drug development. Curr Pharm Des 11:205–225

Rosengren A, Hawken S, Ounpuu S, Sliwa K, Zubaid M, Almahmeed WA, Blackett KN, Sitthi-amorn C, Sato H, Yusuf S (2004) Association of psychosocial risk factors with risk of acute myocardial infarction in 11119 cases and 13648 controls from 52 countries (the INTERHEART study): case-control study. Lancet 364:953–962

Ryckmans T (2010) Modulation of the vasopressin system for the treatment of CNS diseases. Curr Opin Drug Discov Devel 13:538–547

Samson WK, Schell DA (1995) Oxytocin and the anterior pituitary gland. Adv Exp Med Biol 395:355–364

Sawchenko PE, Imaki T, Potter E, Kovacs K, Imaki J, Vale W (1993) The functional neuroanatomy of corticotropin-releasing factor. In: Ciba Foundation symposium, vol 172, pp 5–21, discussion 21–29

Schmidt MV (2011) Animal models for depression and the mismatch hypothesis of disease. Psychoneuroendocrinology 36:330–338

Schmidt MV, Scharf SH, Sterlemann V, Ganea K, Liebl C, Holsboer F, Muller MB (2010) High susceptibility to chronic social stress is associated with a depression-like phenotype. Psychoneuroendocrinology 35:635–643

Shalev I, Israel S, Uzefovsky F, Gritsenko I, Kaitz M, Ebstein RP (2011) Vasopressin needs an audience: neuropeptide elicited stress responses are contingent upon perceived social evaluative threats. Horm Behav 60:121–127

Shimazaki T, Iijima M, Chaki S (2006) The pituitary mediates the anxiolytic-like effects of the vasopressin V1B receptor antagonist, SSR149415, in a social interaction test in rats. Eur J Pharmacol 543:63–67

Simon NG, Guillon C, Fabio K, Heindel ND, Lu SF, Miller M, Ferris CF, Brownstein MJ, Garripa C, Koppel GA (2008) Vasopressin antagonists as anxiolytics and antidepressants: recent developments. Recent Pat CNS Drug Discovery 3:77–93

Smith GW, Aubry JM, Dellu F, Contarino A, Bilezikjian LM, Gold LH, Chen R, Marchuk Y, Hauser C, Bentley CA et al (1998) Corticotropin releasing factor receptor 1-deficient mice display decreased anxiety, impaired stress response, and aberrant neuroendocrine development. Neuron 20:1093–1102

Steckler T, Holsboer F (1999) Corticotropin-releasing hormone receptor subtypes and emotion. Biol Psychiatry 46:1480–1508

Stedenfeld KA, Clinton SM, Kerman IA, Akil H, Watson SJ, Sved AF (2011) Novelty-seeking behavior predicts vulnerability in a rodent model of depression. Physiol Behav 103:210–216

Stemmelin J, Lukovic L, Salome N, Griebel G (2005) Evidence that the lateral septum is involved in the antidepressant-like effects of the vasopressin V1b receptor antagonist, SSR149415. Neuropsychopharmacology 30:35–42

Stetler C, Miller GE (2011) Depression and hypothalamic-pituitary-adrenal activation: a quantitative summary of four decades of research. Psychosom Med 73:114–126

Sutton RE, Koob GF, Le Moal M, Rivier J, Vale W (1982) Corticotropin releasing factor produces behavioural activation in rats. Nature 297:331–333

Swanson LW, Sawchenko PE, Rivier J, Vale WW (1983) Organization of ovine corticotropin-releasing factor immunoreactive cells and fibers in the rat brain: an immunohistochemical study. Neuroendocrinology 36:165–186

Thompson RR, George K, Walton JC, Orr SP, Benson J (2006) Sex-specific influences of vasopressin on human social communication. Proc Natl Acad Sci USA 103:7889–7894

Timpl P, Spanagel R, Sillaber I, Kresse A, Reul JM, Stalla GK, Blanquet V, Steckler T, Holsboer F, Wurst W (1998) Impaired stress response and reduced anxiety in mice lacking a functional corticotropin-releasing hormone receptor 1. Nat Genet 19:162–166

Urani A, Philbert J, Cohen C, Griebel G (2011) The corticotropin-releasing factor 1 receptor antagonist, SSR125543, and the vasopressin 1b receptor antagonist, SSR149415, prevent stress-induced cognitive impairment in mice. Pharmacol Biochem Behav 98:425–431

van Gaalen MM, Stenzel-Poore MP, Holsboer F, Steckler T (2002) Effects of transgenic overproduction of CRH on anxiety-like behaviour. Eur J Neurosci 15:2007–2015

Veenema AH, Blume A, Niederle D, Buwalda B, Neumann ID (2006) Effects of early life stress on adult male aggression and hypothalamic vasopressin and serotonin. Eur J Neurosci 24:1711–1720

Veenema AH, Reber SO, Selch S, Obermeier F, Neumann ID (2008) Early life stress enhances the vulnerability to chronic psychosocial stress and experimental colitis in adult mice. Endocrinology 149:2727–2736

Veith RC, Lewis N, Langohr JI, Murburg MM, Ashleigh EA, Castillo S, Peskind ER, Pascualy M, Bissette G, Nemeroff CB et al (1993) Effect of desipramine on cerebrospinal fluid concentrations of corticotropin-releasing factor in human subjects. Psychiatry Res 46:1–8

Wang SS, Kamphuis W, Huitinga I, Zhou JN, Swaab DF (2008) Gene expression analysis in the human hypothalamus in depression by laser microdissection and real-time PCR: the presence of multiple receptor imbalances. Mol Psychiatry 13:786–799

Wang Z, Ferris CF, De Vries GJ (1994). Role of septal vasopressin innervation in paternal behavior in prairie voles (*Microtus ochrogaster*). Proc Nat Acad Sci USA 91:400–404

Winslow JT, Hastings N, Carter CS, Harbaugh CR, Insel TR (1993) A role for central vasopressin in pair bonding in monogamous prairie voles. Nature 365:545–548

Yehuda R, Golier JA, Kaufman S (2005) Circadian rhythm of salivary cortisol in holocaust survivors with and without PTSD. Am J Psychiatry 162:998–1000

Zhao L, Brinton RD (2004) Suppression of proinflammatory cytokines interleukin-1beta and tumor necrosis factor-alpha in astrocytes by a V1 vasopressin receptor agonist: a cAMP response element-binding protein-dependent mechanism. J Neurosci 24:2226–2235

Zink CF, Stein JL, Kempf L, Hakimi S, Meyer-Lindenberg A (2010) Vasopressin modulates medial prefrontal cortex-amygdala circuitry during emotion processing in humans. J Neurosci 30:7017–7022

Zobel AW, Nickel T, Kunzel HE, Ackl N, Sonntag A, Ising M, Holsboer F (2000) Effects of the high-affinity corticotropin-releasing hormone receptor 1 antagonist R121919 in major depression: the first 20 patients treated. J Psychiatr Res 34:171–181

Long-lasting Consequences of Early Life Stress on Brain Structure, Emotion and Cognition

Harm J. Krugers and Marian Joëls

Abstract During the perinatal period, the brain undergoes substantial structural changes, synaptic rearrangements, and development of neuronal circuits which ultimately determine brain function and behavior. Environmental factors—such as exposure to adverse experiences—have major impact on brain function and structure during this sensitive period. These alterations can be long-lasting, and have been implicated in psychopathology such as cognitive decline and emotional dysfunction. Here we briefly review how early postnatal adversity determines structure and function of the hippocampus, amygdala, and prefrontal cortex (PFC) areas, which are crucial for proper cognitive and emotional function.

Keywords Early life adversity · Stress · Hippocampus · Amygdala

Contents

1 Introduction	82
2 Early Life Experience and Stress-Responsiveness	83
3 Early Life Adversity and Cognitive and Emotional Function	84
4 Early Life Adversity and Neuronal Structure	86
5 Early Life Adversity and Synaptic Function	87
6 Discussion	87
References	89

H. J. Krugers (✉)
Swammerdam Institute for Life Sciences, Center for Neuroscience,
University of Amsterdam, Science Park 904, 1098 XH Amsterdam, The Netherlands
e-mail: h.krugers@uva.nl

M. Joëls
Department of Translational Neuroscience, Brain Center Rudolf Magnus,
University Medical Center Utrecht, 3584 CG Utrecht, The Netherlands

1 Introduction

Several studies provide evidence that the early postnatal period is a sensitive period for proper development of neuronal function and programming of behavior (Hackman et al. 2010). This postnatal context in which individuals develop is dependent on many social and economic factors and several studies indicate that this socioeconomic status is associated with brain development and cognitive function (Hackman et al. 2010). For example, the socioeconomic context in which brain development occurs has been correlated with effects on language and executive function (Hackman et al. 2010; Noble et al. 2007; Farah et al. 2006; Kishiyama et al. 2009).

During the early postnatal period, individuals are particularly dependent on parental care, which is important for cognitive development in human offspring (Hackman et al. 2010; Baram et al. 2012). For example, higher quality of maternal care has been associated with the development of more secure attachment styles later in life (De Wolff and van IJzendoorn 1997; Egeland and Farber 1984; Sroufe 2005; Belsky and Fearon 2002; Egeland et al. 1983; Baram et al. 2012). At the more extreme end, childhood emotional maltreatment has been associated with a profound and lasting negative impact on behavioral and emotional functioning (Van Harmelen et al. 2010; Teicher et al. 2006) and enhanced risk to develop depressive and anxiety disorders in later life (e.g., Kendler et al. 2000; Van Harmelen et al. 2010). Also, institutionalized children demonstrate high rates of psychiatric symptoms, (Ellis et al. 2004).

An important question that remains to be addressed is what the neurobiological correlates of altered cognitive and emotional sensitivity are in individuals reporting early life adversity. At the structural level, emotional maltreatment has been associated with reductions of medial prefrontal cortex (PFC) volume (Van Harmelen et al. 2010), decreased hippocampal volume (Rao et al. 2010; Teicher et al. 2012; Carballedo et al. 2013), and increased amygdala volume in children which were raised by mothers suffering from depression (Lupien et al. 2011). At the functional level, lower social economic status, maternal deprivation and institutionalization has been correlated with altered activity in the PFC and amygdala activity during regulation of negative emotion (Kim et al. 2013; Gee et al. 2013; Cohen et al. 2013). In addition, maltreatment during childhood has been associated with altered hippocampal-subgenual and amygdala-subgenual resting-state functional connectivity (Herringa et al. 2013). Moreover, emotional maltreatment is associated with enhanced amygdala activity (Van Harmelen et al. 2013).

Taken together, these studies in human subjects suggest that early life adversity is associated with long-lasting changes in brain structure and brain function. Given the important role of the medial PFC, amygdala, and hippocampus in the regulation of cognitive function and emotional behavior, this might provide an important link to increased emotional sensitivity and decreased higher cognitive function in individuals reporting early life adversity. While the link with disease is unique for humans and emphasizes the relevance of the problem, these studies so far have been highly correlational. In addition, these studies cannot give insight into the *mechanism* by which stress during early life alters cognitive performance and emotional memory formation later on. Animal studies are highly relevant to get a better understanding of the mechanisms that underlie these effects. In this brief review we will discuss animal studies describing how early postnatal adversity regulates neuronal structure, neuronal function, as well as cognition and emotional behavior.

2 Early Life Experience and Stress-Responsiveness

Many rodent models which are used to study the effects of adverse early life experiences focus on varying the amount of maternal care during the first week(s) after birth. Pioneering studies by Seymour Levine have demonstrated that changes in the early postnatal environment can have lasting consequences for stress-responsiveness (e.g., Weinberg and Levine 1977; Wiener and Levine 1978, 1983). During this period, the presence of the dam is crucial for controlling activity of the hypothalamus pituitary adrenal (HPA-axis) (De Kloet et al. 1988; Stanton et al. 1988; Levine 1994; Schmidt et al. 2002). From approximately postnatal day (PND) 3 to PND 14 rodent pups show a reduced activity of the HPA-axis in response to mild to moderate stressors (Stress HypoResponsive Period (SHRP); Sapolsky and Meaney 1986; Rosenfeld et al. 1992; Schmidt et al. 2002). Disruption of the early life environment and the SHRP by separating pups from the dam has substantial and long-lasting endocrine and behavioral consequences.

For example, brief periods of handling during the early postnatal period in rats has lasting neuroendocrine consequences. In these experiments, pups which were exposed to early handling (EH) were repeatedly picked up by an experimenter and isolated in a small compartment for several minutes, while non-handled (NH) pups were left undisturbed. As a consequence of this mild procedure, EH rats were more active, more explorative, and showed lower glucocorticoid responses to stressors (Pryce and Feldon 2003; Meaney et al. 1988). EH animals may have increased negative feedback sensitivity when compared to NH animals since hypothalamic CRH mRNA levels are lower while glucocorticoid receptor (GR) levels are enhanced (Meaney et al. 1988).

Alterations in HPA-axis activity can also be observed after a single 24 h separation (Maternal Separation, MS) of pups from the dam. Single 24 h MS at PND 3 results in increased basal corticosterone levels in young (3 months) rats but this effect does not maintain into adulthood (Workel et al. 2001; Lehmann et al. 2002).

In contrast to single 24 h MS, repeated MS for 3–6 h MS has been shown to affect HPA-axis responses into adulthood, (e.g., Plotsky and Meaney 1993).

It is important to note that studies on the effects of handling and MS on stress-responsiveness do involve a role of maternal care. For example, handling of pups results in overall increased levels of maternal licking and grooming of the pups by the mother (Liu et al. 1997). Similar effects are seen with MS: while the dam and pups are separated during MS, the dam displays enhanced attention to the pups upon reunion (Oomen et al. 2009). Pioneering work from the group of Michael Meaney has examined in detail the effects of natural variations in maternal care on stress-responsiveness. They classified mothers as high licking/grooming-arched back nursing (High LG-ABN) or low licking/grooming-arched back nursing (Low LG-ABN) (Liu et al. 1997) and reported that the amount of maternal care has profound effects on the development of the HPA-axis. As adults, offspring reared by High LG-ABN mothers show significantly reduced levels of plasma ACTH and corticosterone in response to restraint stress compared to offspring raised by Low LG-ABN mothers, while basal ACTH and corticosterone are unaffected (Liu et al. 1997). The reduced HPA-response to stress correlated with the amount of maternal care, such that higher levels of licking and grooming were correlated with lower HPA-axis responsiveness after restraint stress. In addition, CRH mRNA levels were lower in offspring of High LG-ABN mothers when compared to Low LG-ABN offspring, whereas GRmRNA levels are higher in offspring of High LG-ABN mothers when compared to Low LG-ABN offspring (Liu et al. 1997). Importantly, these effects could be reversed by cross-fostering animals thereby demonstrating that the effects on stress-responsiveness are mediated by variations in maternal care. In agreement, fragmented maternal care, resulting from unpredictable and erratic interaction between the dam and pups as a consequence of chronic early life stress has substantial but temporal impact on HPA-axis activity in rats (Brunson et al. 2005) and mice (Rice et al. 2008).

Not only between but also within litters there are substantial differences in the amount of licking and grooming received among the offspring (see Fig. 1a; Van Hasselt et al. 2012a). In general, male pups receive more maternal care than females. In agreement with the between-litter studies, *individual* levels of maternal care within litters also correlate with the expression of GRs (Van Hasselt et al. 2012a, b, c).

In conclusion, rodent models have demonstrated that the early postnatal period is a sensitive period for the development of HPA-axis responsiveness.

3 Early Life Adversity and Cognitive and Emotional Function

Early life experience has profound impact on cognitive and emotional function, including anxiety like behavior. For instance, rats subjected to maternal separation respond more anxiously to novelty, as measured by reduced feeding in a novel environment and reduced activity in an open field (Aisa et al. 2007; Macrí et al. 2004).

Maternal care too has profound effects on anxiety like behavior. High LG offspring shows reduced anxiety when compared to Low LG offspring (Caldji et al. 1998, 2000; Weaver et al. 2006). Also animals which are exposed to fragmented maternal care display enhanced anxiety like behavior (Ter Horst, unpublished observations).

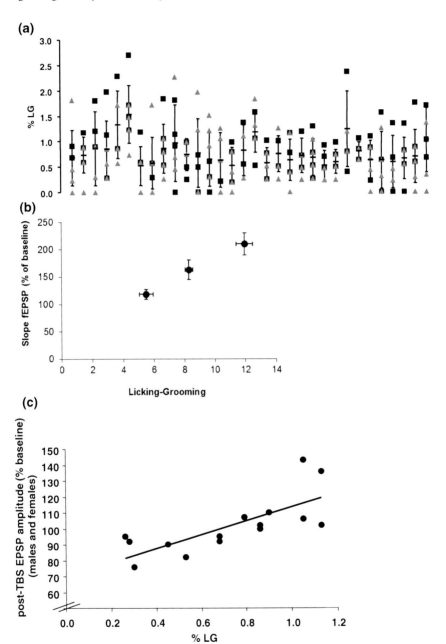

◀ **Fig. 1** The amount of maternal care during the first postnatal week affects adult synaptic plasticity in the hippocampus. **a** The amount of licking-grooming (LG) received by individual pups within a litter varies considerably. The y-axis indicates the time during which individual pups were licked or groomed by the dam, as a percentage of the timeslots during which scoring took place. Each *square* represents one male pup, each *triangle* a female pup and each column represents one litter. The *horizontal stripes* indicate the average amount of care bestowed by the dam on her litter. From Van Hasselt et al. 2012a. **b** CA1 hippocampal synaptic potentiation is enhanced in the adult litters from High LG-ABN animals when compared to Low LG-ABN animals. Mid-LG-ABN animals displays average levels of synaptic potentiation (adapted from Champagne et al. 2008). **c** The degree of synaptic potentiation in the dentate gyrus of adult rats also correlates with the percentage LG received during the first postnatal week when determined at the level of individual animals, examining within-litter variation. From Van Hasselt et al. 2012a

Early life adversity also enhances emotional memory formation. Contextual fear conditioning was found to be enhanced in animals exposed to 24 h MS at PND 3 (Oomen et al. 2010) as well as in Low compared to High LG-ABN animals (Champagne et al. 2008). Moreover, MS enhanced cued-fear conditioning (Oomen et al. 2010). These studies are in agreement with human studies reporting that emotional responsiveness is enhanced in individuals with a history of negative early life experience (Hackman et al. 2010; Van Harmelen et al. 2010, 2013).

While emotional responses are enhanced after exposure to negative early life experience, spatial memory performance and executive function are generally hampered by early life adversity. Exposure to MS, fragmental maternal care, and Low (compared to High) LG-ABN have been associated with impaired Morris Water Maze learning (Aisa et al. 2007; Brunson et al. 2005; Oomen et al. 2010; Liu et al. 2000) and object recognition (Bredy et al. 2003; Rice et al. 2008). MS also hampers PFC-dependent cognitive function by impairing deficits in temporal memory and cognitive flexibility (Lejeune et al. 2013; Baudin et al. 2012). In addition, *individual* levels in maternal care received early in life were reported to correlate with adult decision making in the rat Iowa Gambling task (Van Hasselt et al. 2012c).

4 Early Life Adversity and Neuronal Structure

While animal studies show that early adversity determines emotional responsiveness, relatively little is known about how it affects structure and function of the amygdala. Twenty-four hours of MS does not affect dendritic complexity in the amygdala (Krugers et al. 2012). This contrasts to the effect of chronic early life stress experienced during the entire first postnatal week (caused by drastically limiting the nesting material) which increases dendritic length in the basolateral amygdala (Krugers et al. unpublished observations).

Adverse early experiences substantially affect hippocampal structure in adulthood. Exposure to MS, chronic early life stress, and Low (as opposed to High) levels of LG-ABN have been associated with reduced dendritic complexity, spine number, and markers of synaptic function in various hippocampal subfields (Brunson et al. 2005; Oomen et al. 2010; Champagne et al. 2008; Bagot et al. 2009; Liu et al. 2000). In addition, early life adversity also hampered structural development of the PFC. For instance, repeated neonatal MS in rodents was reported to alter dendritic morphology and spine density in PFC pyramidal neurons (Bock et al. 2005; Monroy et al. 2010; Pascual and Zamora-León 2007).

5 Early Life Adversity and Synaptic Function

Early life adversity not only determines behavior and neuronal structure, but also synaptic function and synaptic plasticity, two critical endpoints for learning and memory (Neves et al. 2008; Kessels and Malinow 2009). Various studies have reported that experiencing MS or chronic early life stress hampers hippocampal synaptic plasticity (Brunson et al. 2005; Oomen et al. 2010). Likewise, long-term potentiation (LTP) can easily be induced in the CA1 area and dentate gyrus in High but not in Low LG-ABN offspring (Champagne et al. 2008; Bagot et al. 2009, 2012), even at the level of individual rats (Fig. 1). Interestingly, NMDA currents are enhanced in offspring of low LG-ABN animals (Bagot et al. 2012), which is in line with altered NMDA receptor composition in maternally deprived animals (Rodenas-Ruano et al. 2012). Surprisingly, application of an NMDA receptor antagonist enabled synaptic plasticity in offspring of low LG-ABN animals, suggesting that the enhanced NMDA receptor function in the hippocampus may underlie the reduced ability to elicit LTP in Low LG-ABN offspring (Bagot et al. 2012).

Importantly, maternal care also determines the sensitivity of synapses for stress hormones. Synaptic plasticity in the CA1 or dentate gyrus of Low LG-ABN animals was found to be facilitated by application of high levels of corticosterone and/or noradrenaline to hippocampal slices (Champagne et al. 2008; Bagot et al. 2009, 2012). This indicates that synapses of low LG-ABN animals do have the ability to express synaptic plasticity and that the sensitivity of synapses is determined by variations in postnatal maternal care. Similar effects have been found after maternal deprivation (Oomen et al. 2010). These observations may indicate that low levels of postnatal maternal care and MS render synapses more efficient under stressful conditions.

6 Discussion

Several studies in humans suggest that negative early life experiences program brain function and behavior into adulthood (for reviews see Hackman et al. 2010; Baram et al. 2012). While such studies are extremely valuable, they are not trivial since they require longitudinal studies over decades. Moreover, control over the environment in these studies is not possible, for obvious reasons, and genetic effects might be an important confounding factor (Korosi et al. 2012).

Also, one should also be cautious in generalizing effects of early life adversity since different studies examine various types of early life adversity (e.g., low socioeconomic status; sexual abuse; emotional maltreatment).

Animal studies have the advantage to make more conclusive correlations between early experience and lasting effects on brain structure and behavior. Moreover, they allow detailed analysis of the underlying molecular and cellular mechanisms. In general, animal studies confirm observations in humans that early life adversity hampers hippocampus-dependent and executive function (Kim et al. 2013; Herringa et al. 2013; Liu et al. 2000; Brunson et al. 2005; Oomen et al. 2010), while emotional responsiveness and emotional memory formation are enhanced (Van Harmelen et al. 2013; Oomen et al. 2010; Champagne et al. 2008).

These studies have also revealed that early life adversity programs activity of the HPA-axis via epigenetic mechanisms (Weaver et al. 2004). In particular, expression of the GR appears to be regulated by maternal care via epigenetic modification, not only in animals (Weaver et al. 2004; Van Hasselt et al. 2012a) but also in humans (Suderman et al. 2012; McGowan et al. 2009). In addition, epigenetic regulation of the FK506 binding protein 5 gene (an important regulator of the stress hormone system) mediates the interaction between gene and childhood trauma (Klengel et al. 2013; Touma et al. 2011). Interestingly, interference with epigenetic modification can normalize the effects of maternal care on behavior (Weaver et al. 2006). Much research has focused on the role of GRs in mediating the long-lasting effects of early life adversity and maternal care. However, other factors may be as important as well. For example, the effects of early life stress during the first postnatal week on neuronal function and neuronal structure are absent in CRH1 knockout mice (Wang et al. 2011, 2013).

Finally, the effects of early life events on brain function, behavior, and adaptation might be context dependent. Individuals exposed to early life adversity might be hampered in performance when exposed to nonstressful conditions but may actually be programmed to adapt well when exposed to stressful conditions (Champagne et al. 2009; Nederhof and Schmidt 2012).

These considerations raise a number of relevant questions that will be important to address in animals in the near future:

(1) How does early life experience determine synaptic function and the sensitivity of synapses for stress hormones?
(2) Is the regulation of glutamatergic synapses (and their sensitivity) relevant for behavioral adaptation?
(3) Is exposure to early life adversity always maladaptive, and if maladaptive, what are resilience factors?

References

Aisa B, Tordera R, Lasheras B, Del Río J, Ramírez MJ (2007) Cognitive impairment associated to HPA axis hyperactivity after maternal separation in rats. Psychoneuroendocrinology. 32:256–266

Bagot RC, van Hasselt FN, Champagne DL, Meaney MJ, Krugers HJ, Joëls M (2009) Maternal care determines rapid effects of stress mediators on synaptic plasticity in adult rat hippocampal dentate gyrus. Neurobiol Learn Mem 92:292–300

Bagot RC, Tse YC, Nguyen HB, Wong AS, Meaney MJ, Wong TP (2012) Maternal care influences hippocampal N-methyl-D-aspartate receptor function and dynamic regulation by corticosterone in adulthood. Biol Psychiatry 72:491–498

Baram TZ, Davis EP, Obenaus A, Sandman CA, Small SL, Solodkin A, Stern H (2012) Fragmentation and unpredictability of early-life experience in mental disorders. Am J Psychiatry 169:907–915

Baudin A, Blot K, Verney C, Estevez L, Santamaria J, Gressens P, Giros B, Otani S, Daugé V, Naudon L (2012) Maternal deprivation induces deficits in temporal memory and cognitive flexibility and exaggerates synaptic plasticity in the rat medial prefrontal cortex. Neurobiol Learn Mem 98:207–214

Belsky J, Fearon RMP (2002) Early attachment security, subsequent maternal sensitivity, and later child development: does continuity in development depend upon continuity of caregiving? Attach Hum Dev 4:361–387

Bock J, Gruss M, Becker S, Braun K (2005) Experience-induced changes of dendritic spine densities in the prefrontal and sensory cortex: correlation with developmental time windows. Cereb Cortex 15:802–808

Bredy TW, Humpartzoomian RA, Cain DP, Meaney MJ (2003) Partial reversal of the effect of maternal care on cognitive function through environmental enrichment. Neuroscience 118:571–576

Brunson KL, Kramár E, Lin B, Chen Y, Colgin LL, Yanagihara TK, Lynch G, Baram TZ (2005) Mechanisms of late-onset cognitive decline after early-life stress. J Neurosci 25:9328–9338

Caldji C, Tannenbaum B, Sharma S, Francis D, Plotsky PM, Meaney MJ (1998) Maternal care during infancy regulates the development of neural systems mediating the expression of fearfulness in the rat. Proc Natl Acad Sci USA 95:5335–5340

Caldji C, Diorio J, Meaney MJ (2000) Variations in maternal care in infancy regulate the development of stress reactivity. Biol Psychiatry 48:1164–1174

Carballedo A, Morris D, Zill P, Fahey C, Reinhold E, Meisenzahl E, Bondy B, Gill M, Möller HJ, Frodl T (2013) Brain-derived neurotrophic factor Val66Met polymorphism and early life adversity affect hippocampal volume. Am J Med Genet B Neuropsychiatr Genet 162:183–190

Champagne DL, Bagot RC, van Hasselt F, Ramakers G, Meaney MJ, de Kloet ER, Joëls M, Krugers H (2008) Maternal care and hippocampal plasticity: evidence for experience-dependent structural plasticity, altered synaptic functioning, and differential responsiveness to glucocorticoids and stress. J Neurosci 28:6037–6045

Champagne DL, de Kloet ER, Joëls M (2009) Fundamental aspects of the impact of glucocorticoids on the (immature) brain. Semin Fetal Neonatal Med 14:136–142

Cohen MM, Jing D, Yang RR, Tottenham N, Lee FS, Casey BJ (2013) Early-life stress has persistent effects on amygdala function and development in mice and humans. Proc Natl Acad Sci USA 110:18274–18278

De Wolff MS, van Ijzendoorn MH (1997) Sensitivity and attachment: a meta-analysis on parental antecedents of infant attachment. Child Dev 68:571–591

De Kloet ER, Rosenfeld P, Van Eekelen JA, Sutanto W, Levine S (1988) Stress, glucocorticoids and development. Prog Brain Res 73:101–120

Egeland B, Sroufe LA, Erickson M (1983) The developmental consequence of different patterns of maltreatment. Child Abuse Negl 7:459–469

Egeland B, Farber EA (1984) Infant-mother attachment: factors related to its development and changes over time. Child Dev 55:753–771

Ellis BH, Fisher PA, Zaharie S (2004) Predictors of disruptive behavior, developmental delays, anxiety, and affective symptomatology among institutionally reared romanian children. J Am Acad Child Adolesc Psychiatry 43:1283–1292

Farah MJ, Shera DM, Savage JH, Betancourt L, Giannetta JM, Brodsky NL, Malmud EK, Hurt H (2006) Childhood poverty: specific associations with neurocognitive development. Brain Res 1110:166–174

Gee DG, Barad-Durnam LJ, Flannery J, Goff B, Humphreys KL, Telzer EH, Hare TA, Bookheimer SY, Tottenham N (2013) Earlt developmental emergence of human amygdala-prefrontal connectivity after maternal deprivation. Proc Natl Acad Sci USA 110:15638–15643

Hackman DA, Farah MJ, Meaney MJ (2010) Socioeconomic status and the brain: mechanistic insights from human and animal research. Nat Rev Neurosci 11:651–659

Herringa RJ, Birn RM, Ruttle PL, Burghy CA, Stodola DE, Davidson RJ, Essex MJ (2013) Childhood maltreatment is associated with altered fear circuitry and increased internalizing symptoms by late adolescence. Proc Natl Acad Sci USA 110:19119–19124

Kendler KS, Thornton LM, Gardner CO (2000) Stressful life events and previous episodes in the etiology of major depression in women: an evaluation of the "kindling" hypothesis. Am J Psychiatry 157:1243–1251

Kessels HW, Malinow R (2009) Synaptic AMPA receptor plasticity and behavior. Neuron 61:340–350

Kim P, Evans GW, Angstadt M, Ho SS, Sripada CS, Swain JE, Liberzon I, Phan KL (2013) Effects of childhood poverty and chronic stress on emotion regulatory brain function in adulthood. Proc Natl Acad Sci USA 110:18442–18447

Kishiyama MM, Boyce WT, Jimenez AM, Perry LM, Knight RT (2009) Socioeconomic disparities affect prefrontal function in children. J Cogn Neurosci 21:1106–1115

Klengel T, Mehta D, Anacker C, Rex-Haffner M, Pruessner JC, Pariante CM, Pace TW, Mercer KB, Mayberg HS, Bradley B, Nemeroff CB, Holsboer F, Heim CM, Ressler KJ, Rein T, Binder EB (2013) Allele-specific FKBP5 DNA demethylation mediates gene-childhood trauma interactions. Nat Neurosci 16:33–41

Korosi A, Naninck EFG, Oomen CA, Schouten M, Krugers H, Fitzsimons C, Lucassen PJ (2012) Early-life stress mediated modulation of adult neurogenesis and behavior. Behav Brain Res 227:400–409

Krugers HJ, Oomen CA, Gumbs M, Li M, Velzing EH, Joels M, Lucassen PJ (2012) Maternal deprivation and dendritic complexity in the basolateral amygdala. Neuropharmacology 62:534–537

Lehmann J, Russig H, Feldon J, Pryce CR (2002) Effect of a single maternal separation at different pup ages on the corticosterone stress response in adult and aged rats. Pharmacol Biochem Behav 73:141–145

Lejeune S, Dourmap N, Martres MP, Giros B, Daugé V, Naudon L (2013) The dopamine D1 receptor agonist SKF 38393 improves temporal order memory performance in maternally deprived rats. Neurobiol Learn Mem 106:268–273

Levine S (1994) The ontogeny of the hypothalamic-pituitary-adrenal axis. The influence of maternal factors. Ann N Y Acad Sci 746:275–288

Liu D, Diorio J, Tannenbaum B, Caldji C, Francis D, Freedman A, Sharma S, Pearson D, Plotsky PM, Meaney MJ (1997) Maternal care, hippocampal glucocorticoid receptors, and hypothalamic-pituitary-adrenal responses to stress. Science 277:1659–1662

Liu D, Diorio J, Day JC, Francis DD, Meaney MJ (2000) Maternal care, hippocampal synaptogenesis and cognitive development in rats. Nat Neurosci 3:799–806

Lupien SJ, Parent S, Evans AC, Tremblay RE, Zelazo PD, Corbo V, Pruessner JC, Séguin JR (2011) Larger amygdala but no change in hippocampal volume in 10-year-old children exposed to maternal depressive symptomatology since birth. Proc Natl Acad Sci USA 108: 14324–14329

Macrí S, Mason GJ, Würbel H (2004) Dissociation in the effects of neonatal maternal separations on maternal care and the offspring's HPA and fear responses in rats. Eur J Neurosci 20: 1017–1024

McGowan PO, Sasaki A, D'Alessio AC, Dymov S, Labonté B, Szyf M, Turecki G, Meaney MJ (2009) Epigenetic regulation of the glucocorticoid receptor in human brain associates with childhood abuse. Nat Neurosci 12:342–348

Meaney MJ, Aitken DH, van Berkel C, Bhatnagar S, Sapolsky RM (1988) Effect of neonatal handling on age-related impairments associated with the hippocampus. Science 239:766–768

Monroy E, Hernández-Torres E, Flores G (2010) Maternal separation disrupts dendritic morphology of neurons in prefrontal cortex, hippocampus, and nucleus accumbens in male rat offspring. J Chem Neuroanat 40:93–101

Nederhof E, Schmidt MV (2012) Mismatch or cumulative stress: toward an integrated hypothesis of programming effects. Physiol Behav 106:691–700

Neves G, Cooke SF, Bliss TV (2008) Synaptic plasticity, memory and the hippocampus: a neural network approach to causality. Nat Rev Neurosci 9:65–75

Noble KG, McCandliss BD, Farah MJ (2007) Socioeconomic gradients predict individual differences in neurocognitive abilities. Dev Sci 10:464–480

Oomen CA, Girardi CE, Cahyadi R, Verbeek EC, Krugers H, Joëls M, Lucassen PJ (2009) Opposite effects of early maternal deprivation on neurogenesis in male versus female rats. PLoS ONE 4:e3675

Oomen CA, Soeters H, Audureau N, Vermunt L, van Hasselt FN, Manders EM, Joëls M, Lucassen PJ, Krugers H (2010) Severe early life stress hampers spatial learning and neurogenesis, but improves hippocampal synaptic plasticity and emotional learning under high-stress conditions in adulthood. J Neurosci 30:6635–6645

Pascual R, Zamora-León SP (2007) Effects of neonatal maternal deprivation and postweaning environmental complexity on dendritic morphology of prefrontal pyramidal neurons in the rat. Acta Neurobiol Exp 67:471–479

Plotsky PM, Meaney MJ (1993) Early, postnatal experience alters hypothalamic corticotropin-releasing factor (CRF) mRNA, median eminence CRF content and stress-induced release in adult rats. Brain Res Mol Brain Res 18:195–200

Pryce CR, Feldon J (2003) Long-term neurobehavioural impact of the postnatal environment in rats: manipulations, effects and mediating mechanisms. Neurosci Biobehav Rev 27:57–71

Rao U, Chen LA, Bidesi AS, Shad MU, Thomas MA, Hammen CL (2010) Hippocampal changes associated with early-life adversity and vulnerability to depression. Biol Psychiatry 67:357–364

Rice CJ, Sandman CA, Lenjavi MR, Baram TZ (2008) A novel mouse model for acute and long-lasting consequences of early life stress. Endocrinology 149:4892–4900

Rodenas-Ruano A, Chávez AE, Cossio MJ, Castillo PE, Zukin RS (2012) REST-dependent epigenetic remodeling promotes the developmental switch in synaptic NMDA receptors. Nat Neurosci 15:1382–1390

Rosenfeld P, Wetmore JB, Levine S (1992) Effects of repeated maternal separations on the adrenocortical response to stress of preweanling rats. Physiol Behav 52:787–791

Sapolsky RM, Meaney MJ (1986) Maturation of the adrenocortical stress response: neuroendocrine control mechanisms and the stress hyporesponsive period. Brain Res 396:64–76

Schmidt MV, Oitzl MS, Levine S, de Kloet ER (2002) The HPA system during the postnatal development of CD1 mice and the effects of maternal deprivation. Brain Res Dev Brain Res 139:39–49

Sroufe LA (2005) Attachment and development: a prospective, longitudinal study from birth to adulthood. Attach Hum Dev 7:349–367

Stanton ME, Gutierrez YR, Levine S (1988) Maternal deprivation potentiates pituitary-adrenal stress responses in infant rats. Behav Neurosci 102:692–700

Suderman M, McGowan PO, Sasaki A, Huang TC, Hallett MT, Meaney MJ, Turecki G, Szyf M (2012) Conserved epigenetic sensitivity to early life experience in the rat and human hippocampus. Proc Natl Acad Sci USA 109:17266–17272

Teicher MH, Samson JA, Polcari A, McGreenery CE (2006) Sticks, stones, and hurtful words: relative effects of various forms of childhood maltreatment. Am J Psychiatry 163:993–1000

Teicher MH, Anderson CM, Polcari A (2012) Childhood maltreatment is associated with reduced volume in the hippocampal subfields CA3, dentate gyrus, and subiculum. Proc Natl Acad Sci USA 109:563–572

Touma C, Gassen NC, Herrmann L, Cheung-Flynn J, Büll DR, Ionescu IA, Heinzmann JM, Knapman A, Siebertz A, Depping AM, Hartmann J, Hausch F, Schmidt MV, Holsboer F, Ising M, Cox MB, Schmidt U, Rein T (2011) FK506 binding protein 5 shapes stress responsiveness: modulation of neuroendocrine reactivity and coping behavior. Biol Psychiatry 70:928–936

van Harmelen AL, de Jong PJ, Glashouwer KA, Spinhoven P, Penninx BW, Elzinga BM (2010) Child abuse and negative explicit and automatic self-associations: the cognitive scars of emotional maltreatment. Behav Res Ther 48:486–494

van Harmelen AL, van Tol MJ, Demenescu LR, van der Wee NJ, Veltman DJ, Aleman A, van Buchem MA, Spinhoven P, Penninx BW, Elzinga BM (2013) Enhanced amygdala reactivity to emotional faces in adults reporting childhood emotional maltreatment. Soc Cogn Affect Neurosci 8:362–369

van Hasselt FN, Cornelisse S, Zhang TY, Meaney MJ, Velzing EH, Krugers HJ, Joëls M (2012a) Adult hippocampal glucocorticoid receptor expression and dentate synaptic plasticity correlate with maternal care received by individuals early in life. Hippocampus 22:255–266

van Hasselt FN, Boudewijns ZS, van der Knaap NJ, Krugers HJ, Joëls M (2012b) Maternal care received by individual pups correlates with adult CA1 dendritic morphology and synaptic plasticity in a sex-dependent manner. J Neuroendocrinol 24:331–340

van Hasselt FN, de Visser L, Tieskens JM, Cornelisse S, Baars AM, Lavrijsen M, Krugers HJ, van den Bos R, Joëls M (2012c) Individual variations in maternal care early in life correlate with later life decision-making and c-fos expression in prefrontal subregions of rats. PLoS ONE 7:e37820

Wang XD, Rammes G, Kraev I, Wolf M, Liebl C, Scharf SH, Rice CJ, Wurst W, Holsboer F, Deussing JM, Baram TZ, Stewart MG, Müller MB, Schmidt MV (2011) Forebrain CRF$_1$ modulates early-life stress-programmed cognitive deficits. J Neurosci 31:13625–13634

Wang XD, Su YA, Wagner KV, Avrabos C, Scharf SH, Hartmann J, Wolf M, Liebl C, Kühne C, Wurst W, Holsboer F, Eder M, Deussing JM, Müller MB, Schmidt MV (2013) Nectin-3 links CRHR1 signaling to stress-induced memory deficits and spine loss. Nat Neurosci 16:706–713

Weaver IC, Cervoni N, Champagne FA, D'Alessio AC, Sharma S, Seckl JR, Dymov S, Szyf M, Meaney MJ (2004) Epigenetic programming by maternal behavior. Nat Neurosci 7:847–854

Weaver IC, Meaney MJ, Szyf M, Weaver IC, Meaney MJ, Szyf M (2006) Maternal care effects on the hippocampal transcriptome and anxiety-mediated behaviors in the offspring that are reversible in adulthood. Proc Natl Acad Sci USA 103:3480–3485

Weinberg J, Levine S (1977) Early handling influences on behavioral and physiological responses during active avoidance. Dev Psychobiol 10:161–169

Wiener SG, Levine S (1978) Perinatal malnutrition and early handling: interactive effects on the development of the pituitary-adrenal system. Dev Psychobiol 11:335–352

Wiener SG, Levine S (1983) Influence of perinatal malnutrition and early handling on the pituitary-adrenal response to noxious stimuli in adult rats. Physiol Behav 31:285–291

Workel JO, Oitzl MS, Fluttert M, Lesscher H, Karssen A, de Kloet ER (2001) Differential and age-dependent effects of maternal deprivation on the hypothalamic-pituitary-adrenal axis of brown norway rats from youth to senescence. J Neuroendocrinol 13:569–580

Mechanisms Linking In Utero Stress to Altered Offspring Behaviour

Theresia H. Mina and Rebecca M. Reynolds

Abstract Development in utero is recognised as a determinant of health in later life, a concept known as early life 'programming'. Several studies in humans have now shown a link between in utero stressors of maternal stress, anxiety and depression and adverse behavioural outcomes for the offspring including poorer cognitive function and behavioural and emotional problems. These behaviours are observed from the very early neonatal period and appear to persist through to adulthood. Underlying mechanisms are not known but overexposure of the developing foetus to glucocorticoids has been proposed. Dysregulation of the maternal and offspring hypothalamic–pituitary–adrenal (HPA) axis has been proposed as a mechanism linking in utero stress with offspring behavioural outcomes. Studies suggest that altered circulating levels of maternal cortisol during pregnancy and/or changes in placental gene expression or methylation, which result in increased glucocorticoid transfer to the developing foetus, are linked to changes in offspring behaviour and in activity of the offspring HPA axis. Further understanding of the underlying pathways and identification of any gestation of vulnerability are needed to help design interventions to reduce in utero stress and improve behavioural outcomes in the offspring.

Keywords Stress · Behaviour · Cortisol · 11 beta hydroxysteroid dehydrogenase · Glucocorticoid receptor

T. H. Mina · R. M. Reynolds (✉)
Endocrinology Unit, Queen's Medical Research Institute,
University/British Heart Foundation Centre for Cardiovascular Science,
47 Little France Crescent, Edinburgh, EH16 4TJ, UK
e-mail: R.Reynolds@ed.ac.uk

Contents

1 Introduction	94
2 Literature Search Strategy	95
3 Definitions of In Utero Stress	95
4 In Utero Stress and Offspring Behaviour	96
5 Mechanisms Linking In Utero Stress with Offspring Behavioural Outcomes	98
6 Changes in Circulating Levels of Maternal Hormones During Pregnancy	100
7 Maternal HPA Axis and Offspring Behaviour	101
8 In Utero Stress and Infant HPA Axis Activity	107
9 Changes in Placental Growth, Gene Expression and Epigenetic Modification in Association with In Utero Stress	110
10 Changes in Placental Growth, Gene Expression and Epigenetic Modification in Association with Offspring Behaviour	113
11 In Utero Stress and Structural Changes in the Offspring Brain	114
12 Can We Intervene to Improve Offspring Outcomes?	115
13 Conclusion	115
References	116

1 Introduction

The early life environment is recognised to be a key time determining the trajectory of future health. Numerous epidemiological studies have shown associations between low birthweight, a marker of an adverse intrauterine environment and subsequent cardiometabolic disease, mental health problems, cognitive decline and other disorders including osteoporosis. This concept is termed 'early life programming' or the 'developmental origins of health and disease' (Barker 1995; Godfrey et al. 2010). The developing foetus is thought to respond to insults in utero with changes in structure, physiology and metabolism which may be initially beneficial, but in later life become maladaptive and predispose to disease. Experimental studies using animal models and translational studies in humans have been carried out to identify possible programming factors and to investigate underlying mechanisms. Maternal mood disorders including 'stress', anxiety and depression have been identified as important factors leading to in utero stress and which are linked to adverse outcomes in the offspring (Cottrell and Seckl 2009). Indeed the World Health Organisation has now highlighted the importance of maternal mental health for well-being of not only the mother, but also her children (http://www.who.int/mental_health/prevention/suicide/MaternalMH/en/).

The child-bearing years are the time that women are most vulnerable to mood disorders. For example, major depressive disorder occurs in 8–13 % of women during pregnancy (Gavin et al. 2005). Accompanying depressive symptoms do not abate during pregnancy and there is a further peak in symptoms in the post-partum period. The most replicated finding in studies investigating the effects of in utero stress on offspring outcomes is of an association between maternal stress and anxiety with preterm birth, shorter gestational age and low birthweight. The evidence supporting these observations has been reviewed in detail elsewhere (for reviews see e.g. Huizink et al. 2004; Mulder et al. 2002; Talge et al. 2007). There is also increasing evidence supporting a link between in utero stress and altered foetal behaviours in utero including increased foetal heart rate and motility assessed using ultrasound and foetal heart rate monitoring (for review see Van den Bergh et al. 2005b). In this review, we focus on the evidence from human studies linking in utero stress secondary to maternal stress, anxiety and/or depressive symptoms in pregnancy with offspring behaviour from the early neonatal period onwards. In particular, we examine the emerging studies that have started to dissect potential underlying mechanisms underlying the links between in utero stress and altered offspring behaviour.

2 Literature Search Strategy

For this narrative review, we carried out a literature search using Pubmed to identify studies linking in utero stress with offspring behaviour. We carried out searches with key MESH terms including: 'maternal anxiety depression offspring behaviour'; 'maternal glucocorticoid offspring behaviour'; 'cortisol pregnancy'; 'maternal anxiety depression epigenetics'; 'placenta offspring behaviour'; 'placenta anxiety depression'; 'umbilical cord behaviour' and 'umbilical anxiety depression'. Titles and abstracts were reviewed and those papers which appeared to meet the pre-set inclusion criteria were selected. We included human studies only. We excluded studies without assessments of maternal mood during pregnancy and also studies investigating the effect of maternal alcoholism, substance abuse, and smoking. Finally, we excluded studies which focused on the effect of stress on obstetric complications, labour and delivery, and birth outcomes and also studies reporting outcomes in twins. The searches also generated a collection of review papers, from which additional relevant studies were identified from reviewing the reference lists. The final literature searches were carried out on 1st December 2013.

3 Definitions of In Utero Stress

The term 'in utero stress' covers several overlapping and related concepts. Stressors that have been shown to be related to adverse outcomes for the offspring include studies of women with a range of maternal mood disorders from clinically diagnosed mental health disorders such as depression (Pawlby et al. 2011) to experiencing symptoms of anxiety and/or depression during pregnancy (reviewed in Nast et al. 2013). Studies also include women involved in natural disasters (e.g. major hurricanes (Kinney et al. 2008), the Canadian Ice Storms (Laplante et al. 2004) or wars (Yehuda et al. 2005; Van Os and Selten 1998; Kleinhaus et al. 2013), women who have experienced a major life event during pregnancy such as bereavement (Khashan et al. 2008), to those experiencing less severe stressors in pregnancy such as work stress (Mozurkewich et al. 2000), or pregnancy-specific and day-to-day hassles (Huizink et al. 2003). A recent systematic review examined the methods applied to assess maternal psychosocial stress during pregnancy in studies which looked at associations with biobehavioural outcomes in the offspring (Nast et al. 2013). Of the 115 identified publications that assessed psychosocial stress using validated methods, there were 43 different instruments which the authors then categorised into assessing one of seven categories: anxiety, depression, daily hassles, aspects of psychological symptomatology (not related to anxiety and depression), life events, specific socio-environmental stressors and stress related to pregnancy and parenting. The authors highlighted that the diverse nature of the stressors and the differences in methods of assessment hamper the comparability of stress research results. For this review, we have considered studies defining in utero stress as maternal stress, anxiety and/or depression.

4 In Utero Stress and Offspring Behaviour

Several studies have now examined the link between maternal stress, anxiety and/or depression and offspring behaviours as a neonate, child, adolescent and young adult. In addition to the previously highlighted problems of studies using different assessments of maternal mood (Nast et al. 2013), there are also several different methods of assessment of infant behaviours. For example, assessment may be carried out by an independent assessor using established rating scales or recordings, or by reporting from either the parent (usually the mother) or a teacher. Nevertheless, despite variations in methodology, the overall findings are consistent with neurodevelopmental and behavioural problems in offspring exposed to in utero stress (reviewed in Van den Bergh et al. 2005b), notably with changes in regulatory and emotional behaviours and in cognitive development. Strikingly, the links between in utero stress and altered behaviour are observable from the very early neonatal period, findings that are consistent with a 'programmed' effect. Even as early as 3–5 days old, infants of mothers with more 'total distress', assessed by combination of a number of validated scales, have more problems with regulatory behaviours (e.g. in alertness and attention) measured using the Brazelton Neonatal Behaviour Assessment scale (Rieger et al. 2004). Likewise, lower scores on this scale have been observed in offspring of anxious mothers at 3 weeks of age (Brouwers et al. 2001). Newborns aged 4–14 days of stressed mothers also score less well on neurological examination (Lou et al. 1994) while maternal depression is associated with poorer newborn scores for orientation and reflexes (Lundy et al. 1999). There are also changes in emotional behaviours as infants of mothers with high antenatal state and trait anxiety show high activity levels at age 2–4 days, cry more at 1 week of age and have a difficult temperament at 10 weeks (Van den Bergh 1990).

Several changes in behaviours are observed in the first 2 years of life in infants exposed to in utero stress. Babies of mothers with high scores of negative life changes during pregnancy exhibit more crying/fussing at 3 and 6 months than infants born to mothers with low negative change scores (Wurmser et al. 2006). Depressed mothers report their babies to be less relaxed, more emotional and more fussy at 3–5 months, behaviours which are associated with poorer maternal interaction with the infant (Field et al. 1985). Maternal depression and anxiety are also related to higher infant negative behavioural reactivity at 4 months (Davis et al. 2004).

A large study of 10,323 children showed an association between higher prenatal anxiety and offspring sleeping difficulties at 18 and 30 weeks (O'Connor et al. 2007) whilst another large study including 2,724 children showed that anxiety during pregnancy was associated with poorer neuromotor development at 4 months (Van Batenburg-Eddes et al. 2009). Maternal prenatal anxiety has also been associated with poorer infant temperament between 4 and 8 months (Vaughn et al. 1987) whilst pregnancy-specific anxiety in mid-pregnancy predicted lower mental and motor developmental scores at 8 months (Buitelaar et al. 2003). In the latter study pregnancy-specific anxiety explained 7 % of the variance of test-affectivity and goal-directedness. Importantly, different types of in utero stress appear to be related to different infant outcomes. For example, infant attention regulation problems at 3 and 8 months were associated with higher maternal perceived stress at 15–17 weeks but not with daily hassles (Huizink 2002; Huizink et al. 2003). Further understanding of these pathways may be useful to guide therapies for maternal stress during pregnancy.

The effects of in utero stress on offspring behaviour persist into childhood, with associations between maternal anxiety and/or depression and behavioural and emotional problems reported in children at 4 and 6–7 years (O'Connor et al. 2002, 2003) and with more attention deficit hyperactivity disorder (ADHD) symptoms at 7–9 years (Rodriguez and Bohlin 2004; Van den Bergh and Marcoen 2004). Maternal pregnancy-specific anxiety is associated with poorer executive function by the child at 6–9 years (Buss et al. 2011). Maternal prenatal stress also appears to affect other aspects of child development such as vulnerability to bullying (Lereya and Wolke 2013), although teasing out pre-and post-natal influences is challenging. One study has suggested that maternal depression has a wider impact on different types of child maladjustment than maternal anxiety, the latter appearing to be more specific to internalising difficulties in the child (Barker et al. 2011) and this may have implications for targeting mental health care in pregnancy. Consistent with this, whilst maternal antenatal depression associates with depression in adolescence (Pawlby et al. 2009), maternal symptoms of anxiety do not predict adolescent psychopathology once maternal depression is taken into account (Hay et al. 2008). Other behavioural changes reported in adolescents of anxious mothers include increased impulsivity (Van den Bergh et al. 2005a) and an increased risk of anxiety (Davis and Sandman 2012), though neither of these studies included adjustment for maternal depression.

Few studies examining the influence of in utero stress on offspring behaviours have extended observations to adulthood. A very recent study with data from more than 4,500 parents and their children showed that antenatal depression was an independent risk factor for depression in the offspring at aged 18 years (Pearson et al. 2013). Offspring were 1.28 times (95 % CI, 1.08–1.51, $p = 0.003$) more likely to have depression at aged 18 years for each standard deviation increase in maternal depression score antenatally, independent of later post-partum maternal depression. These findings suggest persisting effects of an in utero insult on offspring behaviour, highlighting the need to understand more about the potential underlying pathways and whether there are gestations of increased vulnerability.

The evidence described above is thus strongly supportive of an association between in utero stress and altered offspring behaviours. However, in human studies it is harder to establish whether the relationship is causal and it is very hard to distinguish between in utero stressors and post-natal environmental and/or genetic influences. This is particularly relevant for studies of maternal mood as, for example, antenatal depression is a major risk factor for post-partum depression. Therefore, many studies control for as many confounding factors as possible, including maternal psychological state at the time of infant assessment, to try and account for any post-partum influences. Another approach has been to examine behaviour in children born after in vitro fertilisation who are either genetically related or unrelated to the mother (Rice et al. 2010). Using this paradigm in a follow-up study of nearly 800 children, maternal stress in pregnancy was associated with child symptoms of conduct disorder and ADHD. Of note, the conduct disorder occurred in children who were unrelated to their mothers, suggesting the association with in utero stress was independent of genetic factors and so supportive of an environmental influence. In contrast, ADHD also occurred in children who were related to the mothers. Further studies are needed to disentangle genetic from environmental influences.

5 Mechanisms Linking In Utero Stress with Offspring Behavioural Outcomes

Several mechanisms have been proposed to explain the link between in utero stress and adverse offspring behaviours (See Fig. 1). Overexposure of the developing foetus to glucocorticoids has been proposed as one of the key mechanism linking an adverse intrauterine environment with adverse offspring outcomes including behavioural outcomes (Cottrell and Seckl 2009; Reynolds 2013). The hypothalamic–pituitary–adrenal (HPA) axis is one of the major hormonal systems underlying the normal 'stress' response and mood disorders are often attributed to altered HPA axis activity (Gotlib et al. 2008). Therefore, dysregulation of the maternal HPA axis and/or offspring HPA axis is a key candidate pathway linking

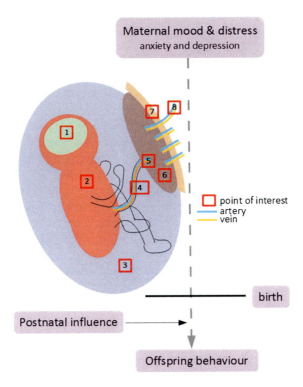

Fig. 1 Potential sites for examining mechanisms underlying the link between in utero stress and offspring behaviour. *Cartoon* shows the sites within the maternal-placental-foetal unit that have been studied to understand the mechanisms linking in utero stress with offspring behaviour. There may be additional post-natal influences. Sites include (*1*) Foetal brain via Magnetic Resonance Imaging (MRI). (*2*) Foetal heart, foetal movement. (*3*) Metabolites in amniotic fluid via amniocentesis. (*4*) Umbilical cord (artery and vein) pulsatility index, cord blood. (*5*) Umbilical blood flow, resistance index. (*6*) Placental blood flow, placental tissue (including studies on metabolites, genetics and epigenetics, gene expression and protein abundance). (*7*) Uterine pulsatility index. (*8*) Hormones and metabolites in maternal peripheral blood

in utero stress with offspring behavioural outcomes. Moreover, the developing brain of the infant is particularly susceptible to the adverse effects of glucocorticoids (McEwen 1999). Human brain development begins early in gestation, with key genes involved in neurodevelopmental processes expressed in the early embryonic period and continuing during foetal and infant development (Kang et al. 2011). Foetal glucocorticoid overexposure is thought to arise either from increased availability of glucocorticoids in the maternal circulation, or from increased transfer of glucocorticoids to the foetus via the placenta. In addition, in utero stressors may effect blood supply to the foetal/placental unit and have direct or indirect effects on offspring brain development (Fig. 1).

6 Changes in Circulating Levels of Maternal Hormones During Pregnancy

Cortisol is secreted by the adrenal gland under regulation of the HPA axis in response to both physiological and psychological stress. Cortisol secretion follows a diurnal pattern with levels peaking in the morning, prior to activity, and declining through the day. Circadian secretion of ACTH (adrenocorticotrophic hormone) from the pituitary is stimulated by the action of CRH (corticotropin releasing hormone) and AVP (vasopressin) from the parvicellular neurons of the hypothalamus under control of the suprachiasmatic nucleus of the hypothalamus. This diurnal cycle of cortisol release can be interrupted by stressors, which cause a premature secretory burst of glucocorticoids. In pregnancy, the activity of the maternal HPA axis undergoes significant changes. (Lindsay and Nieman 2005). Cortisol levels rise dramatically, peaking in the third trimester at three times non-pregnant levels (Jung et al. 2011). This is partly due to placental release of large quantities of CRH into the maternal blood stream from the early second trimester (Campbell et al. 1987); this stimulates the maternal pituitary to secrete ACTH, with consequent increase in maternal cortisol levels (Goland et al. 1994). Placental CRH also acts directly on the maternal adrenal to stimulate cortisol secretion and maternal cortisol can also stimulate placental production of CRH, thus further increasing cortisol levels. Corticosteroid binding globulin (CBG) levels also rise during pregnancy under the influence of oestrogen stimulation and bind some of the available cortisol, yet there are also progressive rises in 24 h urinary free cortisol and plasma free cortisol, the latter increasing by 1.6 fold by the third trimester (Jung et al. 2011). Despite the increasing circulating levels of cortisol, the diurnal pattern of cortisol secretion is maintained (Entringer et al. 2011). However, as pregnancy progresses, the normal physiological responses to stressors, and the cortisol awakening response, a marker of basal HPA activity, are attenuated (Lindsay and Nieman 2005). These dramatic changes in HPA axis activity, and the fact that pregnancy induces other changes that may alter HPA axis activity, such as changes in the immune system and in levels of progesterone (Robinson and Klein 2012), mean that the associations between cortisol levels and mood described in the non-pregnant state may not be the same in pregnancy.

There is some evidence that the maternal HPA axis is dysregulated in pregnancies associated with maternal depression. Maternal depression during pregnancy has been associated with elevated mid-pregnancy corticotrophin releasing hormone (Rich-Edwards et al. 2008; O'Keane et al. 2011) and cortisol levels (Field et al. 2006, 2008), with altered diurnal pattern of cortisol secretion in the second trimester (O'Keane et al. 2011; O'Connor et al. 2013b), and with increased urinary cortisol at the middle to end of the third trimester (27–35 weeks) (Lundy et al. 1999). Evidence is less clear when examining the relationship between maternal anxiety and/or symptoms of stress and changes in the maternal HPA axis. Several studies show that stress and anxiety levels assessed by questionnaire do not necessarily correlate with maternal cortisol levels (Voegtline et al. 2012;

Harville et al. 2009). This is likely to be related to differences in study design, such as the gestation of sample collection for cortisol analysis. The type of maternal stressor also appears to influence the cortisol response in pregnancy. For example, a history of prior major stress (child abuse) increases the cortisol awakening response during pregnancy (Bublitz and Stroud 2012) whilst experience of chronic stressful life events during early pregnancy has been associated with blunted peak salivary cortisol levels in the morning (Obel et al. 2005).

7 Maternal HPA Axis and Offspring Behaviour

Despite the relative lack of correlation between maternal mood and measurements of cortisol in pregnancy, a number of studies have examined the relationship between maternal HPA axis activity and behavioural changes in her offspring. The majority of these studies have used salivary samples to measure free cortisol during pregnancy as samples can be collected at home, i.e. in normal living conditions, without the stress of a clinic visit. The most robust studies include samples collected over 2 days to allow an average cortisol level to be calculated. Salivary cortisol measurements are typically used to examine the cortisol awakening response, or include measurements throughout the day, in a 'day curve', to examine the diurnal rhythm of cortisol secretion. Other investigators have measured cortisol in amniotic fluid. While this may represent cortisol secreted from the maternal circulation and/or from foetal urine, cortisol levels in amniotic fluid generally correlate with maternal serum cortisol levels at the time of sampling (Sarkar et al. 2007) and are thus thought to represent a marker of foetal glucocorticoid overexposure. Only a handful of studies have linked maternal urine or serum cortisol measurements with offspring outcomes. Results of studies assessing the maternal HPA axis and offspring behavioural outcomes are summarised in Table 1. The most consistent findings showing significant associations between cortisol measurements and offspring behaviour are with studies with measurements of morning cortisol levels. Higher maternal morning cortisol concentrations in saliva have been associated with altered infant behaviour in the first year of life including negative infant temperament (De Weerth et al. 2003; Davis et al. 2007) as well as poorer infant cognitive development as assessed by mental and motor development (Buitelaar et al. 2003) and less habituation to repeated stress (De Weerth et al. 2013). The finding of poorer infant cognitive development with greater glucocorticoid exposure was also observed in a study measuring cortisol in amniotic fluid (Bergman et al. 2010b). In contrast, no associations were found between cortisol in amniotic fluid and infant fear reactivity (Bergman et al. 2010a) or between the maternal salivary cortisol awakening response and emotional regulation (Bolten et al. 2013). There is some evidence that the relationship of higher morning cortisol levels and offspring behaviour persists to childhood with higher maternal cortisol levels during pregnancy associating with higher anxiety in childhood (Davis and Sandman 2012). Children with anxiety ratings within the

Table 1 Assessment of maternal HPA axis during pregnancy and links with offspring behaviour

Study*	Country	n	Gestation mother assessed (weeks)	Maternal HPA axis assessment	Infant age at follow-up	Infant assessment	Main Findings	Confounders and covariates included in statistical analysis
Salivary cortisol								
Buitelaar et al. 2003	Netherlands	170	15–17 ('early'), 27–28 ('mid'), 37–38 ('late')	7 saliva samples collected 2 hourly from 8 am–8 pm	3, 8 months	Bayley scales of infant development, temperamental questionnaire	Early morning 8 am cortisol in late pregnancy was negatively related to both mental and motor development at 3 months and motor development at 8 months	Gestational age at birth, birthweight, maternal postnatal stress and depression
De Weerth et al. 2003	Netherlands	17	36	Awakening salivary cortisol	1, 3, 5, 7, 18 and 20 weeks	Behaviour during bathing and infant characteristics questionnaire	Higher prenatal cortisol associated more crying, fussing and negative facial expressions and difficult behaviour with higher scores on emotion and activity. Most marked in young aged 1–7 weeks	
Davis et al. 2007	US	247	18–20, 24–26, 30–32	early afternoon, at least 1 h after eating (mean 14:20 h)	2 months	Negative reactivity assessed using the fear scale of the infant temperament questionnaire	Higher maternal cortisol level in late but not early pregnancy associated with negative infant reactivity	Time of sample collection, maternal postnatal psychological state

(continued)

Table 1 (continued)

Study*	Country	n	Gestation mother assessed (weeks)	Maternal HPA axis assessment	Infant age at follow-up	Infant assessment	Main Findings	Confounders and covariates included in statistical analysis
Davis and Sandman 2012	US	178	19, 25, 31	early afternoon, at least 1 h after eating, (range 14:17–14:46 h)	6–9 years	Achenbach system of empirically based assessment and child behaviour checklist	Elevated average maternal gestational cortisol was associated with higher child anxiety	Age, income, education, marital status, race, maternal psychological stress at time of assessment
Bolten et al. 2013	Switzerland	48	32–34	Salivary cortisol collected at 0, 30, 45 and 60 min after awakening	10 and 14 days 6 months	Neonatal intensive care unit network neurobehavioral scale still face paradigm	Cortisol levels after awakening in mid and late pregnancy were not associated with emotion regulation	Prepartum depressive symptoms
De Weerth et al. 2013	Netherlands	107	32.8	Circadian cortisol curves over 2 days	9 months	Crying and fussing after repeated maternal separation	High maternal cortisol awakening response associated with less infant habituation to repeated stress	Education, smoking, gestation length, birthweight, sex, duration of breastfeeding, non-parental care, time of day

(continued)

Table 1 (continued)

Study*	Country	n	Gestation mother assessed (weeks)	Maternal HPA axis assessment	Infant age at follow-up	Infant assessment	Main Findings	Confounders and covariates included in statistical analysis
Urine cortisol								
Lundy et al. 1999	US	43	27–35 (mean 32.3)	Morning sample collected after completing questionnaires. Cortisol measurements corrected for creatinine	1 week	Brazelton Neonatal Behaviour Assessment	Higher urinary cortisol associated with abnormal reflexes in neonates	Gender, ethnicity, obstetric complication
Amniotic fluid								
Bergman et al. 2010a	UK	108	Mean 17.2 (range 15–32)		17 months	Laboratory temperament assessment battery locomotor version	No relationship between cortisol and fear reactivity	Prenatal, obstetric and socioeconomic factors including age, social class, ethnicity, parity alcohol, smoking birthweight, gestational age, gender

(continued)

Table 1 (continued)

Study*	Country	n	Gestation mother assessed (weeks)	Maternal HPA axis assessment	Infant age at follow-up	Infant assessment	Main Findings	Confounders and covariates included in statistical analysis
Bergman et al. 2010b	UK	125	mean 17.2 (range 15–32)		17 months	Bayley scales of infant development, Ainsworth strange situation assessment	Inverse association between amniotic fluid cortisol and infant cognitive ability. Results moderated by infant parent attachment (remained significant in insecure mother–infant pair bonding)	Prenatal, obstetric and socioeconomic factors including age, social class, ethnicity, parity alcohol, smoking birthweight, gestational age, gender
Baibazarova et al. 2013	Netherlands	158	15.3–18.2		3 months	Infant behaviour questionnaire-revised (Dutch translation)	Trend for an indirect effect of amniotic fluid cortisol on infant distress to limitation and fear via birthweight	Social class, obstetric history, age, education
Maternal blood								
Buitelaar et al. 2003	Netherlands	43	24	ACTH	3, 8 months	Bayley scales of infant development, temperamental questionnaire	High ACTH associated with difficulty and inadaptability of infants of 3 and 8 months of age	Gestational age at birth, birthweight, maternal postnatal stress and depression

(continued)

Table 1 (continued)

Study*	Country	n	Gestation mother assessed (weeks)	Maternal HPA axis assessment	Infant age at follow-up	Infant assessment	Main Findings	Confounders and covariates included in statistical analysis
Bolten et al. 2013	Switzerland	48	32–34	Cortisol measured during laboratory stress test (trier social stress test)	10 and 14 days and 6 months	Neonatal intensive care unit network neurobehavioral scale still face paradigm	Maternal cortisol reactivity to stress during pregnancy was associated with infant's emotion regulation at the age of 6 months	Maternal depressive symptoms
Baibazarova et al. 2013	Netherlands	158	15.3–18.2	Serum cortisol collected after amniocentesis	3 months	Infant behaviour questionnaire-revised (Dutch translation)	Maternal cortisol was associated with amniotic fluid cortisol. Trend for an indirect effect of amniotic fluid cortisol on infant distress to limitation and fear via birthweight	Social class, obstetric history, age, education

*Arranged in chronological order

borderline/clinically significant range were twice as likely to have been exposed to higher maternal cortisol levels compared to children with ratings in the normal range (odds ratio 2.1, 95 % confidence interval 1.1–3.9, $p < 0.05$). Most of these studies have included adjustment for several important confounding factors such as social class and many also include adjustment for current maternal psychological status in an attempt to distinguish between in utero and postnatal effects. Some studies have found that the relationship between higher cortisol levels and altered offspring behaviour is only seen with cortisol levels measured in late gestation (Buitelaar et al. 2003; Davis et al. 2007) suggesting there may be a window of susceptibility at which the developing foetus is most vulnerable to high cortisol levels. However, others have shown no gestation-specific effects (Davis and Sandman 2012) and more research is needed.

8 In Utero Stress and Infant HPA Axis Activity

There is increasing evidence that offspring exposed to in utero stress have altered activity of their own HPA axis (See Table 2). Infants of mothers with symptoms of depression have been noted to have higher urinary cortisol within the first week of life, consistent with overall increased HPA axis activity (Lundy et al. 1999). Several studies have measured cortisol in infant saliva to avoid invasive sample collection. The diurnal pattern of cortisol secretion appears to be altered in children exposed to in utero stress with high morning cortisol levels and then flattening of the day curve (Van den Bergh et al. 2008). The responsiveness of the HPA axis is also altered in relation to maternal mood with increased salivary cortisol response to separation/union stress (O'Connor et al. 2013a) in infants of mothers who were anxious in pregnancy. Intriguingly, the cortisol responses to stress appear to vary according to infant age and also according to the type of stressor with increased or decreased responsivity depending on age and stressor (Tollenaar et al. 2011). Likewise when cortisol responses were assessed in adolescence using a laboratory stress test (carbon dioxide inhalation), there was a blunted response in the children exposed to prenatal stress (Vedhara et al. 2012) suggesting that earlier observed increased responsiveness of the HPA axis may change over time.

Increased foetal glucocorticoid exposure has been suggested as a mechanism linking in utero stress with altered infant HPA axis activity. It is known that increased glucocorticoid exposure alters the set-point of the offspring HPA axis. Maternal and foetal/newborn cortisol levels are correlated (Gitau et al. 2004; Smith et al. 2011) and maternal cortisol levels are associated with reactivity of the newborn HPA axis as demonstrated by a study showing correlations between higher maternal cortisol levels in mid-late gestation and increased cortisol response in the newborn to the stress of a heel prick test (Davis et al. 2011). Higher maternal cortisol levels are also associated by increased cortisol responses in young childhood including responses to the physical stress of vaccination (Gutteling et al. 2004), or the psychological stress of the first day at school

Table 2 Assessment of infant hypothalamic–pituitary–adrenal axis in relation to in utero stress

Study*	Country	n	Infant HPA assessment	Infant age	Maternal assessment during pregnancy	Main findings
Lundy et al. 1999	US	35	Urinary cortisol first morning urine within 24 h of birth	1 week	CES-D, diagnostic interview schedule, foetal attachment scale, maternal stress interview and STAI at 32.3 weeks gestation	Urine cortisol higher in infants of depressed mothers
Kaplan et al. 2008	US	33	Baseline salivary cortisol (collected in the afternoon and prior to the assessment)	4 months	Psychiatric interview in 2nd trimester, CES-D and STAI in 3rd trimester	In women with an antenatal psychiatric diagnosis, infant cortisol levels were higher if the infant had insensitive versus sensitive caregiving, infant cortisol was low in healthy women regardless of sensitivity
Tollenaar et al. 2011	Netherlands	173	Infant salivary cortisol reactivity to bathing, vaccination, still face procedure and strange situation procedure	5 weeks, 8 weeks, 5 months and 12 months	General and pregnancy-related feelings of stress and anxiety, as well as circadian cortisol levels, measured in the last trimester of pregnancy	Maternal prenatal fear of bearing a handicapped child was a consistent predictor of infant cortisol reactivity. Higher fear was significantly related to higher salivary cortisol reactivity to the bathing session and to decreased cortisol reactivity to vaccination and maternal separation
O'Connor et al. 2013a	UK	125	Salivary cortisol before and 20 min after separation-union procedure	17 months	STAI, stressful life events questionnaire, EPDS at 17.2 weeks gestation; amniotic fluid cortisol	Amniotic fluid cortisol predicted infant cortisol response to separation–reunion stress: infants who were exposed to higher levels of cortisol in utero showed higher pre-stress cortisol values and blunted response to stress exposure

(continued)

Table 2 (continued)

Study*	Country	n	Infant HPA assessment	Infant age	Maternal assessment during pregnancy	Main findings
Gutteling et al. 2004	Netherlands	24	Salivary cortisol 1 day in weekend before vaccination, five samples every 3 h 8 am–8 pm. One sample before vaccination, 15–20–25–30 min after vaccination	4.9 years	Every day problem list, pregnancy-related anxiety questionnaire-revised, perceived stress scale and a salivary cortisol day curve (samples collected every 2 h between 8 am to 8 pm) at 16 weeks gestation	Children whose mother had higher morning cortisol levels in pregnancy had higher cortisol following vaccination. Daily hassle and fear of having handicapped child was also associated with significantly higher infant cortisol levels
Gutteling et al. 2005	Netherlands	29	Salivary cortisol after awakening, before lunch, after school (14:50–17:40 h), before bedtime at least 1 h after dinner, for 2 days	5.31 years	Every day problem list, pregnancy-related anxiety questionnaire-revised, perceived stress scale and a salivary cortisol day curve (samples collected every 2 h between 8 am to 8 pm) at 16 weeks gestation	Children whose mother had higher morning cortisol levels during pregnancy and fear of bearing a handicapped child had higher cortisol levels on first school day
O'Connor et al. 2005	UK	74	Salivary cortisol awakening, 30 min after awakening, 4, 9 pm, three consecutive school days with parents' help	10 years	Crown-Crisp index and EPDS at 18 and 32 weeks gestation	Prenatal anxiety (particularly at 32 weeks gestation) associated with increased wakening cortisol in child
Van den Bergh et al. 2008	Belgium	86	Salivary cortisol awakening, 4 h after awakening, 12 h after awakening, in weekend	14–15 years	STAI at 12–22, 23–31, 32–40 weeks gestation	Exposure to antenatal anxiety at 12–22 weeks was associated with a high, flat cortisol day profile in adolescents
Vedhara et al. 2012	UK	139	2 min before CO_2 inhalation, 10, 20, 30 min after inhalation	15.12 years	Crown-Crisp index and EPDS at 18 and 30 weeks gestation	Increased prenatal anxiety related to blunted cortisol responses

*Studies are ordered according to infant age at time of assessment
Abbreviations Centre for Epidemiological Studies Depression Scale (CES-D), Edinburgh Postnatal Depression Scale (EPDS)

(Gutteling et al. 2005). The changes in offspring HPA axis activity associated with glucocorticoid overexposure in utero may persist into adult life as low birth weight is associated with higher fasting cortisol levels (Phillips et al. 1998, 2000; Van Montfoort et al. 2005) and with activation of the HPA axis (Reynolds et al. 2001, 2005). Increased HPA axis activity has also been reported in 6–11 year olds born at term but who were exposed to antenatal glucocorticoids in utero (Alexander et al. 2012).

9 Changes in Placental Growth, Gene Expression and Epigenetic Modification in Association with In Utero Stress

The placenta is located at the interface between mother and foetus and is the key conduit of nutrient supply for the developing foetus. The placenta is capable of responding to changes in the maternal environment with a range of structural and functional adaptations (Lewis et al. 2006). Preliminary data suggests changes in placental growth and development can occur in response to maternal mood from early in pregnancy. For example, a very small study including only 18 women showed a correlation between maternal depressive symptoms and first trimester serum levels of vascular endothelial growth factor (VEGF), soluble fms-like tyrosine kinase 1 (sFlt-1) and placental growth factor, biomarkers of placental development (Fowles et al. 2011). Evidence that maternal mood is associated with altered placental growth also comes from a large study ($N = 78,017$) of singleton pregnancies which demonstrated an association between maternal life stress and increased placental weight at birth (Tegethoff et al. 2010). Emotional symptoms were not related to placental weight. Underlying mechanisms are unknown but are likely to include changes in growth factors within the placenta and altered blood supply with attendant changes in oxygen and nutrient supply. For example, a very small study suggested maternal stress may influence placental DEPTOR (DEP-domain containing and mTOR (mammalian target of rapamycin)-interacting protein), which may act as nutrient sensor in placenta (Mparmpakas et al. 2012), whilst another demonstrated that intrusive thought and emotional distress were associated with reduced foetoplacental blood flow in the third trimester (Helbig et al. 2013). Much more work is needed to understand the mechanisms linking maternal mood to placental development and growth.

A number of studies have started to examine whether in utero stress is associated with changes in placental gene expression with focus on a series of candidate genes involved in glucocorticoid metabolism and serotonin transfer. The placental enzyme 11-beta hydroxysteroid dehydrogenase type 2 (HSD2) metabolises 80–90 % of active maternal cortisol to inactive cortisone, protecting the foetus from excess glucocorticoid exposure (Edwards et al. 1993). Other key players in regulating glucocorticoid action, including glucocorticoid and mineralocorticoid receptors (NR3C1 and NR3C2, respectively) and 11-beta hydroxysteroid dehydrogenase type 1 (HSD1), which catalyses the regeneration of active glucocorticoids, are expressed in placental tissue (Murphy et al. 2006). The serotonin receptor (SLC6A4) mediates placental uptake of serotonin and the placenta is also able to synthesise serotonin from maternal tryptophan precursors (Bonnin et al. 2011). As with the studies examining maternal stress and offspring outcomes, there are variations in the methods used to assess in utero stress, e.g. retrospective/prospective data collection, use of different scales, etc., which means comparison of studies may not be straightforward. In addition, other factors such as mode of delivery, timing between delivery and placental biopsy may lead to differences in outcomes between studies.

O'Donnell et al. collected placental samples from 56 women at the time of elective caesarean section (O'Donnell et al. 2012). Symptoms of maternal anxiety and depression were assessed on the morning prior to the caesarean section using the Spielberger state and trait anxiety scale. This scale has been validated in pregnancy (Gunning et al. 2010) and extensively used to assess mood. They found increased anxiety on both scales (i.e. 'state' indicating anxiety at the time of the questionnaire and 'trait' indicating general levels of mood) was associated with down-regulation of HSD2 mRNA in the placenta. Decreased enzyme activity was confirmed in a smaller subset of the samples. The finding of decreased HSD2 mRNA coupled with decreased enzyme activity would potentially allow increased glucocorticoid exposure to the foetus in women with anxiety symptoms. In contrast, maternal depressive symptoms were not associated with any significant changes in placental HSD2 expression in this study (O'Donnell et al. 2012) or in a larger study (Ponder et al. 2011). However, the history of depression during pregnancy in both of these studies was obtained retrospectively, either through review of delivery and/or antenatal health records (Ponder et al. 2011) which may be subject to recall-bias or by questionnaire at time of elective caesarean section which may be highly influenced by the current situation (O'Donnell et al. 2012).

In addition to studies on genes regulating foetal glucocorticoid exposure, there has been interest in other signalling pathways including monoamines, such as serotonin as mood disorders are commonly attributed to disordered signalling in these pathways (Ressler and Nemeroff 2000). Extensive studies in animal models have demonstrated the important role of serotonin in foetal neurodevelopment (Velasquez et al. 2013). Studies in pregnant women diagnosed with mood disorders, or receiving treatment with selective serotonin reuptake inhibitors (SSRIs), are also consistent with a key role for serotonin pathways in foetal development (Oberlander 2012; Olivier et al. 2013), with effects persisting at least to childhood. The serotonin and glucocorticoid signalling systems also interact, with glucocorticoids regulating serotonin synthesis, transport, re-uptake and neuronal receptor expression, whilst serotonin controls glucocorticoid and mineralocorticoid receptor expression in the central nervous system (Wyrwoll and Holmes 2012). Recent data suggest maternal depression may influence placental 5-hydroxytriptamine (5-HT). mRNA levels of SLC6A4, the transmembrane serotonin transporter were increased in placentas from women with untreated mood disorders and from women treated with SSRIs, compared to controls (Ponder et al. 2011). A major mechanisms for removing 5HT is its metabolism into inactive 5-hydroxyindoleacetic acid by the enzyme monoamine oxidase A (MAO-A). This enzyme is present in the placenta (Zhang et al. 2010) and located in the syncytiotrophoblast (Blakeley et al. 2013) and maternal depression is associated with reduced MAO-A expression in term 'placenta' (Blakeley et al. 2013).

Mechanisms leading to the changes in placental gene expression in response to in utero stress are unknown but there has been much interest in whether epigenetic modifications (i.e. alterations in gene function in the absence of changes in the DNA sequence) could be an underlying mechanism. These epigenetic modifications include DNA methylation, post-translational modification of histones and non-coding RNAs. DNA methylation has been most studied in the context of early life programming. There is increasing evidence to suggest that DNA methylation is influenced by environmental cues (Jaenisch and Bird 2003) and thus can link in utero stress with permanent changes in the epigenome and life-long phenotypic consequences (Weaver et al. 2004).

A number of studies have started to investigate whether maternal mood in pregnancy influences foetal DNA methylation. A study of 82 women showed that prenatal exposure to maternal depression was associated with decreased methylation in umbilical cord leukocytes of the SLC6A4 gene (Devlin et al. 2010). In contrast, another small study including infants of depressed mothers treated with a serotonin reuptake inhibitor antidepressant ($n = 33$), infants of depressed mothers who were not treated ($n = 13$) and infants of normal control mothers ($n = 36$) found that maternal depression/anxiety in the third trimester was associated with increased methylation at the exon 1F promoter of the human GR gene (NR3C1) in cord blood (Oberlander et al. 2008). Variation in DNA methylation in cord blood at this and other loci of the NR3C1 gene was also associated with maternal emotional well-being (particularly anxiety) in another small study of 83 women (Hompes et al. 2013). Another study used a genome-wide approach to examine methylation patterns of >27,000 CpG sites across the genome in umbilical cord blood-derived DNA from the offspring of women undergoing treatment for a mood disorder during the neonatal period ($n = 201$) (Schroeder et al. 2012). There was no association between neonatal umbilical cord blood DNA methylation and maternal psychiatric diagnosis or clinically significant depressive symptoms. The largest study to date investigating maternal mood and offspring early life parameters included 508 infants whose mothers had completed a validated depression questionnaire. DNA methylation in cord blood at the regulatory sequence of the imprinted gene MEG3 differed significantly by maternal mood. Compared with infants of women without depressed mood, infants born to women with severe depressed mood had a 2.4 % higher methylation at the MEG3 DMR. There were no associations of maternal mood with the other imprinted genes studied including methylation of IGF2 and the linked gene H19 (Liu et al. 2012). There is some evidence that DNA methylation may be altered by antidepressant therapy. A study of 436 newborns in cord blood/placenta found a race-dependent change in methylation of the H19 DMRs in those whose mothers used antidepressant drugs during pregnancy (Soubry et al. 2011). Likewise, exposure to an antidepressant medication was associated with differential methylation of CpG sites in TNFRSF21 and CHRNA2 (Schroeder et al. 2012). TNFRSF21 is also known as death receptor 6 (DR6) and is expressed in developing neurons and is involved in refinement of neuronal connections during development. CHRNA2 is a broadly expressed subunit of nicotinic acetylcholine receptors and has been linked to neurocognitive functioning. In this study, the average difference in methylation for both CpG sites was less than 3 % between each group which may be of questionable clinical relevance.

10 Changes in Placental Growth, Gene Expression and Epigenetic Modification in Association with Offspring Behaviour

So are the changes in placental growth, gene expression and epigenetic modifications associated with altered offspring behaviour? There is some evidence that the growth of the placenta is associated with subsequent behavioural problems in the offspring. In the North Finland birth cohort 1986, placental weight and surface area was recorded and then the children ($n = 8101$) were assessed at 8 years for ADHD symptoms, probable psychiatric disturbance, antisocial disorder and neurotic disorder, and were assessed again at 16 years ($n = 6607$) for ADHD symptoms (Khalife et al. 2013). There was a positive association between placental size (placental weight, surface area and placental-to-birth weight ratio) and mental health problems in boys. Increased placental size was linked to overall probable psychiatric disturbance at age 8 years (OR 1.14, [95 % CI 1.04–1.25]), antisocial behaviour at age 8 years (OR 1.14 [95 % CI 1.03–1.27]) and ADHD symptoms (inattention-hyperactivity) at 16 years (OR 1.19 [95 % CI 1.02–1.38]), after adjusting for known confounders. This finding contrasts with another study including 4,976 participants in the North Finland birth cohort 1966 where small placental weight was associated with schizotypal traits in women at 31 years of age (Lahti et al. 2009). More work is needed to understand how the placenta responds to an in utero stress, either through diminished growth (Lahti et al. 2009) or by increasing growth as a compensatory mechanism to an adverse maternal environment as speculated by Khalife et al.

Although no studies have examined whether changes in placental gene expression are associated with later offspring behaviour, supportive evidence that changes in levels of placental HSD2 associate with altered offspring behaviour comes from observations of decreased verbal and visuo-spatial abilities and narrative memory in the 8-year-old children of women who consumed large quantities of liquorice during pregnancy (Räikkönen et al. 2009). Liquorice contains glycyrrhizin, an HSD inhibitor, and so these children are potentially exposed to more glucocorticoids in utero. Liquorice exposure is also associated with significant increases in externalising symptoms, attention, rule breaking and aggression problems with notably a 2.26 fold increase in attention deficit hyperactivity disorder (Räikkönen et al. 2009). In addition, there is some evidence linking DNA methylation changes at birth with neonatal stress responses and behaviour. A small study linked increased methylation at the NR3C1 promoter in cord blood of neonates to increased response to stress (Oberlander et al. 2008). In another study increased placental HSD2 methylation was associated with altered newborn behaviour with reduced scores of quality of movement using the NICU Network Neurobehavioural Scales (Marsit et al. 2012). The same researchers have subsequently shown that infants of mothers who were depressed during pregnancy and had greater methylation of placental NR3C1 CpG2 had poorer self-regulation, more hypotonia and more lethargy that whose mothers were not depressed (Conradt et al. 2013). Infants whose mothers were more anxious during pregnancy and had greater methylation of placental HSD2 CpG4 were more hypotonic compared with infants of mothers who were not anxious during pregnancy. These results are intriguing but do suggest the possibility of identifying biomarkers of future risk at birth by identifying epigenetic modifications in placenta.

11 In Utero Stress and Structural Changes in the Offspring Brain

Abnormalities identified in the brain using magnetic resonance imaging (MRI) correlate with a variety of neurological disorders including schizophrenia, autism, anxiety, depression and attention deficit hyperactivity disorder (Broyd et al. 2009; Philip et al. 2012). Recently, investigators have started to examine whether in utero stress is also associated with alterations in brain development using MRI. The relation between maternal antenatal depression and neural development in newborns (Rifkin-Graboi et al. 2013) was examined using a prospective birth cohort study (Growing up in Singapore Towards Healthy Outcomes [GUSTO]). The investigators focused on changes in the amygdala, a brain region that is important for emotional memory processing and regulates a variety of emotions including fear, depression and anxiety. They used structural MRI to examine the size of the amygdala, and diffusion tensor imaging (DTI) to derive fractional anisotropy and axial diffusivity, measures for characterising the microstructure of the amygdala in 157 newborns aged 6–14 days. Maternal depression was assessed at 26 weeks gestation using the Edinburgh Postnatal Depression Scale. The main finding was that there were changes in the neonatal microstructure (significantly lower fractional anisotropy and axial diffusivity but not volume) of the right amygdala in infants of mothers who had higher scores on the depression scale compared with lower scores. Underlying mechanisms are unknown, but in another study higher maternal cortisol levels measured in earlier but not later gestation were associated with a larger right amygdala volume measured by MRI in girls at age 7 years, but not in boys (Buss et al. 2012). The magnitude of effect was substantial, with a 1 SD increase in maternal cortisol being associated with an approximately 6.4 % increase in the size of the right amygdala. The higher maternal cortisol levels in early gestation were associated with more affective problems in girls, and this association was mediated in part by amygdala volume. New imaging techniques with 3 Tesla magnetic resonance imaging (3T MRI) (Anblagan et al. 2013) may allow measurement of brain volume and sulcal determination in the developing foetus in utero.

12 Can We Intervene to Improve Offspring Outcomes?

The strength of the association between in utero stress and adverse offspring behaviours implies that any intervention that could reduce in utero stress could have huge clinical implications for improving health in the offspring. A number of studies have started to look at ways to alter maternal behaviour during pregnancy. For example, deficiencies of key micronutrients such as folate, vitamin B12, calcium, iron, selenium, zinc and $n - 3$ fatty acids have been associated with low maternal mood in pregnancy (reviewed in Leung and Kaplan 2009), but to date dietary interventions, at least in the post-partum period, have been inconclusive in preventing post-partum depression (Miller et al. 2013). Cognitive behavioural therapy has been shown to be effective in improving maternal mood postnatally (O'Mahen et al. 2013) but this approach has not been tested during pregnancy. One of the largest studies with longest follow-up of the offspring suggesting that behavioural interventions in the mother are effective in improving offspring behaviour comes from follow-up studies from the Nurse Family Partnership randomised controlled trial (Olds et al. 1998; Olds 2002, 2008) Trained nurses visited single, poor, deprived mothers during pregnancy and in the first 2 years of the babies' life and gave the women advice about diet, health and education. The follow-up studies have shown a number of benefits for the offspring including better vocabulary at age 6 years (Olds et al. 2004), reading performance and maths test results at age 9 years (Olds et al. 2007), less substance abuse at 12 years (Olds et al. 2010) and less criminal behaviour at age 19 years (Eckenrode et al. 2011). Whilst the intervention was not specifically designed to target maternal mood or stress, these women were clearly from a group known to be at high risk of stress and this study does suggest that teaching the women about coping strategies and life skills and the supportive environment created had long-lasting beneficial effects on offspring behaviour.

13 Conclusion

An accumulating body of evidence supports a link between in utero stress and adverse offspring behaviours. Alterations in the activity of the maternal HPA axis and/or function of the placenta resulting in foetal glucocorticoid overexposure is a plausible underlying mechanism but more research is needed to understand the pathological pathways and to identify the gestation at which any intervention to prevent in utero stress may be most effective. Lifestyle and behavioural interventions to mothers during pregnancy have been shown to be effective at improving offspring behavioural outcomes, at least in deprived mothers, and more research is needed to design interventions that will be effective across a range of in utero stressors. The accumulating evidence of a link between in utero stress and adverse offspring behaviour suggests that health care professionals should be aware that maternal mental health can potentially impact on the health of both the mother and her child.

Acknowledgements We acknowledge the support of the British Heart Foundation and Tommys. THM is supported by a University of Edinburgh Principal's Award and Charles Darwin Fellowship.

References

Alexander N, Rosenlöcher F, Stalder T et al (2012) Impact of antenatal synthetic glucocorticoid exposure on endocrine stress reactivity in term-born children. J Clin Endocrinol Metab 97:3538–3544
Anblagan D, Yin K, Reynolds R et al (2013) Reliable measurement techniques for motion corrected Fetal brain volume. In: The international society for magnetic resonance in medicine, Salt Lake City, Utah, USA abstract
Baibazarova E, van de Beek C, Cohen-Kettenis PT et al (2013) Influence of prenatal maternal stress, maternal plasma cortisol and cortisol in the amniotic fluid on birth outcomes and child temperament at 3 months. Psychoneuroendocrinology 38:907–915
Barker DJP (1995) Fetal origins of coronary heart disease. Br Med J 311:171–174
Barker ED, Jaffe SR, Uher R et al (2011) The contribution of prenatal and postnatal maternal anxiety and depression to child maladjustment. Depress Anx 28:696–702
Bergman K, Glover V, Sarkar P et al (2010a) In utero cortisol and testosterone exposure and fear reactivity in infancy. Horm Behav 57:306–312
Bergman K, Sarkar P, Glover V et al (2010b) Maternal prenatal cortisol and infant cognitive development: moderation by infant–mother attachment. Biol Psychiatry 67:1026–1032
Blakeley PM, Capron LE, Jensen AB et al (2013) Maternal prenatal symptoms of depression and down regulation of placental monoamine oxidase A expression. J Psychosom Res 75:341–345
Bolten M, Nast I, Skrundz M et al (2013) Prenatal programming of emotion regulation: neonatal reactivity as a differential susceptibility factor moderating the outcome of prenatal cortisol levels. J Pyschosom Res 75:351–357
Bonnin A, Goeden N, Chen K et al (2011) A transient placental source of serotonin for the fetal forebrain. Nature 472:347–350
Brouwers EPM, Van Baar AL, Pop VJM (2001) Maternal anxiety during pregnancy and subsequent infant development. Infant Behavior Dev 24:95–106
Broyd SJ, Demanuele C, Debener S et al (2009) Default-mode brain dysfunction in mental disorders: a systematic review. Neurosci Biobehav Rev 33:279–296
Bublitz MH, Stroud LR (2012) Childhood sexual abuse is associated with cortisol awakening response over pregnancy: preliminary findings. Psychoneuroendocrinology 37:1425–1430
Buitelaar JK, Huizink AC, Mulder EJ et al (2003) Prenatal stress and cognitive development and temperament in infants. Neurobiol Aging 24:S53–S60
Buss C, Davis EP, Hobel CJ et al (2011) Maternal pregnancy-specific anxiety is associated with child executive function at 6–9 years age. Stress 14:665–676

Buss C, Davis EP, Shahbaba B et al (2012) Maternal cortisol over the course of pregnancy and subsequent child amygdala and hippocampus volumes and affective problems. Proc Natl Acad Sci USA 109:E1312–E1319

Campbell EA, Linton EA, Wolfe CD et al (1987) Plasma corticotropin-releasing hormone concentrations during pregnancy and parturition. J Clin Endocrinol Metab 64:1054–1059

Conradt E, Lester BM, Appleton AA et al (2013) The role of DNA methylation of NR3C1 and 11βHSD2 and exposure to maternal mood disorder in utero on newborn neurobehaviour. Epigenetics 8(12):1321–1329 (epub ahead of print)

Cottrell EC, Seckl JR (2009) Prenatal stress, glucocorticoids and the programming of adult disease. Front Behav Neurosci 3:19

Davis EP, Sandman C (2012) Prenatal psychobiological predictors of anxiety risk in preadolescent children. Psychoneuroendocrinol 37:1224–1233

Davis EP, Snidman N, Wadhwa PD et al (2004) Prenatal maternal anxiety and depression predict negative behavioral reactivity in infancy. Infancy 6:319–331

Davis EP, Glynn LM, Schetter CD et al (2007) Prenatal exposure to maternal depression and cortisol influences infant temperament. J Am Acad Child Adolesc Psychiatry 46:737–746

Davis EP, Glynn LM, Waffarn F et al (2011) Prenatal maternal stress programs infant stress regulation. J Child Psychol Psychiatry 52:119–129

De Weerth C, van Hees Y, Buitelaar JK (2003) Prenatal maternal cortisol levels and infant behaviour during the first 5 months. Early Hum Dev 74:139–151

De Weerth C, Buitelaar JK, Beijers R (2013) Infant cortisol and behavioural habituation to weekly maternal separations: links with maternal prenatal cortisol and psychosocial stress. Psychoneuroendocrinology 38:2863–2874

Devlin AM, Brain U, Austin J et al (2010) Prenatal exposure to maternal depressed mood and the MTHFR C677T variant affect SLC6A4 methylation in infants at birth. PLoS One 5:e12201

Eckenrode J, Campa M, Luckey DW et al (2011) Long-term effects of prenatal and infancy nurse home visitation on the life course of youths: 19-year follow-up of a randomized trial. Arch Paedtr Adolesc Med 164:9–15

Edwards CRW, Benediktsson R, Lindsay R et al (1993) Dysfunction of the placental glucocorticoid barrier: a link between the foetal environment and adult hypertension? Lancet 341:355–357

Entringer S, Buss C, Andersen J et al (2011) Ecological momentary assessment of maternal cortisol profiles over a multiple-day period predicts the length of human gestation. Psychosom Med 73:469–474

Field T, Sandberg D, Garcia R et al (1985) Pregnancy problems, postpartum depression and early mother–infant interactions. Dev Psychol 21:1152–1156

Field T, Diego M, Dieter J et al (2006) Prenatal depression effects on the fetus and the newborn: a review. Inf Behav Develop 29:445–455

Field T, Diego MA, Hernandez-Reif M et al (2008) Prenatal dysthymia versus major depression effects on maternal cortisol and fetal growth. Depress. Anx 25:E1

Fowles ER, Murphey C, Ruiz RJ (2011) Exploring relationships among psychosocial status, dietary quality, and measures of placental development during the first trimester in low-income women. Biol Res Nurs 13:70–79

Gavin NI, Gaynes BN, Lohr KN et al (2005) Perinatal depression: a systematic review of prevalence and incidence. Obstet Gynecol. 106:1071–1083

Gitau R, Fisk NM, Glover V (2004) Human fetal and maternal corticotrophin releasing hormone responses to acute stress. Arch Dis Child Fetal Neonatal Ed 89(1):F29–F32

Godfrey KM, Gluckman PD, Hanson MA (2010) Developmental origins of metabolic disease: life course and intergenerational perspectives. Trends Endocrinol Metab 21:199–205

Goland RS, Jozak S, Conwell I (1994) Placental corticotropin-releasing hormone and the hypercortisolism of pregnancy. Am J Obstet Gynecol 171:1287–1291

Gotlib IH, Joorman J, Minor KL et al (2008) HPA axis reactivity: a mechanism underlying the associations among serotonin TLPR stress, and depression. Biol Psychiatry 63(9):847–851

Gunning MD, Denison FC, Stockley CJ et al (2010) Assessing maternal anxiety in pregnancy with the State-Trait Anxiety Inventory (STAI): issues of validity, location and participation. J Reprod Infant Psyc 28(3):266–273

Gutteling BM, de Weerth C, Buitelaar JK (2004) Maternal prenatal stress and 4–6 year old children's salivary cortisol concentrations pre- and post-vaccination. Stress 7:257–260

Gutteling BM, de Weerth C, Buitelaar JK (2005) Prenatal stress and children's cortisol reaction to the first day of school. Psychoneuroendocrinology 30:541–549

Harville EW, Savitz DA, Dole N et al (2009) Stress questionnaires and stress biomarkers during pregnancy. J Women's Health 18:1425–1433

Hay DF, Pawlby S, Waters CS et al (2008) Antepartum and postpartum exposure to maternal depression: different effects on different adolescent outcomes. J Child Psychol Psychiatry 49:1079–1088

Helbig A, Kaasen A, Malt UF et al (2013) Does antenatal maternal psychological distress affect placental circulation in the third trimester? PloS One 8:e57071

Hompes T, Izzi B, Gellens E et al (2013) Investigating the influence of maternal cortisol and emotional state during pregnancy on the DNA methylation status of the glucocorticoid receptor gene (NR3C1) promoter region in cord blood. J Psychiatr Res 47:880–891

Huizink AC (2002) Psychological measures of prenatal stress as predictor of infant temperament. J Am Acad Child Adolesc Psychiatry 41:1078–1085

Huizink AC, Robles de Medina PG, Mulder EJ et al (2003) Stress during pregnancy is associated with developmental outcome in infancy. J Child Psychol Psychiatry 44:1025–1036

Huizink AC, Mulder EJ, Buitelarr JK (2004) Prenatal stress and risk for psychopathology: specific effects of induction of general susceptibility? Psychol Bull 130(1):115–142

Jaenisch R, Bird A (2003) Epigenetic regulation of gene expression: how the genome integrates intrinsic and environmental signals. Nature Genetics 33:245–254

Jung C, Ho JT, Torpy DJ et al (2011) A longitudinal study of plasma and urinary cortisol in pregnancy and postpartum. J Clin Endocrinol Metab 96:1533–1540

Kang HJ, Kawasawa YI, Cheng F et al (2011) Spatio-temporal transcriptome of the human brain. Nature 478:483–489

Kaplan LA, Evans L, Monk C (2008) Effects of mothers' prenatal psychiatric status and postnatal caregiving on infant biobehavioral regulation: can prenatal programming be modified? Early Hum Develop 84:249–256

Khalife N, Glover V, Hartikainen AL et al (2013) Prenatal glucocorticoid treatment and later mental health in children and adolescents. PLoS ONE 8:e81394

Khashan AS, Abel KM, McNamee R et al (2008) Higher risk of offspring schizophrenia following antenatal maternal exposure to severe adverse life events. Arch Gen Psychiatry 65:146–152

Kinnney DK, Miller AM, Crowley DJ et al (2008) Autism prevalence following prenatal exposure to hurricanes and tropical storms in Louisiana. J Autism Dev Disord 38:481–488

Kleinhaus K, Harlap S, Perrin M et al (2013) Prenatal stress and affective disorders in a population birth cohort. Bipolar Disord 15:92–99

Lahti J, Räikkönen K, Sovio U et al (2009) Early-life origins of schizotypal traits in adulthood. Br J Psychiatry 195:132–137

Laplante DP, Barr RG, Brunet A et al (2004) Stress during pregnancy affects general intellectual and language functioning in human toddlers. Pediatr Res 56:400–410

Lereya ST, Wolke D (2013) Prenatal family adversity and maternal mental health and vulnerability to peer victimisation at school. J Child Psychol Psychiatry 54:644–652

Leung BMY, Kaplan BJ (2009) Perinatal depression: prevalence, risks, and the nutrition link—a review of the literature. J Am Diet Assoc 109:1566–1575

Lewis RM, Poore KR, Godfrey KM (2006) The role of the placenta in the developmental origins of health and disease–implications for practice. Rev Gynaecol Perinatal Pract 6:70–79

Lindsay JR, Nieman LK (2005) The hypothalamic–pituitary–adrenal axis in pregnancy: challenges in disease detection and treatment. Endocr Rev 26:775–799

Liu Y, Murphy SK, Murtha AP et al (2012) Depression in pregnancy, infant birth weight and DNA methylation of imprint regulatory elements. Epigenetics 7:735–746

Lou H, Hansen D, Nordentoft M, Pryds O et al (1994) Prenatal stressors of human life affect fetal brain development. Dev Med Child Neurol 36:826–832

Lundy BL, Jones NA, Field T et al (1999) Prenatal depression effects on neonates. Infant Behav Develop 22:119–129

Marsit CJ, Maccani MA, Padbury JF et al (2012) Placental 11-Beta Hydroxysteroid Dehydrogenase Methylation Is Associated with Newborn Growth and a Measure of Neurobehavioral Outcome. Plos One 7(3):e33794

McEwen BS (1999) Stress and the aging hippocampus. Front Neuroendocrinol 20:49–70

Miller BJ, Murray L, Beckmann MM et al (2013) Dietary supplements for preventing postnatal depression. Cochrane Database Syst Rev 10:CD009104

Mozurkewich EL, Luke B, Avni M et al (2000) Working conditions and adverse pregnancy outcome: a meta-analysis. Obstetr Gynecol 95:623–635

Mparmpakas D, Zachariades E, Goumenou A et al (2012) Placental DEPTOR as a stress sensor during pregnancy. Clin Sci 122:349–359

Mulder EJ, Robles de Medina PG, Huizink AC et al (2002) Prenatal maternal stress: effects on pregnancy and the (unborn) child. Early Hum Dev 70:3–14

Murphy VE, Smith R, Giles WB et al (2006) Endocrine regulation of human fetal growth: the role of the mother, placenta and fetus. Endocr Rev 27:141–169

Nast I, Bolten M, Meinlschmidt G et al (2013) How to measure prenatal stress? A systematic review of psychometric instruments to assess psychosocial stress during pregnancy. Paediatr Perinat Epidemiol 27:313–322

O'Connor TG, Heron J, Golding J et al (2002) Maternal antenatal anxiety and children's behavioural/emotional problems at 4 years: report from the Avon longitudinal study of parents and children. Br J Psychiatry 180:502–508

O'Connor TG, Heron J, Golding J et al (2003) Maternal antenatal anxiety and behavioural/emotional problems in children: a test of a programming hypothesis. J Child Psychol Psychiatry 44:1025–1036

O'Connor TG, Ben-Shlomo Y, Heron J et al (2005) Prenatal anxiety predicts individual differences in cortisol in pre-adolescent children. Biol Psychiatry 58:211–217

O'Connor TG, Caprariello P, Blackmore ER et al (2007) Prenatal mood disturbance predicts sleep problems in infancy and toddlerhood. Early Hum Dev 83:451–458

O'Connor TG, Bergman K, Sarkar P et al (2013a) Prenatal cortisol exposure predicts infant cortisol response to acute stress. Dev Psychobiol 55:145–155

O'Connor TG, Tang W, Gilchrist MA et al (2013b) Diurnal cortisol patterns and psychiatric symptoms in pregnancy: short-term longitudinal study. Biol Psychol 96C:35–41

O'Donnell KJ, Bugge Jensen A, Freeman L et al (2012) Maternal prenatal anxiety and downregulation of placental 11β-HSD2. Psychoneuroendocrinology 37:818–826

O'Keane V, Lightman S, Marsh M et al (2011) Increased pituitary-adrenal activation and shortened gestation in a sample of depressed pregnant women: A pilot study. J Affect Disord 130:300–305

O'Mahen H, Himle JA, Fedock G et al (2013) A Pilot randomized controlled trial of cognitive behavioural therapy for perinatal depression adapted for women with low incomes. Depress Anx 30:679–687

Obel C, Hedegaard M, Henriksen TB et al (2005) Stress and salivary cortisol during pregnancy. Psychoneuroendocrinology 30:647–656

Oberlander TF (2012) Fetal serotonin signaling: setting pathways for early childhood development and behavior. J Adolesc Health 52(S2):S9–S16

Oberlander TF, Weinberg J, Papsdorf M et al (2008) Prenatal exposure to maternal depression, neonatal methylation of human glucocorticoid receptor gene (NR3C1) and infant cortisol stress responses. Epigenetics 3:97–106

Olds DL (2002) Prenatal and infancy home visiting by nurses: from randomized trials to community replication. Prev Sci 3:153–172

Olds DL (2008) Preventing child maltreatment and crime with prenatal and infancy support of parents: the nurse-family partnership. J Scand Stud Criminol Crime Prev 9:2–24

Olds D, Henderson CR Jr, Cole R et al (1998) Long-term effects of nurse home visitation on children's criminal and antisocial behaviour: 15 year follow-up of a randomized controlled trial. JAMA 280:1238–1244

Olds DL, Kitzman H, Cole R et al (2004) Effects of nurse home-visiting on maternal life course and child development: age 6 follow-up results of a randomized trial. Pediatrics 114:1550–1559

Olds DL, Kitzman H, Hanks C et al (2007) Effects of nurse home visiting on maternal and child functioning: age-9 follow-up of a randomized trial. Pediatrics 120:e832–e845

Olds DL, Kitzman HJ, Cole RE et al (2010) Enduring effects of prenatal and infancy home visiting by nurses on maternal life course and government spending: follow-up of a randomized trial among children at age 12 years. Arch Paediatr Adolesc Med 164:419–424

Olivier JD, Akerud H, Kaihola H et al (2013) The effects of maternal depression and maternal selective serotonin reuptake inhibitor exposure on offspring. Front Cell Neurosci 7:73

Pawlby S, Hay DF, Sharp D et al (2009) Antenatal depression predicts depression in adolescent offspring: prospective longitudinal community-based study. J Affect Disord 113:236–343

Pawlby S, Hay D, Sharp D et al (2011) Antenatal depression and offspring psychopathology: the influence of childhood maltreatment. Br J Psychiatry 199:106–112

Pearson RM, Evans J, Kounali D et al (2013) Maternal depression during pregnancy and the postnatal period: risks and possible mechanisms for offspring depression at age 18 years. JAMA Psychiatry 70:1312–1319

Philip RC, Dauvermann MR, Whalley HC et al (2012) A systematic review and meta-analysis of the fMRI investigation of autism spectrum disorders. Neurosci Biobehav Rev 36:901–942

Phillips DI, Barker DJ, Fall CH et al (1998) Elevated plasma cortisol concentrations: a link between low birth weight and the insulin resistance syndrome? J Clin Endocrinol Metab 83:757–760

Phillips DI, Walker BR, Reynolds RM et al (2000) Low birth weight predicts elevated plasma cortisol concentrations in adults from 3 populations. Hypertension 35:1301–1306

Ponder KL, Salisbury A, McGonnigal B et al (2011) Maternal depression and anxiety are associated with altered gene expression in the human placenta without modification by antidepressant use: implications for fetal programming. Dev Psychobiol 53:711–723

Räikkönen K, Pesonen AK, Heinonen K et al (2009) Maternal licorice consumption and detrimental cognitive and psychiatric outcomes in children. Am J Epidemiol 170:1137–1146

Ressler KJ, Nemeroff CB (2000) Role of serotonergic and noradrenergic systems in the pathophysiology of depression and anxiety disorders. Depress Anxiety 12(S1):2–19

Reynolds RM (2013) Glucocorticoid excess and the developmental origins of disease: two decades of testing the hypothesis—2012 Curt Richter Award Winner. Psychoneuroendocrinology 38(1):1–11

Reynolds RM, Walker BR, Syddall HE et al (2001) Altered control of cortisol secretion in adult men with low birth weight and cardiovascular risk factors. J Clin Endocrinol Metab 86:245–250

Reynolds RM, Walker BR, Syddall HE et al (2005) Is there a gender difference in the associations of birthweight and adult hypothalamic–pituitary–adrenal axis activity? Eur J Endocrinol 152:249–253

Rice F, Harold GT, Bolvin J et al (2010) The links between prenatal stress and offspring development and psychopathology: disentangling environmental and inherited influences. Psychol Med 40:345–355

Rich-Edwards JW, Mohllajee AP, Kleinman K et al (2008) Elevated midpregnancy corticotropin-releasing hormone is associated with prenatal, but not postpartum, maternal depression. J Clin Endocrinol Metab 93:1946–1951

Rieger M, Pirke K-M, Buske-Kirschbaum A (2004) Influence of stress during pregnancy on HPA activity and neonatal behavior. Ann NY Acad Sci 1032:228–230

Rifkin-Graboi A, Bai J, Chen H et al (2013) Prenatal maternal depression associates with microstructure of right amygdala in neonates at birth. Biol Psychiatry 74:837–844

Robinson DP, Klein SL (2012) Pregnancy and pregnancy-associated hormones alter immune responses and disease pathogenesis. Horm Behav 62:263–271

Rodriguez A, Bohlin G (2004) Are maternal smoking and stress during pregnancy related to ADHD symptoms in children? J Child Psychol Psychiatry 46:246–254

Sarkar P, Bergman K, Fisk NM et al (2007) Ontogeny of foetal exposure to maternal cortisol using midtrimester amniotic fluid as a biomarker. Clin Endocrinol 66:636–640

Schroeder JW, Smith AK, Brennan PA et al (2012) DNA methylation in neonates born to women receiving psychiatric care. Epigenetics 7:409–414

Smith AK, Jeffrey Newport D, Ashe MP (2011) Predictors of neonatal hypothalamic-pituitary-adrenal axis activity at delivery. Clin Endocrinol. doi:10.1111/j.1365-2265.2011.03998.x

Soubry A, Murphy S, Huang Z et al (2011) The effects of depression and use of antidepressive medicines during pregnancy on the methylation status of the IGF2 imprinted control regions in the offspring. Clinical Epigenetics 3:2

Talge NM, Neal C, Glover V (2007) Antenatal maternal stress and long-term effects on child neurodevelopment: how and why? J Child Psychol Psychiatry 48:245–261

Tegethoff M, Greene N, Olsen J et al (2010) Maternal psychosocial stress during pregnancy and placenta weight: evidence from a national cohort study. PLoS ONE 5:e14478

Tollenaar MS, Beijers R, Jansen J et al (2011) Maternal prenatal stress and cortisol reactivity to stressors in human infants. Stress 14:53–65

Van Batenburg-Eddes T, de Groot L, Huizink AC et al (2009) Maternal symptoms of anxiety during pregnancy affect infant neuromotor development: the generation R study. Dev Neuropsychol 34:476–943

Van den Bergh BRH (1990) The influence of maternal emotions during pregnancy on fetal and neonatal behaviour. Pre Perinat Psychol J 5:119–130

Van den Bergh BRH, Marcoen A (2004) High antenatal maternal anxiety is related to ADHD symptoms, externalizing problems and anxiety in 8/9 year olds. Child Dev 75:1085–1097

Van den Bergh BRH, Mennes M, Oosterlaan J et al (2005a) High antenatal maternal anxiety is related to impulsivity during performance on cognitive tasks in 14- and 15-year-olds. Neurosci Biobehav Rev 29:259–269

Van den Bergh BRH, Mulder EJH, Mennes M et al (2005b) Antenatal maternal anxiety and stress and the neurobehavioural development of the fetus and child: links and possible mechanisms. A review. Neurosci Biobehav Rev 29:237–258

Van den Bergh BRH, Van Calster B, Smits T et al (2008) Antenatal maternal anxiety is related to HPA-axis dysregulation and self-reported depressive symptoms in adolescence: a prospective study on the fetal origins of depressed mood. Neuropsychopharmacology 33:536–545

Van Montfoort N, Finken MJ, le Cessie S et al (2005) Could cortisol explain the association between birth weight and cardiovascular disease in later life? A meta-analysis. Eur J Endocrinol 153:811–817

Van Os J, Selten JP (1998) Prenatal exposure to maternal stress and subsequent schizophrenia. The May 1940 invasion of the Netherlands. Br J Psychiatry 172:324–326

Vaughn BE, Bradley C, Joffe L et al (1987) Maternal characteristics measured prenatally are predictive of ratings of temperamental "difficulty" on the Carey infant temperament questionnaire. Dev Psychol 23:152–161

Vedhara K, Metcalfe C, Brant H et al (2012) Maternal mood and neuroendocrine programming: effects of time of exposure and sex. J Neuroendocrinol 24:999–1011

Velasquez JC, Goeden N, Bonnin A (2013) Placental serotonin: implications for the developmental effects of SSRIs and maternal depression. Front Cell Neurosci 7:47

Voegtline KM, Costigan KA, Kivlighan KT et al (2012) Concurrent levels of maternal salivary cortisol are unrelated to self-reported psychological measures in low-risk pregnant women. Arch Womens Ment Health 16:101–108

Weaver IC, Cervoni N, Champagne FA et al (2004) Epigenetic programming by maternal behavior. Nature Neuroscience 7:847–854

Wurmser H, Rieger M, Domogalla C et al (2006) Association between life stress during pregnancy and infant crying in the first six months postpartum: a prospective longitudinal study. Early Human Dev 82:341–349

Wyrwoll CS, Holmes MC (2012) Prenatal excess glucocorticoid exposure and adult affective disorders: a role for serotonergic and catecholamine pathways. Neuroendocrinol 95:47–55

Yehuda R, Engel SM, Brand SR et al (2005) Transgenerational effects of posttraumatic stress disorder in babies of mothers exposed to the World Trade Center attacks during pregnancy. J Clin Endocrinol Metab 90:4115–4118

Zhang H, Smith GN, Liu X et al (2010) Association of MAOA, 5-HTT, and NET promoter polymorphisms with gene expression and protein activity in human placentas. Physiol Genomics 42:85–92

Does Stress Elicit Depression? Evidence From Clinical and Preclinical Studies

Helle M. Sickmann, Yan Li, Arne Mørk, Connie Sanchez and Maria Gulinello

Abstract Exposure to stressful situations may induce or deteriorate an already existing depression. Stress-related depression can be elicited at an adolescent/adult age but evidence also shows that early adverse experiences even at the fetal stage may predispose the offspring for later development of depression. The hypothalamus–pituitary–adrenal axis (HPA-axis) plays a key role in regulating the stress response and dysregulation in the system has been linked to depression both in humans and in animal models. This chapter critically reviews clinical and preclinical findings that may explain how stress can cause depression, including HPA-axis changes and alterations beyond the HPA-axis. As stress does not elicit depression in the majority of the population, this motivated research to focus on understanding the biology underlying resilient versus sensitive subjects. Animal models of depression have contributed to a deeper understanding of these mechanisms. Findings from these models will be presented.

Keywords Depression · Clinical studies · Animal models · Stress resilience · Vulnerability factors

---

H. M. Sickmann
Department of Drug Design and Pharmacology, Faculty of Health and Medical Sciences, University of Copenhagen, Copenhagen, Denmark

H. M. Sickmann · A. Mørk
Lundbeck Research DK, Ottiliavej 9, Copenhagen DK-2500, Denmark

Y. Li · C. Sanchez (✉)
Lundbeck Research USA, 215 College Road, Paramus, NJ 07652, USA
e-mail: cs@lundbeck.com

M. Gulinello
Albert Einstein College of Medicine, Yeshiva University,
1410 Pelham Pkwy S, Bronx, NY 10461, USA

Contents

1	Introduction	124
2	The Stress System	124
3	The Stress System and Depression: Is There a Link?	127
	3.1 Stressors and Risk of Depression	127
	3.2 HPA-Axis Dysregulation in Depression	128
4	Stress-Based Animal Models of Depression	130
	4.1 Chronic Single Stress Models of Depression in Adult Animals	136
	4.2 Multiple Chronic Stressors in Adult Animals	137
	4.3 Early Life Stress	139
5	Beyond the HPA-Axis: How Might Stress Cause Depression?	140
	5.1 Mechanisms of Stress and Depression in Humans	140
	5.2 Mechanisms of Stress and Depression in Animal Models	141
	5.3 Susceptibility and Resistance to Depression	142
6	Summary and Future Perspectives	143
References		144

1 Introduction

Stressful events may provoke depressive symptoms or exacerbate an already existing depression. It is currently under debate whether affective illness results from the negative effects of stressors, or rather if the differential physiological and cognitive processes that underlie depression cause dysfunctional response to stressors, or some combination of the two. While it is clear that the risk of depression is correlated with increased incidence and duration of stressors, the majority of persons undergoing traumatic and stressful events do not develop depression and, furthermore, no specific trauma or set of stressors can adequately account for the majority of instances of depression. Here, we review and discuss the current clinical and preclinical evidence relating to the disease biology underlying stress and its relationship to depression. Furthermore, we will also review evidence that questions the link between stress and depression.

2 The Stress System

A normal stress response involves a switch from basal activity to a stress reactivity phase in which plasma and brain stress hormones increase. This is followed by a stress recovery phase in which stress hormones return to baseline levels following the removal of the stimulus (McEwen 1998) primarily via direct negative feedback but also potentially via various other peptides that have all been related to depression (Dubrovsky 2000; Swaab et al. 2005).

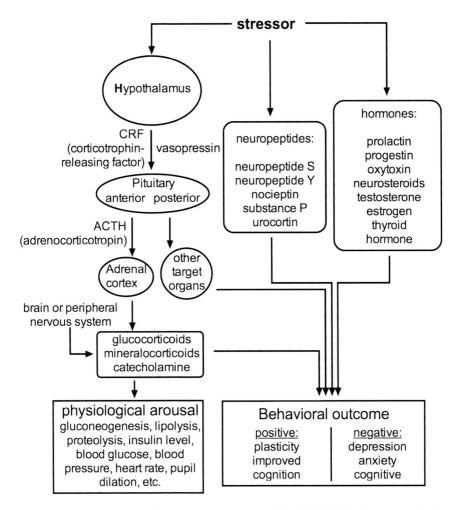

Fig. 1 Simplified summary of the stress response network. Both the HPA-axis system and other effectors are involved in stress response, inducing various physical and affective changes. Not included in the diagram are feedback controls occurring on various levels of this network

Glucocorticoids and their nuclear receptors have become almost synonymous with stress and clearly play an important role in regulation of stress outcomes. However, the stress response is more complex, involving layers of regulation and crosstalk between different systems. The most rapid physiological responses to stressors involve activation of the sympathetic nervous system and the release of adrenaline which mediate sympathomimetic changes such as cardiovascular changes and pupil dilation. This is followed by activation of the hypothalamic–pituitary–adrenal axis (HPA-axis) which is the most common operationally defined outcome of stress. In response to stress, the paraventricular nucleus (PVN) of the hypothalamus secretes corticotropin-releasing factor (CRF), which stimulates the synthesis and release of adrenocorticotropin (ACTH) from the anterior pituitary. ACTH then stimulates the synthesis and release of glucocorticoids (primarily cortisol in humans and corticosterone in rodents) from the adrenal cortex [Fig. 1; (Nestler et al. 2002; McEwen 1998; Aguilera et al. 2008)].

Even at this early level of regulation to the stress response, there may be critical differences between the response to acute and chronic stressors that may have implications for the susceptibility to the negative effects of stressors. During acute stress CRF and vasopressin (released from the supraoptic nuclei of the hypothalamus) seem to induce the release of ACTH from the anterior pituitary in a synergistic manner but during repeated stress there is evidence of a switch from CRF-ergic to vasopressinergic drive of the PVN (Aguilera et al. 2008; Lightman 2008) [however see (Chen et al. 2008; Bergström et al. 2008)]. Vasopressin deficient rats show reduced depression-like behavior, though these have normal basal and stress-induced corticosterone levels (Mlynarik et al. 2007). Vasopressin is also one of the peptide hormones important in regulation of social behavior, and may thus be important in this aspect of depression (Litvin et al. 2011; Egashira et al. 2007).

The activity of the HPA-axis is controlled by several other brain regions, including the hippocampus, which exerts an inhibitory influence on hypothalamic CRF-containing neurons, and the amygdala, that appears to have an excitatory influence (McEwen 2000; Sapolsky 2000; Nestler et al. 2002). High levels of glucocorticoids are normally self-limiting, as these exert negative feedback, via regulation of the hypothalamus and hippocampus.

There are two types of intracellular nuclear receptors mediating the effects of circulating glucocorticoids,—the mineralocorticoid receptor (MR, type I) and the glucocorticoid receptor (GR, type II) (McEwen 2000; de Kloet et al. 1999). The MR has equal affinity (high) for mineralocorticoids (e.g., aldosterone) and glucocorticoids and is thought to be critical in regulating activity during low glucocorticoid levels, such as circadian variation. The GR is a low-affinity receptor, and appears to be involved in the modulation of actions during high levels of glucocorticoids and in the negative feedback response (Heuser et al. 1994). Glucocorticoids also have rapid effects via interaction with membrane bound receptors (french-Mullen 1995; Towle and Sze 1983).

In addition to glucocorticoids, stressors reliably alter numerous other factors that should also be considered an integral part of the stress response. These include catecholamines, growth hormone, prolactin, various peptides, progesterone, testosterone, and thyroid hormones, the levels of which are reliably altered in response to stressors (Torner and Neumann 2002; Wuttke et al. 1987; Kant et al. 1987). The role of several of these "forgotten" stress hormones and peptides has been inadequately studied, despite the fact that these can profoundly affect mood. Progesterone levels, for example, are increased within 5 min of a stressful stimuli in both sexes and is primarily of adrenal origin (Deis et al. 1989). Progestins can also bind to both the MR and the GR (Myles and Funder 1996; Quinkler et al. 2002). Furthermore, estrogens regulate cortisol binding proteins which can, in turn, affect cortisol levels (Swaab et al. 2005). Cortisol levels can also be regulated by other steroid hormones, such as estrogens and androgens, in other ways. Gonadal steroid hormones receptors are present in CRH containing neurons in the PVN (Swaab et al. 2005) and the CRH promoter contains both androgen and estrogen response elements (Vamvakopoulos and Chrousos 1993; Bao et al. 2005). It is thus not surprising that estrogens may regulate susceptibility to depression-like outcomes after stressors (Li et al. 2010a). Elucidating the crosstalk between these hormones and the glucocorticoids may help close some of the gaps in our understanding of depression and may be useful to parse out some of the apparently idiosyncratic symptoms of depression, such as weight gain, sleep disruption, and changes in appetite in addition to sex differences in the response to stressors and in the susceptibility to depression.

3 The Stress System and Depression: Is There a Link?

The evidence for a relationship between stress and depression comes mainly from two lines of research: first, the observation that exposure to stressful events is correlated with the incidence of depression and second, the evidence of HPA-axis dysregulation in a subset of depressed patients. There are elsewhere excellent reviews detailing the evidence and history of the stress-induced depression hypothesis (Holsboer 2000, 2001; van Praag 2004; Kendler et al. 1999; Kessler 1997; Stetler and Miller 2011).

3.1 Stressors and Risk of Depression

Numerous studies suggest that exposure to stressful life events is a major correlate with depression (Schmidt et al. 2010b; Kendler et al. 1999; Tao et al. 2011; Brown et al. 1987; Strauss et al. 2011; Liu and Alloy 2010) and for subsequent HPA-axis dysfunction (Nemeroff et al. 1984; Frodl et al. 2010; Risch et al. 2009). Long-term exposure to particularly uncontrollable and unpredictable life stressors is often said to be a major factor in the development of depression (Kessler 1997; Kendler et al. 1999). Stressors are associated with the onset, symptom severity, and relapse of depressive disorders (Kendler et al. 1999; Lewinsohn et al. 1999; Hammen et al. 1992; Burke et al. 2005). However, there are several limitations of this proposed relationship. While some studies report changes in basal or stimulated stress responses in depressed patients, others have failed to replicate some of these phenomena (Croes et al. 1993). In many cases stressful events do not elicit depressive episodes and conversely, many episodes of depression occur in the absence of considerable life stress (Bonanno 2004).

In fact, the relationship between stress and depression is primarily based on correlational data, thus there are alternative explanations that do not infer stress as a causal factor in depression. These include the proposal that higher numbers of stressful events occur as the result of depression (Hamilton et al. 2013; Lyubomirsky et al. 1998). There may also be an increased propensity in depressed patients to perceive and report events as stressful (Hammen 1991; Lyubomirsky et al. 1998; Hamilton et al. 2013; Power et al. 2013). A prolonged physiological response to stressors evident in depression (Siegle et al. 2001) in concert with activation in brain areas (such as the amygdala) responsible for encoding emotional features would be consistent with this possibility. Thus, although there seems to be a link between stress and depression, stress is neither necessary nor sufficient to cause depression in the majority of the population.

3.2 HPA-Axis Dysregulation in Depression

The dysregulation of the HPA-axis evident in depressed patients is one foundation of the stress-depression hypothesis. Deregulated HPA-axis function can include altered basal HPA activity (Halbreich et al. 1985), altered negative feedback mechanisms in the HPA-axis, such as dexamethasone nonsuppression, and/or hyper-reactivity of the stress response (see Fig. 2).

Some studies indicate that depressed patients have elevated basal cortisol levels (Halbreich et al. 1985; Young et al. 1994), but other studies have reported no effect (Burke et al. 2005). Generally, when more specific inclusion criteria are used (i.e., diagnostic groups, hospitalization status, age), higher numbers of depressed individuals with greater median cortisol levels compared to nondepressed individuals are reported. Moreover, a recent meta-analysis demonstrated that HPA-axis activity may vary between diagnostic groups as patients with psychotic and melancholic depression are more likely to display HPA-axis hyperactivity (Stetler and Miller 2011), substantiating the hypothesis that some individuals may be susceptible to HPA-axis dysregulation during depression.

Alternatively to basal stress hyperactivity, failure of negative feedback may result also from exposure to chronic stress and can be evident in depressed patients. Administration of low doses of the exogenous steroid, dexamethasone, normally inhibits endogenous cortisol secretion, but in depressed patients dexamethasone fails to lower endogenous cortisol (a.k.a. dexamethasone nonsuppression). However, dexamethasone nonsuppression occurs in less than half of depressed subjects (Rubin et al. 1987; Holsboer et al. 1980). Other assays for HPA-axis negative feedback (combined dexamethasone/CRH test) may increase the sensitivity for detecting failure of negative feedback in depressed patients. In studies where this test was employed, as many as 90 % of depressed patients had some failure of feedback when the subjects were also grouped into different age ranges (Heuser et al. 1994; Watson et al. 2006). It should also be noted that about 30 % of subjects without depression also exhibit dexamethasone nonsuppression (Silver 1986).

It has also been suggested that the acute stress response may be hyperactive in depressed subjects (Schmidt et al. 2010b; Nestler et al. 2002; Trestman et al. 1991).

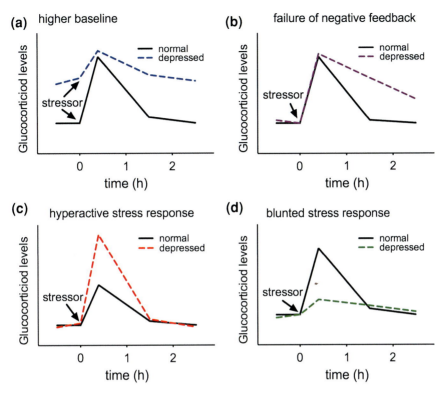

Fig. 2 HPA-axis dysfunction in depression. Using plasma glucocorticoid levels as an example we here illustrate the several types of abnormal stress responses displayed in subsets of depressed patients. **a** Some patients may have increased basal cortisol levels. **b** A prolonged elevation of the hormone levels after the cessation of stressful stimuli may result from a failure of feedback inhibition. **c** Some patients may have higher stress hormone levels during exposure to stressful stimuli. **d** Finally, a blunted response to stressors is also evident in some patients, though this can be confounded by circadian dysregulation and/or higher basal responses. (McEwen 2000; Kehne and Cain 2010; Young et al. 2000)

Higher than normal activation of the HPA-axis occurs in a subset of patients, is ameliorated by antidepressant treatment and is significantly lower during remission (Trestman et al. 1991; Holsboer 2001). Roughly half of the patients exhibit some evidence of HPA-axis hyperactivity, though not all necessarily in exactly the same measures (Nestler et al. 2002). However, it should be noted that hypercortisolemia is not specific to depression, but is also found in other psychiatric disorders.

Although many studies report some type of HPA-axis hyperactivity in depression, there are also conflicting studies. For example, there is evidence that the stress response may be blunted in depression and depressed patients may have lower cortisol and decreased CRF levels in response to a stressor (Brindle et al. 2013; Burke et al. 2005; Geracioti et al. 1992). Cortisol response to stressful stimuli can also be "flatter" in depressed patients (Young et al. 2000; Trestman et al. 1991). This would not be inconsistent with the blunted affective response to both aversive and appetitive stimuli evident in depressed patients. Other factors known to affect HPA-axis measurements include age and gender, depression subtype and/or diagnosis, and depression severity (Burke et al. 2005; Maes et al. 1994), complicating interpretation of the studies. Furthermore, although increased reactivity of the HPA-axis is evident for some people at high-risk for depression (Schmidt et al. 2010b), studies such as the Munich Vulnerability Study indicate that only about 20 % of healthy people who have a high familial risk of depression display abnormal dexamethasone suppression (Holsboer 2000). Furthermore, stress hyperactivity is more likely to be a feature of hospitalization status (Maes et al. 1994) than of depression per se, highlighting the variability in the association of HPA-axis measures with the risk of depression (Kendler et al. 2002, 2006).

Pharmacotherapies reducing stress hormone levels can be associated with a biologically well-defined subset of depressed patients—particularly those with adrenal disorders. Inhibitors of glucocorticoid synthesis, such as ketoconazole, metyrapone, and aminogluthetimide have shown antidepressant activity in patients with Cushing's syndrome (Reus and Wolkowitz 2001). There have also been attempts to reduce HPA activity in patients without a concomitant adrenal disease via inhibition of the CRF receptor. Despite promising results from the first CRF1 receptor antagonist trial in depressed patients, subsequent trials yielded negative results (Griebel and Holsboer 2012).

4 Stress-Based Animal Models of Depression

Stress can be modeled in animals by exogenous stress hormone administration or by exposing animals to aversive stimuli (e.g., restrainers, shock), and these are among the most commonly used models for inducing depression-like behavior in rodents. While these procedures often result in many characteristics similar to depression (Casarotto and Andreatini 2007; Moreau et al. 1992; Forbes et al. 1996) they also often result in contradictory results see Table 1. Chronic corticosterone administration, for example, does not always impair HPA-axis function (Young 1995) and can sometimes even reduce depression-like behavior (Xu et al. 2011). Animal models can be used to examine causal relationships between stressors and the development of depression-related outcomes and also to clarify the mechanisms of the HPA-axis dysfunction that may be evident in depressed patients.

Animal models of stress-induced depression include prenatal and early postnatal stress, social stressors (isolation, subordination and defeat), the application of single stressors (e.g., restraint, shock) and long-term administration of a combination of various unpredictable stressors, known as uncontrollable chronic stress or chronic mild stress (CMS). These models of chronic stress generally simulate aspects of the core symptoms of depression, including behavioral despair in the forced swim (FST) and learned helplessness tests, social withdrawal in a variety of social tests and anhedonia in the saccharin (or sucrose) preference test, operant tasks, and self-stimulation (Blanchard et al. 2001; Koolhaas et al. 1997; Willner

Table 1 Variability of endocrine and behavioral outcomes in animal models of depression

Model	Species	Body weight	Plasma corticosterone levels	Brain GR levels	Forced swim	Anhedonia (saccharin/sucrose preference)
Chronic mild stress	Mice	No change (Schweizer et al. 2009; Pothion et al. 2004; Zhou et al. 2011), decrease (Taksande et al. 2013; Strekalova et al. 2004),	Increase (Taksande et al. 2013; Zhao et al. 2012b; Detanico et al. 2009; Couch et al. 2013; Zhou et al. 2011), no change (Zhao et al. 2012b)	Decrease (Froger et al. 2004; Mao et al. 2009; Zhou et al. 2011)	Increase (Willner 2005; Venzala et al. 2013; Strekalova et al. 2004; Strekalova and Steinbusch 2010; Taksande et al. 2013; Zhou et al. 2011), no change (Schweizer et al. 2009; Willner 2005)	Increase (Venzala et al. 2013; Strekalova et al. 2004; Strekalova and Steinbusch 2010; Taksande et al. 2013; Zhou et al. 2011) no change (Ducottet and Belzung 2005); depend on strain (Schweizer et al. 2009; Pothion et al. 2004)
	Rats	Increase (Zhao et al. 2012a), no change (Patermain et al. 2011; Grippo et al. 2005; Gronli et al. 2004; Dalla et al. 2005) or decrease (Dalla et al. 2005; Bielajew et al. 2002; Li et al. 2010b; Matthews et al. 1995),	Increase (Grippo et al. 2005; Castro et al. 2012; Dalla et al. 2005; Herman et al. 1995; Herman and Spencer 1998; Kim et al. 1999; Wu and Wang 2010), no change (Dalla et al. 2005; Patermain et al. 2005; Zhao et al. 2012a; Gronli et al. 2004; Bielajew et al. 2002; Wu and Wang 2010), decrease (Bielajew et al. 2002)	No change (Lopez et al. 1998; van Riel et al. 2003; Herman and Spencer 1998) or decrease (Pan et al. 2010; Xu et al. 2006; Zheng et al. 2006; Kim et al. 1999; Herman et al. 1995; Yau et al. 2001)	Increase (Castro et al. 2012; Crema et al. 2013), female higher than male (Dalla et al. 2008), decrease (Suo et al. 2013)	Increase (Bergström et al. 2008; Dalla et al. 2005, 2008; Grippo et al. 2005; D'Aquila et al. 1994; Crema et al. 2013; Gronli et al. 2004; Li et al. 2010b), no change (Nielsen et al. 2000; Bielajew et al. 2002; Matthews et al. 1995; Baker et al. 2006; Forbes et al. 1996; Bergström et al. 2008)

(continued)

Table 1 (continued)

Model	Species	Body weight	Plasma corticosterone levels	Brain GR levels	Forced swim	Anhedonia (saccharin/sucrose preference)
Chronic social stress	Mice	Increase (Savignac et al. 2011; Razzoli et al. 2011b; Dadomo et al. 2011), no change (Slattery et al. 2012; Schmidt et al. 2010a) or decrease (Savignac et al. 2011; Razzoli et al. 2011b; Dadomo et al. 2011; Bartolomucci et al. 2005),	Increase (Lehmann et al. 2013; Schmidt et al. 2010a, b; Bhatnagar and Vining 2003; Bowens et al. 2012; Dadomo et al. 2011; Bartolomucci et al. 2005; Gomez-Lazaro et al. 2012), no change (Uschold-Schmidt et al. 2012; Bhatnagar and Vining 2003; Gomez-Lazaro et al. 2012) or decrease (Savignac et al. 2011; Martin and Brown 2010; Gomez-Lazaro et al. 2012)	No change (Bowens et al. 2012; Schmidt et al. 2010a, b) or decrease (Bartolomucci et al. 2005)	Increase (Lehmann et al. 2013) or no change (Venzala et al. 2013; Slattery et al. 2012; Covington et al. 2009; Razzoli et al. 2011b),	Increase (Venzala et al. 2013; Krishnan et al. 2007; Wilkinson et al. 2009; Covington et al. 2009) or no change (Slattery et al. 2012),
	Rats	No change (Calvo et al. 2011; Hayashida et al. 2010) decrease (Kanarik et al. 2011; Rygula et al. 2005; Iio et al. 2011; Carnevali et al. 2012).	Increase or no change depend on strain (Wu and Wang 2010; Calvo et al. 2011), no change in female, decrease in male (McCormick et al. 2008; Djordjevic et al. 2009)	No change or decrease (Calvo et al. 2011), translocate to nucleus (Djordjevic et al. 2009)	Increase (Rygula et al. 2005; Hayashida et al. 2010; Iio et al. 2011; Caldarone et al. 2000; Kanarik et al. 2011; Calvo et al. 2011) or no change (Calvo et al. 2011; Castro et al. 2012; Kanarik et al. 2011)	Increase (Rygula et al. 2005; Kanarik et al. 2011; Carnevali et al. 2012)

(continued)

Table 1 (continued)

Model	Species	Body weight	Plasma corticosterone levels	Brain GR levels	Forced swim	Anhedonia (saccharin/sucrose preference)
Chronic single stress	Mice	No change (Mozhui et al. 2010) or decrease (Mozhui et al. 2010; Li et al. 2010a)	Increase or no change depend on strain (Mozhui et al. 2010; Wilson and Weber 2013; Bowers et al. 2008; Kim et al. 2011)	No change (Kim et al. 2011)	Increase (Lehmann et al. 2013; Li et al. 2010a; Swiergiel et al. 2008), decrease or no change (Mozhui et al. 2010; Wilson and Weber 2013; Hata et al. 1999)	Increase (Wilson and Weber 2013)
	Rats	Decrease (Chiba et al. 2012; O'Mahony et al. 2011; Lee et al. 2012), less weight gain (Rabasa et al. 2013; Naert et al. 2011; Ulloa et al. 2010),	Increase (Chiba et al. 2012; Rabasa et al. 2013; Lee et al. 2012; Ulloa et al. 2010; Makino et al. 1995), faster response to stress (Naert et al. 2011), no change (O'Mahony et al. 2011; Zhang et al. 2011)	Increase (Zhang et al. 2011), decrease (Makino et al. 1995)	Increase (Chiba et al. 2012; Ulloa et al. 2010; Rabasa et al. 2010; Lee et al. 2012; Zhang et al. 2011), no change (Cancela et al. 1991; Crema et al. 2013; Gregus et al. 2005; Huynh et al. 2011) or decrease (Suvrathan et al. 2010; Swiergiel et al. 2007; Platt and Stone 1982).	Increase (Chiba et al. 2012; Naert et al. 2011; Zhang et al. 2011), no change (Huynh et al. 2011) or decrease (Ely et al. 1997),

(continued)

Table 1 (continued)

Model	Species	Body weight	Plasma corticosterone levels	Brain GR levels	Forced swim	Anhedonia (saccharin/sucrose preference)
Chronic corticosterone	Mice	No change (Gourley et al. 2008a), slower weight gain (Wu et al. 2013)	Increase (Murray et al. 2008; Hodes et al. 2012; Gourley et al. 2008b), decrease (Howell et al. 2011; Xu et al. 2011)	No change (Wu et al. 2013), decrease (Hodes et al. 2012; Zhou et al. 2011)	Increase (Gourley et al. 2008a; Wu et al. 2013; Zhou et al. 2011), no change (David et al. 2009; Murray et al. 2008)	Increase (Gourley et al. 2008a; Wu et al. 2013; Zhou et al. 2011)
	Rats	No change (Ulloa et al. 2010), decrease (Barr et al. 2000)	Increase (Ulloa et al. 2010; Lee et al. 2009; Herman and Spencer 1998; Fan et al. 2013)	Decrease (Wang and Wang 2009), decrease or no change depend on subregion (Herman and Spencer 1998)	Increase (Ulloa et al. 2010; Lee et al. 2009; Gregus et al. 2005), decrease in female (Brotto et al. 2001)	Increase (Fan et al. 2013; Gourley et al. 2009),
Acute stress (restraint, swim, shock, or social defeat)	Mice	No change or decrease (Razzoli et al. 2011a),	Increase (Lehmann et al. 2013; Bowers et al. 2008; Spencer et al. 2012; Swiergiel et al. 2008; Krishnan et al. 2007), decrease (Stone et al. 1997)	No change (Spencer et al. 2012)	Increase (Hayase 2011), no change (Kinsey et al. 2007; Couch et al. 2013).	No change (Razzoli et al. 2011a; Stone et al. 1997), decrease (Hayase 2011)
	Rats	Increase (Zelena et al. 1999), decrease (Razzoli et al. 2009; Zelena et al. 1999)	Increase (Razzoli et al. 2009; Armario et al. 1991; Daniels et al. 2004; Djordjevic et al. 2009; Brunton and Russell 2010), less increase in female (Ver Hoeve et al. 2013), no change (Zelena et al. 1999; Daniels et al. 2004)	Change cellular distribution (Djordjevic et al. 2009), decrease (Yau et al. 2001)	Increase (Suvrathan et al. 2010; Chiba et al. 2012; Cancela et al. 1991; Dalla et al. 2008), no change (Razzoli et al. 2009; Ver Hoeve et al. 2013) or decrease (Armario et al. 1991)	No change (Hollis et al. 2010; Razzoli et al. 2009; Ely et al. 1997)

(continued)

Table 1 (continued)

Model	Species	Body weight	Plasma corticosterone levels	Brain GR levels	Forced swim	Anhedonia (saccharin/sucrose preference)
Prenatal or early postnatal stress	Mice	No change (Mueller and Bale 2008; Pechnick et al. 2006)	Increase (Mueller and Bale 2008) or no change (Pechnick et al. 2006; Mueller and Bale 2008)	Decrease (Lee et al. 2011; Mueller and Bale 2008)	Increase (Pechnick et al. 2006; Mueller and Bale 2008) or no change (Mueller and Bale 2008)	Increase (Mueller and Bale 2008)
	Rats	No change (Keshet and Weinstock 1995) or decrease (Borsonelo et al. 2011)	Increase (Brunton and Russell 2010; Dugovic et al. 1999; Green et al. 2011), less negative feedback (Morley-Fletcher et al. 2003), less response (Van den Hove et al. 2013; Borsonelo et al. 2011) or no change (Holson et al. 1995; Cory-Slechta et al. 2013; Borsonelo et al. 2011).	No difference or decrease depends on sex and subregion (Brunton and Russell 2010), decrease (Green et al. 2011)	Increase (Morley-Fletcher et al. 2003, 2011), no change (Borsonelo et al. 2011) or decrease (Cory-Slechta et al. 2013).	Increase (Keshet and Weinstock 1995), no change (Van den Hove et al. 2013) or decrease (Keshet and Weinstock 1995),

Note 1. In forced swim test: increase = increase of immobility
2. In anhedonia test: increase = less preference for sweetened water

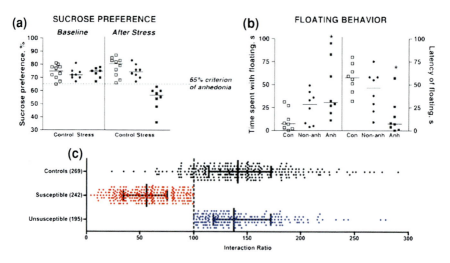

Fig. 3 Variability of stress-induced outcomes in rodents assays of depression-like behavior. Exposure to chronic stress tends to induce anhedonia (less preference for sweetened water) a, increase immobility in the forced swim test b and reduce social interaction c. However, typically only a subset of animals display negative outcomes and the performances of the same animal in different tasks may not be related, i.e., animal having most severe anhedonia may not be the one that display significant immobility. It should also be noted that several studies have failed to find significant effects. Figures are reprinted with permission—a and b from (Strekalova and Steinbusch 2010) and c from (Krishnan et al. 2007)

and Mitchell 2002; Nestler et al. 2002; Strekalova et al. 2004; Strekalova and Steinbusch 2010). It should be noted that the depression-like behaviors resulting from stress exposure are variable (Willner 2005; Strekalova et al. 2004; Strekalova and Steinbusch 2010) and while there may be a mean difference between stressed and control groups, this is typically due to increased depression-like behaviors in a subset of the animals (see Fig. 3). For an excellent review on stress-based animal models [see (Willner et al. 2013)].

4.1 Chronic Single Stress Models of Depression in Adult Animals

Among the single stressors used to investigate the relationship between chronic stress and depression, social stressors are increasingly being investigated, and the most commonly used social stressors include social defeat and social isolation. Social defeat is induced in rodents by repeated exposure to a dominant conspecific (as in the resident-intruder paradigm) (Blanchard et al. 2001; Dadomo et al. 2011; Razzoli et al. 2011b; Albonetti and Farabollini 1994). This paradigm is based on the fact that adult male rodents will establish a territory and the resident will attack or threaten unfamiliar males intruding in its home cage (Mitchell and Fletcher 1993; Kemble 1993; Koolhaas et al. 2013). Allowing the resident to threaten the intruder repeatedly produces defeated subjects, which elicits depression-like behavior (Carnevali et al. 2012; Hammack et al. 2012; Iio et al. 2011; Razzoli et al. 2009; Bhatnagar and Vining 2003) much as in humans (Björkqvist 2001), however see (van der Staay et al. 2008; Paul et al. 2011; Blanchard et al. 1995; Koolhaas et al. 1997). Social isolation can also cause negative behavioral outcomes, but it is unclear if they reliably do so in the current accepted assays of depression-like behavior in rodents. Some studies indicate an increase in behavioral despair, social impairment, and impaired hedonic responses (Carnevali et al. 2012; Grippo et al. 2007; Martin and Brown 2010), however, others rather demonstrate anxiety-like or OCD-like deficits (Kim and Kirkpatrick 1996; Thorsell et al. 2006).

Chronic restraint is also a common way of inducing stress and examining the relationship between stress and depression-like behavior in rodents. Typically, repeated restraint causes anxiety-, despair-, and depression-like behaviors, alters sleep and alters glucocorticoid receptor expression (Chiba et al. 2012; Hammack et al. 2012; Hayase 2011; Suvrathan et al. 2010; Wood et al. 2008; Hegde et al. 2008), some of which can be replicated by chronic, sustained corticosterone administration (Gregus et al. 2005). However, the extent and nature of the behavioral outcomes are, highly variable. Furthermore, the exact nature of the HPA-axis response is dependent on the brain region, sex, age, and the time after stressor cessation at which the specific molecular target is assessed.

4.2 Multiple Chronic Stressors in Adult Animals

Chronic mild stress (CMS) is one of the most commonly used rodent models of depression. CMS consists of the application of a combination of various stressful stimuli over a long period of time (generally several weeks), including circadian disruption, food and/or water restriction, aversive cage conditions, social stress, intermittent strong/aversive stimuli, changes in room temperature, electrical shock, forced exposure to an elevated platform, and forced swimming among others (Castro et al. 2012; Baker et al. 2006; Casarotto and Andreatini 2007; Dalla et al. 2005; D'Aquila et al. 1994; Lin et al. 2002; Matthews et al. 1995; Moreau et al. 1992; Gregus et al. 2005; Wu and Wang 2010; Strekalova and Steinbusch 2010). CMS generally results in a subset of animals which display anhedonia in the saccharin/sucrose preference test or intra cranial self-stimulation (Lin et al. 2002; Willner et al. 1992), or immobility in the forced swim and/or tail suspension tests or social deficits. This is consistent with other stress models, such as chronic social stress, that also result in about half the subjects with depression-like behavior while the other half remain resistant (Krishnan et al. 2007). This circumstance is also similar to clinical stress literature in which only a subset of subjects are affected.

In addition to the fact that there is a subset of vulnerable subjects in studies that find negative effects of stressors on subsequent behavior, in some cases exposure to stressors can also have positive behavioral outcomes such as increasing hedonic behavior (Barr et al. 2000) and reducing behavioral despair (Platt and Stone 1982; Brotto et al. 2001; Xu et al. 2011). This highlights the relevance of adopting a "screen versus mean" approach wherein subjects with a known behavioral outcome after stress are compared to subjects who do not show a depressive phenotype to these stimuli, as it is evident that the question is not truly how does stress cause depression, but in whom does stress cause depression (Ducottet and Belzung 2005; Castro et al. 2012; Strekalova and Steinbusch 2010; Krishnan et al. 2007).

The hypotheses of a chronically deregulated HPA-axis as the cause of depression-like behavior following stressful events are not unequivocally supported by preclinical studies. For example, when mice are subjected to chronic social stress only about half of the animals demonstrated depression-like behavior, whereas both resistant and susceptible mice showed increased corticosterone response to acute stressors compared to unstressed control mice (Krishnan et al. 2007). Also, although male rats most robustly show depression-like outcomes (lower body weight and robust anhedonia) compared to females, there is no difference in corticosterone levels in males undergoing CMS compared to controls, while females undergoing CMS did have higher levels of corticosterone than controls (Dalla et al. 2005). Thus, there is no clear-cut relation between depression-like behavior and HPA-axis response in numerous studies that have looked at these multiple measures in the same subjects. Thus, other factors than the absolute corticosterone, GR, and MR levels may regulate the susceptibility to stress-induced depression in rodents. For reviews with different points of view see (Hill et al. 2012; Willner 2005; Wiborg 2013) and Psychopharmacology vol. 134, issue 4.

Consistent with this view, the assessment of GR expression, negative feedback, and other HPA-axis measures in animal models of stress-induced depression has likewise resulted in variable and equivocal data. Clearly, the role of the GR and GR levels are a function of brain region, so we will focus on the hippocampus here (see Table 1). Several studies have reported downregulation of GR after chronic stress (Zheng et al. 2006; Xu et al. 2006; Kim et al. 1999; Herman et al. 1995; Yau et al. 2001). Other studies however, failed to find any effect of chronic stress on hippocampal GR expression levels (Lopez et al. 1998; van Riel et al. 2003). Chronic corticosterone administration does not always impair negative feedback (Young 1995). Results are also conflicting with respect to CRF. Predictably, mice lacking CRF receptors show a marked impairment of HPA response to stressors but have either no behavioral outcomes or possibly an anxiety-like phenotype (Kormos and Gaszner 2013; Muglia et al. 1995; Weninger et al. 1999), indicating that none of the core symptoms of depression are observed when manipulating this aspect of the HPA-axis. On the other hand, rodents overexpressing CRF have blunted stress responses (Yu et al. 2008) and reduced depression-like behavior (Regev et al. 2011; Kormos and Gaszner 2013), which is not consistent with the idea that a higher stress reactivity would lead to depression-like behavior. Finally, CRF antagonists show reliably antidepressant-like effects only when endogenous

CRF levels are high (Griebel et al. 2002; Maciag et al. 2002; Nielsen et al. 2004; Okuyama et al. 1999; Zorrilla et al. 2002). This latter observation may explain the inadequate outcome in CRF receptor 1 antagonist trials in mixed populations of depressed patients (Griebel and Holsboer 2012).

4.3 Early Life Stress

It has become increasingly evident that detrimental early life experiences, including prenatal stress, can permanently alter physiology and behavior and increase the risk of psychiatric disorders (Davis and Sandman 2012; Khashan et al. 2008). Children born to women who were stressed while pregnant are prone to depression (Kleinhaus et al. 2013; Van den Bergh et al. 2008). The transgenerational changes in depression risk after exposure to prenatal stress are not limited to severe trauma but also occur after exposure to milder stressors (Huizink et al. 2003). Prospective studies have shown that children up to 10 years old may suffer from sleep problems, reduced cognitive performance, and increased fearfulness as a consequence of exposure to prenatal stress (Austin et al. 2005; Bergman et al. 2007). However, exposure to early postnatal stress also increases the risk for developing depression (Nemeroff 2004; Mullen et al. 1996) and many studies of prenatal stress do not address the confounding effects of postnatal stress exposure. Animal studies are useful models in this field, as the environment and stress intensity can be better controlled and/or systematically manipulated. Prenatally stressed offspring may be derived by maternal exposure to the same type of stressor repeatedly (e.g., restraint stress) or to multiple types of stressors (CMS). Maternal separation is a widely used model for studying effects of early postnatal stress.

Animals exposed to pre- or early postnatal stress show depression-like behavioral changes including behavioral despair in the FST, diminished pleasure-seeking, greater acquisition of learned helplessness behavior, and increased REM sleep and sleep fragmentation (Abdul Aziz et al. 2012; Brunton and Russell 2010; Morley-Fletcher et al. 2003; Keshet and Weinstock 1995; Secoli and Teixeira 1998; Dugovic et al. 1999). Similar to adult stress models, depressive-like behavior after exposure to early life stress varies between studies. In some studies pre- or early postnatal stress did not impair behavior or did so in just one sex (Brunton and Russell 2010; Van den Hove et al. 2013). Different stress paradigms (timing, duration, degree and types of stressors) and age and strain of the animals when used for testing may explain some of the inconsistencies across studies. Similar to studies in humans, it is difficult to exclude the impact of differences in postnatal maternal care. Cross-fostering studies, where prenatally stressed pups were raised by nonstressed mothers, address these issues and indeed it has been indicated that maternal care is affected by exposure to stress during gestation (Anisman et al. 1998; Del Cerro et al. 2010) and that maternal behavior impacts offspring behavior (Uchida et al. 2010; Anisman et al. 1998).

There are various mechanisms by which the fetal brain may be impacted by early life stress. Studies have indicated that maternal stress may restrict the blood flow to the fetus (Teixeira et al. 1999) thus reducing nutrient availability and compromise cell development. Glucocorticoids are important for proper brain development, however, exposure to chronically increased levels may increase the risk of diseases later in life (Harris and Seckl 2011). Some studies indicate that maternal corticosterone may be transported through the placenta (Weinstock 2005) however, others suggest that factors other than glucocorticoids are mediating the effects of prenatal stress (Salomon et al. 2011). This is not surprising in light of the fact that the placenta contains "barriers" to the effects of fluctuating maternal glucocorticoids, including enzymes and protein chaperones (Harris and Seckl 2011) and indicate that a complex interplay of mechanisms are involved in determining the actual outcome of stress exposure in utero. Despite these complexities, there are intriguing reports of epigenetic changes (e.g., DNA methylation or histone modifications) that may contribute to the underlying effects of early life stress (Monteleone et al. 2013; Szyf 2013; Anisman et al. 1998). For instance, methylation changes were observed in genes involved in synapse formation in the hippocampus after exposure to prenatal stress (Monteleone et al. 2013). Also, there are methylation changes in the DNA coding for CRF and GR in male offspring following exposure to prenatal stress (Mueller and Bale 2008; Szyf 2013) similar to methylation changes in the GR genes evident in human studies (Radtke et al. 2011).

5 Beyond the HPA-Axis: How Might Stress Cause Depression?

5.1 Mechanisms of Stress and Depression in Humans

There are several theories beyond the HPA-axis as to how stressors may induce depression which are not mutually exclusive. These include increased cell death, reduced neurogenesis, and other forms of reduced structural plasticity and neurotrophic activity (Zunszain et al. 2011). There is understandably little human data, most of which is derived from either postmortem analysis or from imaging technologies (i.e., MRI).

The neurodegeneration hypothesis is essentially based on two lines of reasoning. First, depressed patients may have smaller hippocampal volumes (Frodl et al. 2002) and this may be restored by remission in depressive state and antidepressant treatment (Detanico et al. 2009; Henn et al. 2004). Stressors may also reduce hippocampal volume indicating a mechanism by which stress could elicit depression. Human postmortem studies tend to indicate that there may be decreased cell numbers in specific brain regions (Fuchs et al. 2004).

The neurogenesis hypothesis of major depression suggests that stress-induced decreases of neurogenesis may underlie the long-term decreases in hippocampal volume (Jacobs 2002). While the hypothesis is appealing, the support is inconsistent (Hanson et al. 2011). Intriguing evidence indicates that activation of GR induces pro-apoptotic pathways in the hippocampus whereas activation of MR induces anti-apoptotic pathways, particularly by changing the ratio of proapoptotic molecules, such as Bax, relative to the anti-apoptotic molecules Bcl-2 or Bcl-x(L) (Almeida et al. 2000), but these are not consistent with the putative antidepressant effects of the MR antagonist, spironolactone as shown in animal studies (Wu et al. 2013).

The neurotrophin hypothesis of depression proposes that chronic stress associated with depression decreases neurotrophins. This hypothesis is based on the observation that antidepressants tend to increase neurotrophic factors (Duman and Monteggia 2006; Hodes et al. 2010) although this may be related to the efficacy of antidepressants rather than correlated to depression per se (Adachi et al. 2008; Greenwood et al. 2007; Ibarguen-Vargas et al. 2009; Marais et al. 2009).

Though often quoted, not all studies are consistent with data linking decreased hippocampal volume to depression. A recent study suggests different short- and long-term effects of negative life stressors on hippocampal volumes in older adults (Zannas et al. 2013). Furthermore, increased cell death has not been replicated in studies with exogenous steroid administration (Lucassen et al. 2001a; Muller et al. 2001; Swaab et al. 2005). In addition, dexamethasone suppressors and non-suppressors do not have different hippocampal volume (Axelson et al. 1993). Finally, PTSD patients also have smaller hippocampal volume demonstrating that changes in volume is not specific to depression (Yehuda et al. 1995),

The focus on the hippocampus is also of arguable utility as other brain regions such as amygdala and prefrontal cortex are also involved in depression-like symptoms. Although some human postmortem studies indicate neurodegeneration in specific brain regions (Fuchs et al. 2004) the association between neurodegeneration as a result of glucocorticoid signaling is not robustly supported by studies in humans (Swaab et al. 2005).

5.2 Mechanisms of Stress and Depression in Animal Models

Animal models can more clearly elucidate the role of neurodegeneration, neurogenesis, and neurotrophic factors that may accompany depression-like behavior. Chronic stress models that induce depression-like behavior are also typically accompanied by decreased neurogenesis in rodents (Fuchs et al. 2004). This decreased cell proliferation is ameliorated by antidepressants, but is not necessarily related to improvement in depression-like behavior (Jayatissa et al. 2008, 2009; Henn and Vollmayr 2004).

It has been suggested that chronic stress may induce depression by increasing cell death and/or by reducing structural plasticity, such as retraction or simplification of dendritic arbors (Lee et al. 2002; Swaab et al. 2005). In support of such ideas, chronic stress or chronic administration of corticosterone causes a loss of pyramidal neurons in the hippocampus, reduced dendritic branching, and reactive glial cell proliferation (Sapolsky 1999; Lee et al. 2002; Haynes et al. 2001). Chronic stress in rodents is also accompanied by increased markers of apoptosis (Fuchs et al. 2004), though these markers are not always indicative of actually decreased cell number as the aforementioned data have not always been replicated (Fuchs et al. 2004; Vollmann-Honsdorf et al. 1997; Lucassen et al. 2001b).

5.3 Susceptibility and Resistance to Depression

Results from humans and animal models indicate that stress may induce depression but only in a susceptible population. Recent studies comparing depression-resistant and depression-susceptible subjects exposed to the same stressors elucidate the mechanisms that may underlie susceptibility to depression. Gene-profiling studies demonstrate that behavioral resistance to stress-induced mood dysfunction is an active neurobiological process that is not simply the absence of vulnerability (Krishnan et al. 2007). Resistance likely depends on an increased degree of molecular plasticity. For example, a considerably larger number of genes related to synaptic plasticity are regulated in resistant subgroups compared to susceptible groups who underwent the same exposure to chronic stress. This is consistent with psychological studies that suggest that coping is an active process. The link between susceptibility to depression and reduced degree of plasticity is supported by several lines of evidence—first, stressors that induce mood dysfunction also impair cognitive function, structural plasticity, and long-term potentiation (LTP, a form of neuronal plasticity), and second, the primary factors differentiating resistant and susceptible subjects arguably fall into the global category of plasticity-related moieties.

If active plasticity in response to stress were to be a critical factor distinguishing susceptible from resistant subjects, one would predict that stressors that induce depression should also modulate cognition (Gourley et al. 2008a) and LTP, as is indeed the case in the amygdala (Kulisch and Albrecht 2013) and the hippocampus (Kim et al. 2006). Similarly to depression-like outcomes, LTP deficits following stress only occur in rats unable to control their stress exposure (Shors et al. 1989). In this context, it is noteworthy that antidepressant drugs and treatments that restore normal emotional behavior also restore cognitive deficits and improve the reduced LTP found in animals with depression-like behavior (Cui et al. 2006; Kim et al. 2006; Marais et al. 2009). Thus, the mechanism of plasticity underlying cognition may not be dissimilar to the mechanisms which underlie coping and stress resistance.

The mechanisms of shared cognitive and emotional vulnerability include many plasticity-related targets. Among these are neurotropic factors, the best studied of which in this context is brain-derived neurotrophic factor (BDNF). BDNF is increased in the hippocampus, nucleus accumbens, and ventral tegmental area of resistant animals. BDNF levels are associated with cognitive function, immobility in the FST, and anhedonia demonstrated after chronic corticosterone administration

(Gourley et al. 2008a; Bergström et al. 2008). In humans, BDNF levels are reduced in plasma and postmortem brain tissue of depressed patients (Castren and Rantamaki 2008). In stress-induced depression models BDNF is also mostly reported to be decreased (Berry et al. 2012; De Vry et al. 2012; Duric and McCarson 2005; Maniam and Morris 2010), and thus in line with human studies. Deletion of BDNF renders female, but not male mice, more susceptible to stress-induced depression-like behavior (Autry et al. 2009; Ibarguen-Vargas et al. 2009; Monteggia et al. 2007), supporting that neurotropic factors may be important in determining resilience. Furthermore, the efficacy of antidepressant treatments may be related to activation of the BDNF receptor, TrkB (Saarelainen et al. 2003; Castren and Rantamaki 2010; Rantamaki et al. 2007; Razzoli et al. 2011c). It should be noted however, that direct brain administration of BDNF can produce anxiety-like behavior (Casarotto et al. 2012), so it is not a simple case of more is better.

In addition to growth factors and their downstream pathways, other factors regulating neuronal function and plasticity may also play a role in stress-induced susceptibility to depression, including synaptic proteins, neuronal receptors, and ion channels and second messenger systems. Expression of neuronal cell adhesion molecules is influenced by stressors, regulates synaptic structure, and influences negative behavioral outcomes of stressors (Sandi 2004; Aonurm-Helm et al. 2008). Expression of specific potassium channels in association with increased firing in the ventral tegmental area differentiates susceptible from resistant animals (Krishnan et al. 2007). Other second messengers, such as calcium-sensitive adenylyl cyclase, may also be common pathways mediating the effects of susceptibility to stressors on depression, cognition, and plasticity (Krishnan et al. 2008; Razzoli et al. 2010). These avenues not only elucidate new molecular targets by which to manipulate the response to stressors, but also may be related to the specific behavioral profiles that are evident in stress-susceptible subjects (Padilla et al. 2010; Krishnan et al. 2007; Castro et al. 2012; Aonurm-Helm et al. 2008).

6 Summary and Future Perspectives

Whereas it is a commonly held and plausible belief that stress can lead to negative behavioral sequalae, including depression, results from humans and animals rather support that this is likely to be the case only in a susceptible subset of subjects. Susceptibility to stress-related depression is determined by a complex system of interacting factors, including many hormones, peptides, neurosteroids, and growth factors. Elucidating the active coping and plasticity mechanisms that underlie susceptibility to stressors is necessary to prevent negative behavioral outcomes of stress exposure and can lead to development of rational pharmacological treatments founded on an understanding of the disease biology.

References

Abdul Aziz NH, Kendall DA, Pardon MC (2012) Prenatal exposure to chronic mild stress increases corticosterone levels in the amniotic fluid and induces cognitive deficits in female offspring, improved by treatment with the antidepressant drug amitriptyline. Behav Brain Res 231(1):29–39

Adachi M, Barrot M, Autry AE, Theobald D, Monteggia LM (2008) Selective loss of brain-derived neurotrophic factor in the dentate gyrus attenuates antidepressant efficacy. Biol Psychiatry 63(7):642–649

Aguilera G, Subburaju S, Young S, Chen J (2008) The parvocellular vasopressinergic system and responsiveness of the hypothalamic pituitary adrenal axis during chronic stress. Prog Brain Res 170:29–39

Albonetti ME, Farabollini F (1994) Social stress by repeated defeat: effects on social behaviour and emotionality. Behav Brain Res 62(2):187–193

Almeida OF, Conde GL, Crochemore C, Demeneix BA, Fischer D, Hassan AH, Meyer M, Holsboer F, Michaelidis TM (2000) Subtle shifts in the ratio between pro- and antiapoptotic molecules after activation of corticosteroid receptors decide neuronal fate. FASEB J 14(5):779–790

Anisman H, Zaharia MD, Meaney MJ, Merali Z (1998) Do early-life events permanently alter behavioral and hormonal responses to stressors? Int J Dev Neurosci 16(3–4):149–164

Aonurm-Helm A, Jurgenson M, Zharkovsky T, Sonn K, Berezin V, Bock E, Zharkovsky A (2008) Depression-like behaviour in neural cell adhesion molecule (NCAM)-deficient mice and its reversal by an NCAM-derived peptide, FGL. Eur J Neurosci 28(8):1618–1628

Armario A, Gil M, Marti J, Pol O, Balasch J (1991) Influence of various acute stressors on the activity of adult male rats in a holeboard and in the forced swim test. Pharmacol Biochem Behav 39(2):373–377

Austin MP, Hadzi-Pavlovic D, Leader L, Saint K, Parker G (2005) Maternal trait anxiety, depression and life event stress in pregnancy: relationships with infant temperament. Early Hum Dev 81(2):183–190

Autry AE, Adachi M, Cheng P, Monteggia LM (2009) Gender-specific impact of brain-derived neurotrophic factor signaling on stress-induced depression-like behavior. Biol Psychiatry 66(1):84–90

Axelson DA, Doraiswamy PM, McDonald WM, Boyko OB, Tupler LA, Patterson LJ, Nemeroff CB, Ellinwood EH Jr, Krishnan KR (1993) Hypercortisolemia and hippocampal changes in depression. Psychiatry Res 47(2):163–173

Baker SL, Kentner AC, Konkle ATM, Barbagallo LS-M, Bielajew C (2006) Behavioral and physiological effects of chronic mild stress in female rats. Physiol Behav 87(2):314–322

Bao AM, Hestiantoro A, Van Someren EJ, Swaab DF, Zhou JN (2005) Colocalization of corticotropin-releasing hormone and oestrogen receptor-alpha in the paraventricular nucleus of the hypothalamus in mood disorders. Brain 128(Pt 6):1301–1313

Barr AM, Brotto LA, Phillips AG (2000) Chronic corticosterone enhances the rewarding effect of hypothalamic self-stimulation in rats. Brain Res 875(1–2):196–201

Bartolomucci A, Palanza P, Sacerdote P, Panerai AE, Sgoifo A, Dantzer R, Parmigiani S (2005) Social factors and individual vulnerability to chronic stress exposure. Neurosci Biobehav Rev 29(1):67–81

Bergman K, Sarkar P, O'Connor TG, Modi N, Glover V (2007) Maternal stress during pregnancy predicts cognitive ability and fearfulness in infancy. J Am Acad Child Adolesc Psychiatry 46(11):1454–1463

Bergström A, Jayatissa MN, Mørk A, Wiborg O (2008) Stress sensitivity and resilience in the chronic mild stress rat model of depression; an in situ hybridization study. Brain Res 1196:41–52

Berry A, Bellisario V, Capoccia S, Tirassa P, Calza A, Alleva E, Cirulli F (2012) Social deprivation stress is a triggering factor for the emergence of anxiety- and depression-like behaviours and leads to reduced brain BDNF levels in C57BL/6J mice. Psychoneuroendocrinology 37(6):762–772

Bhatnagar S, Vining C (2003) Facilitation of hypothalamic-pituitary-adrenal responses to novel stress following repeated social stress using the resident/intruder paradigm. Horm Behav 43(1):158–165

Bielajew C, Konkle AT, Merali Z (2002) The effects of chronic mild stress on male Sprague-Dawley and Long Evans rats: I. Biochemical and physiological analyses. Behav Brain Res 136(2):583–592

Björkqvist K (2001) Social defeat as a stressor in humans. Physiol Behav 73(3):435–442

Blanchard DC, Spencer RL, Weiss SM, Blanchard RJ, McEwen B, Sakai RR (1995) Visible burrow system as a model of chronic social stress: behavioral and neuroendocrine correlates. Psychoneuroendocrinology 20(2):117–134

Blanchard RJ, McKittrick CR, Blanchard DC (2001) Animal models of social stress: effects on behavior and brain neurochemical systems. Physiol Behav 73(3):261–271

Bonanno GA (2004) Loss, trauma, and human resilience: have we underestimated the human capacity to thrive after extremely aversive events? Am psychol 59(1):20–28

Borsonelo EC, Suchecki D, Galduroz JC (2011) Effect of fish oil and coconut fat supplementation on depressive-type behavior and corticosterone levels of prenatally stressed male rats. Brain Res 1385:144–150

Bowens N, Heydendael W, Bhatnagar S, Jacobson L (2012) Lack of elevations in glucocorticoids correlates with dysphoria-like behavior after repeated social defeat. Physiol Behav 105(4):958–965

Bowers SL, Bilbo SD, Dhabhar FS, Nelson RJ (2008) Stressor-specific alterations in corticosterone and immune responses in mice. Brain Behav Immun 22(1):105–113

Brindle RC, Ginty AT, Conklin SM (2013) Is the association between depression and blunted cardiovascular stress reactions mediated by perceptions of stress? Int J psychophysiol: Off J Int Org Psychophysiol 90(1):66–72

Brotto LA, Gorzalka BB, Barr AM (2001) Paradoxical effects of chronic corticosterone on forced swim behaviours in aged male and female rats. Eur J Pharmacol 424(3):203–209

Brown GW, Bifulco A, Harris TO (1987) Life events, vulnerability and onset of depression: some refinements. Br J Psychiatry 150(1):30–42

Brunton PJ, Russell JA (2010) Prenatal social stress in the rat programmes neuroendocrine and behavioural responses to stress in the adult offspring: sex-specific effects. J Neuroendocrinol 22(4):258–271

Burke HM, Davis MC, Otte C, Mohr DC (2005) Depression and cortisol responses to psychological stress: a meta-analysis. Psychoneuroendocrinology 30(9):846–856

Caldarone BJ, George TP, Zachariou V, Picciotto MR (2000) Gender differences in learned helplessness behavior are influenced by genetic background. Pharmacol Biochem Behav 66(4):811–817

Calvo N, Cecchi M, Kabbaj M, Watson SJ, Akil H (2011) Differential effects of social defeat in rats with high and low locomotor response to novelty. Neuroscience 183:81–89

Cancela LM, Rossi S, Molina VA (1991) Effect of different restraint schedules on the immobility in the forced swim test: modulation by an opiate mechanism. Brain Res Bull 26(5):671–675

Carnevali L, Mastorci F, Graiani G, Razzoli M, Trombini M, Pico-Alfonso MA, Arban R, Grippo AJ, Quaini F, Sgoifo A (2012) Social defeat and isolation induce clear signs of a depression-like state, but modest cardiac alterations in wild-type rats. Physiol Behav 106(2):142–150

Casarotto PC, Andreatini R (2007) Repeated paroxetine treatment reverses anhedonia induced in rats by chronic mild stress or dexamethasone. Eur Neuropsychopharmacol 17(11):735–742

Casarotto PC, de Bortoli VC, Zangrossi H Jr (2012) Intrahippocampal injection of brain-derived neurotrophic factor increases anxiety-related, but not panic-related defensive responses: involvement of serotonin. Behav Pharmacol 23(1):80–88

Castren E, Rantamaki T (2008) Neurotrophins in depression and antidepressant effects. Novartis Found symp 289:43–52; discussion 53–49, 87–93

Castren E, Rantamaki T (2010) The role of BDNF and its receptors in depression and antidepressant drug action: reactivation of developmental plasticity. Dev Neurobiol 70(5):289–297

Castro JE, Diessler S, Varea E, Marquez C, Larsen MH, Cordero MI, Sandi C (2012) Personality traits in rats predict vulnerability and resilience to developing stress-induced depression-like behaviors, HPA axis hyper-reactivity and brain changes in pERK1/2 activity. Psychoneuroendocrinology 37(8):1209–1223

Chen J, Young S, Subburaju S, Sheppard J, Kiss A, Atkinson H, Wood S, Lightman S, Serradeil-Le Gal C, Aguilera G (2008) Vasopressin does not mediate hypersensitivity of the hypothalamic pituitary adrenal axis during chronic stress. Ann N Y Acad Sci 1148:349–359

Chiba S, Numakawa T, Ninomiya M, Richards MC, Wakabayashi C, Kunugi H (2012) Chronic restraint stress causes anxiety- and depression-like behaviors, downregulates glucocorticoid receptor expression, and attenuates glutamate release induced by brain-derived neurotrophic factor in the prefrontal cortex. Prog Neuropsychopharmacol Biol Psychiatry 39(1):112–119

Cory-Slechta DA, Merchant-Borna K, Allen JL, Liu S, Weston D, Conrad K (2013) Variations in the nature of behavioral experience can differentially alter the consequences of developmental exposures to lead, prenatal stress, and the combination. Toxicol Sci 131(1):194–205

Couch Y, Anthony DC, Dolgov O, Revischin A, Festoff B, Santos AI, Steinbusch HW, Strekalova T (2013) Microglial activation, increased TNF and SERT expression in the prefrontal cortex define stress-altered behaviour in mice susceptible to anhedonia. Brain Behav Immun 29:136–146

Covington HE 3rd, Maze I, LaPlant QC, Vialou VF, Ohnishi YN, Berton O, Fass DM, Renthal W, Rush AJ 3rd, Wu EY, Ghose S, Krishnan V, Russo SJ, Tamminga C, Haggarty SJ, Nestler EJ (2009) Antidepressant actions of histone deacetylase inhibitors. J Neurosci 29(37):11451–11460

Crema LM, Pettenuzzo LF, Schlabitz M, Diehl L, Hoppe J, Mestriner R, Laureano D, Salbego C, Dalmaz C, Vendite D (2013) The effect of unpredictable chronic mild stress on depressive-like behavior and on hippocampal A1 and striatal A2A adenosine receptors. Physiol Behav 109:1–7

Croes S, Merz P, Netter P (1993) Cortisol reaction in success and failure condition in endogenous depressed patients and controls. Psychoneuroendocrinology 18(1):23–35

Cui M, Yang Y, Yang J, Zhang J, Han H, Ma W, Li H, Mao R, Xu L, Hao W, Cao J (2006) Enriched environment experience overcomes the memory deficits and depressive-like behavior induced by early life stress. Neurosci Lett 404(1–2):208–212

D'Aquila PS, Brain P, Willner P (1994) Effects of chronic mild stress on performance in behavioural tests relevant to anxiety and depression. Physiol Behav 56(5):861–867

Dadomo H, Sanghez V, Di Cristo L, Lori A, Ceresini G, Malinge I, Parmigiani S, Palanza P, Sheardown M, Bartolomucci A (2011) Vulnerability to chronic subordination stress-induced depression-like disorders in adult 129SvEv male mice. Prog Neuropsychopharmacol Biol Psychiatry 35(6):1461–1471

Dalla C, Antoniou K, Drossopoulou G, Xagoraris M, Kokras N, Sfikakis A, Papadopoulou-Daifoti Z (2005) Chronic mild stress impact: are females more vulnerable? Neuroscience 135(3):703–714

Dalla C, Antoniou K, Kokras N, Drossopoulou G, Papathanasiou G, Bekris S, Daskas S, Papadopoulou-Daifoti Z (2008) Sex differences in the effects of two stress paradigms on dopaminergic neurotransmission. Physiol Behav 93(3):595–605

Daniels WM, Richter L, Stein DJ (2004) The effects of repeated intra-amygdala CRF injections on rat behavior and HPA axis function after stress. Metab Brain Dis 19(1–2):15–23

David DJ, Samuels BA, Rainer Q, Wang JW, Marsteller D, Mendez I, Drew M, Craig DA, Guiard BP, Guilloux JP, Artymyshyn RP, Gardier AM, Gerald C, Antonijevic IA, Leonardo ED, Hen R (2009) Neurogenesis-dependent and -independent effects of fluoxetine in an animal model of anxiety/depression. Neuron 62(4):479–493

Davis EP, Sandman CA (2012) Prenatal psychobiological predictors of anxiety risk in preadolescent children. Psychoneuroendocrinology 37(8):1224–1233 de Kloet ER, Oitzl MS, Joëls M (1999) Stress and cognition: are corticosteroids good or bad guys? Trends Neurosci 22(10):422–426

De Vry J, Prickaerts J, Jetten M, Hulst M, Steinbusch HW, van den Hove DL, Schuurman T, van der Staay FJ (2012) Recurrent long-lasting tethering reduces BDNF protein levels in the dorsal hippocampus and frontal cortex in pigs. Horm Behav 62(1):10–17

Deis RP, Leguizamon E, Jahn GA (1989) Feedback regulation by progesterone of stress-induced prolactin release in rats. J Endocrinol 120(1):37–43

Del Cerro MC, Perez-Laso C, Ortega E, Martin JL, Gomez F, Perez-Izquierdo MA, Segovia S (2010) Maternal care counteracts behavioral effects of prenatal environmental stress in female rats. Behav Brain Res 208(2):593–602

Detanico BC, Piato AL, Freitas JJ, Lhullier FL, Hidalgo MP, Caumo W, Elisabetsky E (2009) Antidepressant-like effects of melatonin in the mouse chronic mild stress model. Eur J Pharmacol 607(1–3):121–125

Djordjevic A, Adzic M, Djordjevic J, Radojcic MB (2009) Stress type dependence of expression and cytoplasmic-nuclear partitioning of glucocorticoid receptor, hsp90 and hsp70 in Wistar rat brain. Neuropsychobiology 59(4):213–221

Dubrovsky B (2000) The specificity of stress responses to different nocuous stimuli: neurosteroids and depression. Brain Res Bull 51(6):443–455

Ducottet C, Belzung C (2005) Correlations between behaviours in the elevated plus-maze and sensitivity to unpredictable subchronic mild stress: evidence from inbred strains of mice. Behav Brain Res 156(1):153–162

Dugovic C, Maccari S, Weibel L, Turek FW, Van Reeth O (1999) High corticosterone levels in prenatally stressed rats predict persistent paradoxical sleep alterations. J Neurosci 19(19):8656–8664

Duman RS, Monteggia LM (2006) A neurotrophic model for stress-related mood disorders. Biol Psychiatry 59(12):1116–1127

Duric V, McCarson KE (2005) Hippocampal neurokinin-1 receptor and brain-derived neurotrophic factor gene expression is decreased in rat models of pain and stress. Neuroscience 133(4):999–1006

Egashira N, Tanoue A, Matsuda T, Koushi E, Harada S, Takano Y, Tsujimoto G, Mishima K, Iwasaki K, Fujiwara M (2007) Impaired social interaction and reduced anxiety-related behavior in vasopressin V1a receptor knockout mice. Behav Brain Res 178(1):123–127

Ely DR, Dapper V, Marasca J, Correa JB, Gamaro GD, Xavier MH, Michalowski MB, Catelli D, Rosat R, Ferreira MB, Dalmaz C (1997) Effect of restraint stress on feeding behavior of rats. Physiol Behav 61(3):395–398

Fan Y, Chen P, Li Y, Cui K, Noel DM, Cummins ED, Peterson DJ, Brown RW, Zhu MY (2013) Corticosterone administration up-regulated expression of norepinephrine transporter and dopamine beta-hydroxylase in rat locus coeruleus and its terminal regions. J Neurochem 128(3):445–458

ffrench-Mullen J (1995) Cortisol inhibition of calcium currents in guinea pig hippocampal CA1 neurons via G-protein-coupled activation of protein kinase C. J Neurosci 15(1):903–911

Forbes NF, Stewart CA, Matthews K, Reid IC (1996) Chronic mild stress and sucrose consumption: validity as a model of depression. Physiol Behav 60(6):1481–1484

Frodl T, Meisenzahl EM, Zetzsche T, Born C, Groll C, Jager M, Leinsinger G, Bottlender R, Hahn K, Moller HJ (2002) Hippocampal changes in patients with a first episode of major depression. Am J psychiatry 159(7):1112–1118

Frodl T, Reinhold E, Koutsouleris N, Reiser M, Meisenzahl EM (2010) Interaction of childhood stress with hippocampus and prefrontal cortex volume reduction in major depression. J Psychiatr Res 44(13):799–807

Froger N, Palazzo E, Boni C, Hanoun N, Saurini F, Joubert C, Dutriez-Casteloot I, Enache M, Maccari S, Barden N, Cohen-Salmon C, Hamon M, Lanfumey L (2004) Neurochemical and behavioral alterations in glucocorticoid receptor-impaired transgenic mice after chronic mild stress. J Neurosci 24(11):2787–2796

Fuchs E, Czeh B, Kole MH, Michaelis T, Lucassen PJ (2004) Alterations of neuroplasticity in depression: the hippocampus and beyond. European Neuropsychopharmacol: J Eur Coll Neuropsychopharmacol 14(Suppl 5):S481–S490

Geracioti TD Jr, Loosen PT, Gold PW, Kling MA (1992) Cortisol, thyroid hormone, and mood in atypical depression: a longitudinal case study. Biol Psychiatry 31(5):515–519

Gomez-Lazaro E, Garmendia L, Beitia G, Perez-Tejada J, Azpiroz A, Arregi A (2012) Effects of a putative antidepressant with a rapid onset of action in defeated mice with different coping strategies. Prog Neuropsychopharmacol Biol Psychiatry 38(2):317–327

Gourley SL, Kedves AT, Olausson P, Taylor JR (2009) A history of corticosterone exposure regulates fear extinction and cortical NR2B, GluR2/3, and BDNF. Neuropsychopharmacology 34(3):707–716

Gourley SL, Kiraly DD, Howell JL, Olausson P, Taylor JR (2008a) Acute hippocampal brain-derived neurotrophic factor restores motivational and forced swim performance after corticosterone. Biol Psychiatry 64(10):884–890

Gourley SL, Wu FJ, Kiraly DD, Ploski JE, Kedves AT, Duman RS, Taylor JR (2008b) Regionally specific regulation of ERK MAP kinase in a model of antidepressant-sensitive chronic depression. Biol Psychiatry 63(4):353–359

Green MK, Rani CS, Joshi A, Soto-Pina AE, Martinez PA, Frazer A, Strong R, Morilak DA (2011) Prenatal stress induces long term stress vulnerability, compromising stress response systems in the brain and impairing extinction of conditioned fear after adult stress. Neuroscience 192:438–451

Greenwood BN, Strong PV, Foley TE, Thompson RS, Fleshner M (2007) Learned helplessness is independent of levels of brain-derived neurotrophic factor in the hippocampus. Neuroscience 144(4):1193–1208

Gregus A, Wintink AJ, Davis AC, Kalynchuk LE (2005) Effect of repeated corticosterone injections and restraint stress on anxiety and depression-like behavior in male rats. Behav Brain Res 156(1):105–114

Griebel G, Holsboer F (2012) Neuropeptide receptor ligands as drugs for psychiatric diseases: the end of the beginning? Nat Rev Drug Discov 11(6):462–478

Griebel G, Simiand J, Steinberg R, Jung M, Gully D, Roger P, Geslin M, Scatton B, Maffrand JP, Soubrie P (2002) 4-(2-Chloro-4-methoxy-5-methylphenyl)-N-[(1S)-2-cyclopropyl-1-(3-fluoro-4-methylp henyl)ethyl]5-methyl-N-(2-propynyl)-1, 3-thiazol-2-amine hydrochloride (SSR125543A), a potent and selective corticotrophin-releasing factor(1) receptor antagonist. II. Characterization in rodent models of stress-related disorders. J Pharmacol Exp Ther 301(1):333–345

Grippo AJ, Francis J, Beltz TG, Felder RB, Johnson AK (2005) Neuroendocrine and cytokine profile of chronic mild stress-induced anhedonia. Physiol Behav 84(5):697–706

Grippo AJ, Gerena D, Huang J, Kumar N, Shah M, Ughreja R, Sue Carter C (2007) Social isolation induces behavioral and neuroendocrine disturbances relevant to depression in female and male prairie voles. Psychoneuroendocrinology 32(8–10):966–980

Gronli J, Murison R, Bjorvatn B, Sorensen E, Portas CM, Ursin R (2004) Chronic mild stress affects sucrose intake and sleep in rats. Behav Brain Res 150(1–2):139–147

Halbreich U, Asnis GM, Shindledecker R, Zumoff B, Nathan R (1985) Cortisol secretion in endogenous depression: I. basal plasma levels. Arch Gen Psychiatry 42(9):904–908

Hamilton JL, Stange JP, Shapero BG, Connolly SL, Abramson LY, Alloy LB (2013) Cognitive vulnerabilities as predictors of stress generation in early adolescence: pathway to depressive symptoms. J Abnorm Child Psychol 41(7):1027–1039

Hammack SE, Cooper MA, Lezak KR (2012) Overlapping neurobiology of learned helplessness and conditioned defeat: implications for PTSD and mood disorders. Neuropharmacology 62(2):565–575

Hammen C (1991) Generation of stress in the course of unipolar depression. J Abnorm Psychol 100(4):555–561

Hammen C, Davila J, Brown G, Ellicott A, Gitlin M (1992) Psychiatric history and stress: predictors of severity of unipolar depression. J Abnorm Psychol 101(1):45–52

Hanson ND, Owens MJ, Nemeroff CB (2011) Depression, antidepressants, and neurogenesis: a critical reappraisal. Neuropsychopharmacology 36(13):2589–2602

Harris A, Seckl J (2011) Glucocorticoids, prenatal stress and the programming of disease. Horm Behav 59(3):279–289

Hata T, Nishikawa H, Itoh E, Watanabe A (1999) Depressive state with anxiety in repeated cold-stressed mice in forced swimming tests. Jpn J Pharmacol 79(2):243–249

Hayase T (2011) Depression-related anhedonic behaviors caused by immobilization stress: a comparison with nicotine-induced depression-like behavioral alterations and effects of nicotine and/or "antidepressant" drugs. J Toxicol Sci 36(1):31–41

Hayashida S, Oka T, Mera T, Tsuji S (2010) Repeated social defeat stress induces chronic hyperthermia in rats. Physiol Behav 101(1):124–131

Haynes LE, Griffiths MR, Hyde RE, Barber DJ, Mitchell IJ (2001) Dexamethasone induces limited apoptosis and extensive sublethal damage to specific subregions of the striatum and hippocampus: implications for mood disorders. Neuroscience 104(1):57–69

Hegde P, Singh K, Chaplot S, Rao BSS, Chattarji S, Kutty BM, Laxmi TR (2008) Stress-induced changes in sleep and associated neuronal activity in rat hippocampus and amygdala. Neuroscience 153(1):20–30

Henn F, Vollmayr B, Sartorius A (2004) Mechanisms of depression: the role of neurogenesis. Drug Discov Today: Dis Mech 1(4):407–411

Henn FA, Vollmayr B (2004) Neurogenesis and depression: etiology or epiphenomenon? Biol Psychiatry 56(3):146–150

Herman JP, Adams D, Prewitt C (1995) Regulatory changes in neuroendocrine stress-integrative circuitry produced by a variable stress paradigm. Neuroendocrinology 61(2):180–190

Herman JP, Spencer R (1998) Regulation of hippocampal glucocorticoid receptor gene transcription and protein expression in vivo. J Neurosci 18(18):7462–7473

Heuser I, Yassouridis A, Holsboer F (1994) The combined dexamethasone/CRH test: a refined laboratory test for psychiatric disorders. J Psychiatr Res 28(4):341–356

Hill MN, Hellemans KG, Verma P, Gorzalka BB, Weinberg J (2012) Neurobiology of chronic mild stress: parallels to major depression. Neurosci Biobehav Rev 36(9):2085–2117

Hodes GE, Brookshire BR, Hill-Smith TE, Teegarden SL, Berton O, Lucki I (2012) Strain differences in the effects of chronic corticosterone exposure in the hippocampus. Neuroscience 222:269–280

Hodes GE, Hill-Smith TE, Suckow RF, Cooper TB, Lucki I (2010) Sex-specific effects of chronic fluoxetine treatment on neuroplasticity and pharmacokinetics in mice. J Pharmacol Exp Ther 332(1):266–273

Hollis F, Wang H, Dietz D, Gunjan A, Kabbaj M (2010) The effects of repeated social defeat on long-term depressive-like behavior and short-term histone modifications in the hippocampus in male Sprague-Dawley rats. Psychopharmacology 211(1):69–77

Holsboer F (2000) The corticosteroid receptor hypothesis of depression. Neuropsychopharmacology 23(5):477–501

Holsboer F (2001) Stress, hypercortisolism and corticosteroid receptors in depression: implicatons for therapy. J Affect Disord 62(1–2):77–91

Holsboer F, Bender W, Benkert O, Klein HE, Schmauss M (1980) Diagnostic value of dexamethasone suppression test in depression. Lancet 2(8196):706

Holson RR, Gough B, Sullivan P, Badger T, Sheehan DM (1995) Prenatal dexamethasone or stress but not ACTH or corticosterone alter sexual behavior in male rats. Neurotoxicol Teratol 17(4):393–401

Howell KR, Kutiyanawalla A, Pillai A (2011) Long-term continuous corticosterone treatment decreases VEGF receptor-2 expression in frontal cortex. PLoS ONE 6(5):e20198

Huizink AC, de Medina PGR, Mulder EJ, Visser GH, Buitelaar JK (2003) Stress during pregnancy is associated with developmental outcome in infancy. J Child Psychol Psychiatry 44(6):810–818

Huynh TN, Krigbaum AM, Hanna JJ, Conrad CD (2011) Sex differences and phase of light cycle modify chronic stress effects on anxiety and depressive-like behavior. Behav Brain Res 222(1):212–222

Ibarguen-Vargas Y, Surget A, Vourc'h P, Leman S, Andres CR, Gardier AM, Belzung C (2009) Deficit in BDNF does not increase vulnerability to stress but dampens antidepressant-like effects in the unpredictable chronic mild stress. Behav Brain Res 202(2):245–251

Iio W, Matsukawa N, Tsukahara T, Kohari D, Toyoda A (2011) Effects of chronic social defeat stress on MAP kinase cascade. Neurosci Lett 504(3):281–284

Jacobs BL (2002) Adult brain neurogenesis and depression. Brain Behav Immun 16(5):602–609

Jayatissa MN, Bisgaard CF, West MJ, Wiborg O (2008) The number of granule cells in rat hippocampus is reduced after chronic mild stress and re-established after chronic escitalopram treatment. Neuropharmacology 54(3):530–541

Jayatissa MN, Henningsen K, West MJ, Wiborg O (2009) Decreased cell proliferation in the dentate gyrus does not associate with development of anhedonic-like symptoms in rats. Brain Res 1290:133–141

Kanarik M, Alttoa A, Matrov D, Kõiv K, Sharp T, Panksepp J, Harro J (2011) Brain responses to chronic social defeat stress: effects on regional oxidative metabolism as a function of a hedonic trait, and gene expression in susceptible and resilient rats. Eur Neuropsychopharmacol 21(1):92–107

Kant GJ, Leu JR, Anderson SM, Mougey EH (1987) Effects of chronic stress on plasma corticosterone ACTH and prolactin. Physiol Behav 40(6):775–779

Kehne JH, Cain CK (2010) Therapeutic utility of non-peptidic CRF1 receptor antagonists in anxiety, depression, and stress-related disorders: evidence from animal models. Pharmacol Ther 128(3):460–487

Kemble ED (1993) 8—Resident–Intruder Paradigms for the Study of Rodent Aggression. In: Conn PM (ed) Methods in neurosciences, vol 14. Academic press, pp 138–150

Kendler KS, Gardner CO, Prescott CA (2002) Toward a comprehensive developmental model for major depression in women. Am J psychiatry 159(7):1133–1145

Kendler KS, Gardner CO, Prescott CA (2006) Toward a comprehensive developmental model for major depression in men. Am J psychiatry 163(1):115–124

Kendler KS, Karkowski LM, Prescott CA (1999) Causal relationship between stressful life events and the onset of major depression. Am J psychiatry 156(6):837–841

Keshet GI, Weinstock M (1995) Maternal naltrexone prevents morphological and behavioral alterations induced in rats by prenatal stress. Pharmacol Biochem Behav 50(3):413–419

Kessler RC (1997) The effects of stressful life events on depression. Annu Rev Psychol 48(1):191–214

Khashan AS, Abel KM, McNamee R, Pedersen MG, Webb RT, Baker PN, Kenny LC, Mortensen PB (2008) Higher risk of offspring schizophrenia following antenatal maternal exposure to severe adverse life events. Arch Gen Psychiatry 65(2):146–152

Kim BS, Kim MY, Leem YH (2011) Hippocampal neuronal death induced by kainic acid and restraint stress is suppressed by exercise. Neuroscience 194:291–301

Kim CK, Yu W, Edin G, Ellis L, Osborn JA, Weinberg J (1999) Chronic intermittent stress does not differentially alter brain corticosteroid receptor densities in rats prenatally exposed to ethanol. Psychoneuroendocrinology 24(6):585–611

Kim EJ, Kim WR, Chi SE, Lee KH, Park EH, Chae JH, Park SK, Kim HT, Choi JS (2006) Repetitive transcranial magnetic stimulation protects hippocampal plasticity in an animal model of depression. Neurosci Lett 405(1–2):79–83

Kim JW, Kirkpatrick B (1996) Social isolation in animal models of relevance to neuropsychiatric disorders. Biol Psychiatry 40(9):918–922

Kinsey SG, Bailey MT, Sheridan JF, Padgett DA, Avitsur R (2007) Repeated social defeat causes increased anxiety-like behavior and alters splenocyte function in C57BL/6 and CD-1 mice. Brain Behav Immun 21(4):458–466

Kleinhaus K, Harlap S, Perrin M, Manor O, Margalit-Calderon R, Opler M, Friedlander Y, Malaspina D (2013) Prenatal stress and affective disorders in a population birth cohort. Bipolar Disord 15(1):92–99

Koolhaas JM, Coppens CM, de Boer SF, Buwalda B, Meerlo P, Timmermans PJ (2013) The resident-intruder paradigm: a standardized test for aggression, violence and social stress. J Vis Exp 77:e4367 (JoVE)

Koolhaas JM, De Boer SF, De Rutter AJ, Meerlo P, Sgoifo A (1997) Social stress in rats and mice. Acta Physiol Scand Suppl 640:69–72

Kormos V, Gaszner B (2013) Role of neuropeptides in anxiety, stress, and depression: from animals to humans. Neuropeptides 47(6):401–419

Krishnan V, Graham A, Mazei-Robison MS, Lagace DC, Kim KS, Birnbaum S, Eisch AJ, Han PL, Storm DR, Zachariou V, Nestler EJ (2008) Calcium-sensitive adenylyl cyclases in depression and anxiety: behavioral and biochemical consequences of isoform targeting. Biol Psychiatry 64(4):336–343

Krishnan V, Han MH, Graham DL, Berton O, Renthal W, Russo SJ, Laplant Q, Graham A, Lutter M, Lagace DC, Ghose S, Reister R, Tannous P, Green TA, Neve RL, Chakravarty S, Kumar A, Eisch AJ, Self DW, Lee FS, Tamminga CA, Cooper DC, Gershenfeld HK, Nestler EJ (2007) Molecular adaptations underlying susceptibility and resistance to social defeat in brain reward regions. Cell 131(2):391–404

Kulisch C, Albrecht D (2013) Effects of single swim stress on changes in TRPV1-mediated plasticity in the amygdala. Behav Brain Res 236(1):344–349

Lee AL, Ogle WO, Sapolsky RM (2002) Stress and depression: possible links to neuron death in the hippocampus. Bipolar Disord 4(2):117–128

Lee B, Shim I, Lee H-J, Yang Y, Hahm D-H (2009) Effects of acupuncture on chronic corticosterone-induced depression-like behavior and expression of neuropeptide Y in the rats. Neurosci Lett 453(3):151–156

Lee B, Yun HY, Shim I, Lee H, Hahm DH (2012) Bupleurum falcatum prevents depression and anxiety-like behaviors in rats exposed to repeated restraint stress. J Microbiol Biotechnol 22(3):422–430

Lee EJ, Son GH, Chung S, Lee S, Kim J, Choi S, Kim K (2011) Impairment of fear memory consolidation in maternally stressed male mouse offspring: evidence for nongenomic glucocorticoid action on the amygdala. J Neurosci 31(19):7131–7140

Lehmann ML, Mustafa T, Eiden AM, Herkenham M, Eiden LE (2013) PACAP-deficient mice show attenuated corticosterone secretion and fail to develop depressive behavior during chronic social defeat stress. Psychoneuroendocrinology 38(5):702–715

Lewinsohn PM, Allen NB, Seeley JR, Gotlib IH (1999) First onset versus recurrence of depression: differential processes of psychosocial risk. J Abnorm Psychol 108(3):483–489

Li W, Li QJ, An SC (2010a) Preventive effect of estrogen on depression-like behavior induced by chronic restraint stress. Neurosci Bull 26(2):140–146

Li Y, Zheng X, Liang J, Peng Y (2010b) Coexistence of anhedonia and anxiety-independent increased novelty-seeking behavior in the chronic mild stress model of depression. Behav Process 83(3):331–339

Lightman SL (2008) The neuroendocrinology of stress: a never ending story. J Neuroendocrinol 20(6):880–884

Lin D, Bruijnzeel AW, Schmidt P, Markou A (2002) Exposure to chronic mild stress alters thresholds for lateral hypothalamic stimulation reward and subsequent responsiveness to amphetamine. Neuroscience 114(4):925–933

Litvin Y, Murakami G, Pfaff DW (2011) Effects of chronic social defeat on behavioral and neural correlates of sociality: Vasopressin, oxytocin and the vasopressinergic V1b receptor. Physiol Behav 103(3–4):393–403

Liu RT, Alloy LB (2010) Stress generation in depression: a systematic review of the empirical literature and recommendations for future study. Clin Psychol Rev 30(5):582–593

Lopez JF, Chalmers DT, Little KY, Watson SJ (1998) A.E. Bennett Research Award. Regulation of serotonin1A, glucocorticoid, and mineralocorticoid receptor in rat and human hippocampus: implications for the neurobiology of depression. Biol Psychiatry 43(8):547–573

Lucassen PJ, Muller MB, Holsboer F, Bauer J, Holtrop A, Wouda J, Hoogendijk WJ, De Kloet ER, Swaab DF (2001a) Hippocampal apoptosis in major depression is a minor event and absent from subareas at risk for glucocorticoid overexposure. Am J Pathol 158(2):453–468

Lucassen PJ, Vollmann-Honsdorf GK, Gleisberg M, Czeh B, De Kloet ER, Fuchs E (2001b) Chronic psychosocial stress differentially affects apoptosis in hippocampal subregions and cortex of the adult tree shrew. Eur J Neurosci 14(1):161–166

Lyubomirsky S, Caldwell ND, Nolen-Hoeksema S (1998) Effects of ruminative and distracting responses to depressed mood on retrieval of autobiographical memories. J Pers Soc Psychol 75(1):166–177

Maciag CM, Dent G, Gilligan P, He L, Dowling K, Ko T, Levine S, Smith MA (2002) Effects of a non-peptide CRF antagonist (DMP696) on the behavioral and endocrine sequelae of maternal separation. Neuropsychopharmacology 26(5):574–582

Maes M, Calabrese J, Meltzer HY (1994) The relevance of the in- versus outpatient status for studies on HPA-axis in depression: Spontaneous hypercortisolism is a feature of major depressed inpatients and not of major depression per se. Prog Neuropsychopharmacol Biol Psychiatry 18(3):503–517

Makino S, Smith MA, Gold PW (1995) Increased expression of corticotropin-releasing hormone and vasopressin messenger ribonucleic acid (mRNA) in the hypothalamic paraventricular nucleus during repeated stress: association with reduction in glucocorticoid receptor mRNA levels. Endocrinology 136(8):3299–3309

Maniam J, Morris MJ (2010) Voluntary exercise and palatable high-fat diet both improve behavioural profile and stress responses in male rats exposed to early life stress: role of hippocampus. Psychoneuroendocrinology 35(10):1553–1564

Mao QQ, Ip SP, Ko KM, Tsai SH, Che CT (2009) Peony glycosides produce antidepressant-like action in mice exposed to chronic unpredictable mild stress: effects on hypothalamic-pituitary-adrenal function and brain-derived neurotrophic factor. Prog Neuropsychopharmacol Biol Psychiatry 33(7):1211–1216

Marais L, Stein DJ, Daniels WM (2009) Exercise increases BDNF levels in the striatum and decreases depressive-like behavior in chronically stressed rats. Metab Brain Dis 24(4):587–597

Martin AL, Brown RE (2010) The lonely mouse: verification of a separation-induced model of depression in female mice. Behav Brain Res 207(1):196–207

Matthews K, Forbes N, Reid IC (1995) Sucrose consumption as an hedonic measure following chronic unpredictable mild stress. Physiol Behav 57(2):241–248

McCormick CM, Smith C, Mathews IZ (2008) Effects of chronic social stress in adolescence on anxiety and neuroendocrine response to mild stress in male and female rats. Behav Brain Res 187(2):228–238

McEwen BS (1998) Protective and damaging effects of stress mediators. N Engl J Med 338(3):171–179

McEwen BS (2000) Allostasis and allostatic load: implications for neuropsychopharmacology. Neuropsychopharmacology 22(2):108–124

Mitchell PJ, Fletcher A (1993) Venlafaxine exhibits pre-clinical antidepressant activity in the resident-intruder social interaction paradigm. Neuropharmacology 32(10):1001–1009

Mlynarik M, Zelena D, Bagdy G, Makara GB, Jezova D (2007) Signs of attenuated depression-like behavior in vasopressin deficient Brattleboro rats. Horm Behav 51(3):395–405

Monteggia LM, Luikart B, Barrot M, Theobold D, Malkovska I, Nef S, Parada LF, Nestler EJ (2007) Brain-derived neurotrophic factor conditional knockouts show gender differences in depression-related behaviors. Biol Psychiatry 61(2):187–197

Monteleone MC, Adrover E, Pallares ME, Antonelli MC, Frasch AC, Brocco MA (2013) Prenatal stress changes the glycoprotein GPM6A gene expression and induces epigenetic changes in rat offspring brain. Epigenetics 9(1):152–160

Moreau JL, Jenck F, Martin JR, Mortas P, Haefely WE (1992) Antidepressant treatment prevents chronic unpredictable mild stress-induced anhedonia as assessed by ventral tegmentum self-stimulation behavior in rats. Eur Neuropsychopharmacol 2(1):43–49

Morley-Fletcher S, Darnaudery M, Koehl M, Casolini P, Van Reeth O, Maccari S (2003) Prenatal stress in rats predicts immobility behavior in the forced swim test. effects of a chronic treatment with tianeptine. Brain Res 989(2):246–251

Morley-Fletcher S, Mairesse J, Soumier A, Banasr M, Fagioli F, Gabriel C, Mocaer E, Daszuta A, McEwen B, Nicoletti F, Maccari S (2011) Chronic agomelatine treatment corrects behavioral, cellular, and biochemical abnormalities induced by prenatal stress in rats. Psychopharmacology 217(3):301–313

Mozhui K, Karlsson RM, Kash TL, Ihne J, Norcross M, Patel S, Farrell MR, Hill EE, Graybeal C, Martin KP, Camp M, Fitzgerald PJ, Ciobanu DC, Sprengel R, Mishina M, Wellman CL, Winder DG, Williams RW, Holmes A (2010) Strain differences in stress responsivity are associated with divergent amygdala gene expression and glutamate-mediated neuronal excitability. J Neurosci 30(15):5357–5367

Mueller BR, Bale TL (2008) Sex-specific programming of offspring emotionality after stress early in pregnancy. J Neurosci 28(36):9055–9065

Muglia L, Jacobson L, Dikkes P, Majzoub JA (1995) Corticotropin-releasing hormone deficiency reveals major fetal but not adult glucocorticoid need. Nature 373(6513):427–432

Mullen PE, Martin JL, Anderson JC, Romans SE, Herbison GP (1996) The long-term impact of the physical, emotional, and sexual abuse of children: a community study. Child Abuse Negl 20(1):7–21

Muller MB, Lucassen PJ, Yassouridis A, Hoogendijk WJ, Holsboer F, Swaab DF (2001) Neither major depression nor glucocorticoid treatment affects the cellular integrity of the human hippocampus. Eur J Neurosci 14(10):1603–1612

Murray F, Smith DW, Hutson PH (2008) Chronic low dose corticosterone exposure decreased hippocampal cell proliferation, volume and induced anxiety and depression like behaviours in mice. Eur J Pharmacol 583(1):115–127

Myles K, Funder JW (1996) Progesterone binding to mineralocorticoid receptors: in vitro and in vivo studies. Am J Physiol—Endocrinol Metab 270(4):E601–E607

Naert G, Ixart G, Maurice T, Tapia-Arancibia L, Givalois L (2011) Brain-derived neurotrophic factor and hypothalamic-pituitary-adrenal axis adaptation processes in a depressive-like state induced by chronic restraint stress. Mol Cell Neurosci 46(1):55–66

Nemeroff CB (2004) Neurobiological consequences of childhood trauma. J Clin Psychiatry 65(1):18–28

Nemeroff CB, Widerlov E, Bissette G, Walleus H, Karlsson I, Eklund K, Kilts CD, Loosen PT, Vale W (1984) Elevated concentrations of CSF corticotropin-releasing factor-like immunoreactivity in depressed patients. Science 226(4680):1342–1344

Nestler EJ, Barrot M, DiLeone RJ, Eisch AJ, Gold SJ, Monteggia LM (2002) Neurobiology of depression. Neuron 34(1):13–25

Nielsen CK, Arnt J, Sanchez C (2000) Intracranial self-stimulation and sucrose intake differ as hedonic measures following chronic mild stress: interstrain and interindividual differences. Behav Brain Res 107(1–2):21–33

Nielsen DM, Carey GJ, Gold LH (2004) Antidepressant-like activity of corticotropin-releasing factor type-1 receptor antagonists in mice. Eur J Pharmacol 499(1–2):135–146

O'Mahony CM, Clarke G, Gibney S, Dinan TG, Cryan JF (2011) Strain differences in the neurochemical response to chronic restraint stress in the rat: relevance to depression. Pharmacol Biochem Behav 97(4):690–699

Okuyama S, Chaki S, Kawashima N, Suzuki Y, Ogawa S, Nakazato A, Kumagai T, Okubo T, Tomisawa K (1999) Receptor binding, behavioral, and electrophysiological profiles of nonpeptide corticotropin-releasing factor subtype 1 receptor antagonists CRA1000 and CRA1001. J Pharmacol Exp Ther 289(2):926–935

Padilla E, Shumake J, Barrett DW, Holmes G, Sheridan EC, Gonzalez-Lima F (2010) Novelty-evoked activity in open field predicts susceptibility to helpless behavior. Physiol Behav 101(5):746–754

Pan Y, Wang FM, Qiang LQ, Zhang DM, Kong LD (2010) Icariin attenuates chronic mild stress-induced dysregulation of the LHPA stress circuit in rats. Psychoneuroendocrinology 35(2):272–283

Paternain L, Garcia-Diaz DF, Milagro FI, Gonzalez-Muniesa P, Martinez JA, Campion J (2011) Regulation by chronic-mild stress of glucocorticoids, monocyte chemoattractant protein-1 and adiposity in rats fed on a high-fat diet. Physiol Behav 103(2):173–180

Paul ED, Hale MW, Lukkes JL, Valentine MJ, Sarchet DM, Lowry CA (2011) Repeated social defeat increases reactive emotional coping behavior and alters functional responses in serotonergic neurons in the rat dorsal raphe nucleus. Physiol Behav 104(2):272–282

Pechnick RN, Kariagina A, Hartvig E, Bresee CJ, Poland RE, Chesnokova VM (2006) Developmental exposure to corticosterone: behavioral changes and differential effects on leukemia inhibitory factor (LIF) and corticotropin-releasing hormone (CRH) gene expression in the mouse. Psychopharmacology 185(1):76–83

Platt JE, Stone EA (1982) Chronic restraint stress elicits a positive antidepressant response on the forced swim test. Eur J Pharmacol 82(3–4):179–181

Pothion S, Bizot J-C, Trovero F, Belzung C (2004) Strain differences in sucrose preference and in the consequences of unpredictable chronic mild stress. Behav Brain Res 155(1):135–146

Power RA, Wingenbach T, Cohen-Woods S, Uher R, Ng MY, Butler AW, Ising M, Craddock N, Owen MJ, Korszun A, Jones L, Jones I, Gill M, Rice JP, Maier W, Zobel A, Mors O, Placentino A, Rietschel M, Lucae S, Holsboer F, Binder EB, Keers R, Tozzi F, Muglia P, Breen G, Craig IW, Muller-Myhsok B, Kennedy JL, Strauss J, Vincent JB, Lewis CM, Farmer AE, McGuffin P (2013) Estimating the heritability of reporting stressful life events captured by common genetic variants. Psychol Med 43(9):1965–1971

Quinkler M, Meyer B, Bumke-Vogt C, Grossmann C, Gruber U, Oelkers W, Diederich S, Bahr V (2002) Agonistic and antagonistic properties of progesterone metabolites at the human mineralocorticoid receptor. Eur J Endocrinol 146(6):789–799

Rabasa C, Delgado-Morales R, Gomez-Roman A, Nadal R, Armario A (2013) Adaptation of the pituitary-adrenal axis to daily repeated forced swim exposure in rats is dependent on the temperature of water. Stress 16(6):698–705

Radtke KM, Ruf M, Gunter HM, Dohrmann K, Schauer M, Meyer A, Elbert T (2011) Transgenerational impact of intimate partner violence on methylation in the promoter of the glucocorticoid receptor. Transl Psychiatry 1:e21

Rantamaki T, Hendolin P, Kankaanpaa A, Mijatovic J, Piepponen P, Domenici E, Chao MV, Mannisto PT, Castren E (2007) Pharmacologically diverse antidepressants rapidly activate brain-derived neurotrophic factor receptor TrkB and induce phospholipase-Cgamma signaling pathways in mouse brain. Neuropsychopharmacology 32(10):2152–2162

Razzoli M, Andreoli M, Maraia G, Di Francesco C, Arban R (2010) Functional role of calcium-stimulated adenylyl cyclase 8 in adaptations to psychological stressors in the mouse: implications for mood disorders. Neuroscience 170(2):429–440

Razzoli M, Carboni L, Andreoli M, Ballottari A, Arban R (2011a) Different susceptibility to social defeat stress of BalbC and C57BL6/J mice. Behav Brain Res 216(1):100–108

Razzoli M, Carboni L, Andreoli M, Michielin F, Ballottari A, Arban R (2011b) Strain-specific outcomes of repeated social defeat and chronic fluoxetine treatment in the mouse. Pharmacol Biochem Behav 97(3):566–576

Razzoli M, Carboni L, Arban R (2009) Alterations of behavioral and endocrinological reactivity induced by 3 brief social defeats in rats: relevance to human psychopathology. Psychoneuroendocrinology 34(9):1405–1416

Razzoli M, Domenici E, Carboni L, Rantamaki T, Lindholm J, Castren E, Arban R (2011c) A role for BDNF/TrkB signaling in behavioral and physiological consequences of social defeat stress. Genes Brain Behav 10(4):424–433

Regev L, Neufeld-Cohen A, Tsoory M, Kuperman Y, Getselter D, Gil S, Chen A (2011) Prolonged and site-specific over-expression of corticotropin-releasing factor reveals differential roles for extended amygdala nuclei in emotional regulation. Mol psychiatry 16(7):714–728

Reus VI, Wolkowitz OM (2001) Antiglucocorticoid drugs in the treatment of depression. Expert Opin Investig Drugs 10(10):1789–1796

Risch N, Herrell R, Lehner T, Liang KY, Eaves L, Hoh J, Griem A, Kovacs M, Ott J, Merikangas KR (2009) Interaction between the serotonin transporter gene (5-HTTLPR), stressful life events, and risk of depression: a meta-analysis. JAMA 301(23):2462–2471

Rubin RT, Poland RE, Lesser IM, Winston RA, Blodgett AL (1987) Neuroendocrine aspects of primary endogenous depression. I. Cortisol secretory dynamics in patients and matched controls. Arch Gen Psychiatry 44(4):328–336

Rygula R, Abumaria N, Flugge G, Fuchs E, Ruther E, Havemann-Reinecke U (2005) Anhedonia and motivational deficits in rats: impact of chronic social stress. Behav Brain Res 162(1):127–134

Saarelainen T, Hendolin P, Lucas G, Koponen E, Sairanen M, MacDonald E, Agerman K, Haapasalo A, Nawa H, Aloyz R, Ernfors P, Castren E (2003) Activation of the TrkB neurotrophin receptor is induced by antidepressant drugs and is required for antidepressant-induced behavioral effects. J Neurosci 23(1):349–357

Salomon S, Bejar C, Schorer-Apelbaum D, Weinstock M (2011) Corticosterone mediates some but not other behavioural changes induced by prenatal stress in rats. J Neuroendocrinol 23(2):118–128

Sandi C (2004) Stress, cognitive impairment and cell adhesion molecules. Nat Rev Neurosci 5(12):917–930

Sapolsky RM (1999) Glucocorticoids, stress, and their adverse neurological effects: relevance to aging. Exp Gerontol 34(6):721–732

Sapolsky RM (2000) Glucocorticoids and hippocampal atrophy in neuropsychiatric disorders. Arch Gen Psychiatry 57(10):925–935

Savignac HM, Finger BC, Pizzo RC, O'Leary OF, Dinan TG, Cryan JF (2011) Increased sensitivity to the effects of chronic social defeat stress in an innately anxious mouse strain. Neuroscience 192:524–536

Schmidt MV, Scharf SH, Liebl C, Harbich D, Mayer B, Holsboer F, Müller MB (2010a) A novel chronic social stress paradigm in female mice. Horm Behav 57(4–5):415–420

Schmidt MV, Scharf SH, Sterlemann V, Ganea K, Liebl C, Holsboer F, Muller MB (2010b) High susceptibility to chronic social stress is associated with a depression-like phenotype. Psychoneuroendocrinology 35(5):635–643

Schweizer MC, Henniger MS, Sillaber I (2009) Chronic mild stress (CMS) in mice: of anhedonia, 'anomalous anxiolysis' and activity. PLoS ONE 4(1):e4326

Secoli SR, Teixeira NA (1998) Chronic prenatal stress affects development and behavioral depression in rats. Stress 2(4):273–280

Shors TJ, Seib TB, Levine S, Thompson RF (1989) Inescapable versus escapable shock modulates long-term potentiation in the rat hippocampus. Science 244(4901):224–226

Siegle GJ, Granholm E, Ingram RE, Matt GE (2001) Pupillary and reaction time measures of sustained processing of negative information in depression. Biol Psychiatry 49(7):624–636

Silver H (1986) Physical complaints correlate better with depression than do dexamethasone suppression test results. J Clin Psychiatry 47(4):179–181

Slattery DA, Uschold N, Magoni M, Bar J, Popoli M, Neumann ID, Reber SO (2012) Behavioural consequences of two chronic psychosocial stress paradigms: anxiety without depression. Psychoneuroendocrinology 37(5):702–714

Spencer SJ, Xu L, Clarke MA, Lemus M, Reichenbach A, Geenen B, Kozicz T, Andrews ZB (2012) Ghrelin regulates the hypothalamic-pituitary-adrenal axis and restricts anxiety after acute stress. Biol Psychiatry 72(6):457–465

Stetler C, Miller GE (2011) Depression and hypothalamic-pituitary-adrenal activation: a quantitative summary of four decades of research. Psychosom Med 73(2):114–126

Stone EA, Zhang Y, Quartermain D (1997) The effect of stress on spontaneous nest leaving behavior in the mouse: an improved model of stress-induced behavioral pathology. Stress 1(3):145–154

Strauss K, Dapp U, Anders J, von Renteln-Kruse W, Schmidt S (2011) Range and specificity of war-related trauma to posttraumatic stress; depression and general health perception: displaced former World War II children in late life. J Affect Disord 128(3):267–276

Strekalova T, Spanagel R, Bartsch D, Henn FA, Gass P (2004) Stress-induced anhedonia in mice is associated with deficits in forced swimming and exploration. Neuropsychopharmacology 29(11):2007–2017

Strekalova T, Steinbusch HW (2010) Measuring behavior in mice with chronic stress depression paradigm. Prog Neuropsychopharmacol Biol Psychiatry 34(2):348–361

Suo L, Zhao L, Si J, Liu J, Zhu W, Chai B, Zhang Y, Feng J, Ding Z, Luo Y, Shi H, Shi J, Lu L (2013) Predictable chronic mild stress in adolescence increases resilience in adulthood. Neuropsychopharmacology 38(8):1387–1400

Suvrathan A, Tomar A, Chattarji S (2010) Effects of chronic and acute stress on rat behaviour in the forced-swim test. Stress 13(6):533–540

Swaab DF, Bao AM, Lucassen PJ (2005) The stress system in the human brain in depression and neurodegeneration. Ageing Res Rev 4(2):141–194

Swiergiel AH, Leskov IL, Dunn AJ (2008) Effects of chronic and acute stressors and CRF on depression-like behavior in mice. Behav Brain Res 186(1):32–40

Swiergiel AH, Zhou Y, Dunn AJ (2007) Effects of chronic footshock, restraint and corticotropin-releasing factor on freezing, ultrasonic vocalization and forced swim behavior in rats. Behav Brain Res 183(2):178–187

Szyf M (2013) DNA methylation, behavior and early life adversity. J Genet Genomics 40(7):331–338

Taksande BG, Faldu DS, Dixit MP, Sakaria JN, Aglawe MM, Umekar MJ, Kotagale NR (2013) Agmatine attenuates chronic unpredictable mild stress induced behavioral alteration in mice. Eur J Pharmacol 720(1–3):115–120

Tao M, Li Y, Xie D, Wang Z, Qiu J, Wu W, Sun J, Wang Z, Tao D, Zhao H, Tian T, Zhang J, Gao C, Niu Q, Li Q, Liu S, Liu J, Zhang Y, He Q, Rong H, Gan Z, Li J, Chen X, Pan J, Li Y, Cui Y, Han W, Ma H, Xie S, Jin G, Li L, Zhang R, Tan Q, Zhang J, Guan J, Shi S, Chen Y, Kendler KS, Flint J, Gao Z (2011) Examining the relationship between lifetime stressful life events and the onset of major depression in Chinese women. J Affect Disord 135(1–3):95–99

Teixeira JM, Fisk NM, Glover V (1999) Association between maternal anxiety in pregnancy and increased uterine artery resistance index: cohort based study. BMJ 318(7177):153–157

Thorsell A, Slawecki CJ, El Khoury A, Mathe AA, Ehlers CL (2006) The effects of social isolation on neuropeptide Y levels, exploratory and anxiety-related behaviors in rats. Pharmacol Biochem Behav 83(1):28–34

Torner L, Neumann ID (2002) The brain prolactin system: involvement in stress response adaptations in lactation. Stress 5(4):249–257

Towle AC, Sze PY (1983) Steroid binding to synaptic plasma membrane: differential binding of glucocorticoids and gonadal steroids. J Steroid Biochem 18(2):135–143

Trestman RL, Coccaro EF, Bernstein D, Lawrence T, Gabriel SM, Horvath TB, Siever LJ (1991) Cortisol responses to mental arithmetic in acute and remitted depression. Biol Psychiatry 29(10):1051–1054

Uchida S, Hara K, Kobayashi A, Otsuki K, Hobara T, Yamagata H, Watanabe Y (2010) Maternal and genetic factors in stress-resilient and -vulnerable rats: a cross-fostering study. Brain Res 1316:43–50

Ulloa JL, Castaneda P, Berrios C, Diaz-Veliz G, Mora S, Bravo JA, Araneda K, Menares C, Morales P, Fiedler JL (2010) Comparison of the antidepressant sertraline on differential depression-like behaviors elicited by restraint stress and repeated corticosterone administration. Pharmacol Biochem Behav 97(2):213–221

Uschold-Schmidt N, Nyuyki KD, Füchsl AM, Neumann ID, Reber SO (2012) Chronic psychosocial stress results in sensitization of the HPA axis to acute heterotypic stressors despite a reduction of adrenal in vitro ACTH responsiveness. Psychoneuroendocrinology 37(10):1676–1687

Vamvakopoulos NC, Chrousos GP (1993) Evidence of direct estrogenic regulation of human corticotropin-releasing hormone gene expression. Potential implications for the sexual dimophism of the stress response and immune/inflammatory reaction. J Clin Investig 92(4):1896–1902

Van den Bergh BR, Van Calster B, Smits T, Van Huffel S, Lagae L (2008) Antenatal maternal anxiety is related to HPA-axis dysregulation and self-reported depressive symptoms in adolescence: a prospective study on the fetal origins of depressed mood. Neuropsychopharmacology 33(3):536–545

Van den Hove DL, Kenis G, Brass A, Opstelten R, Rutten BP, Bruschettini M, Blanco CE, Lesch KP, Steinbusch HW, Prickaerts J (2013) Vulnerability versus resilience to prenatal stress in male and female rats; Implications from gene expression profiles in the hippocampus and frontal cortex. Eur Neuropsychopharmacol 23(10):1226–1246

van der Staay FJ, de Groot J, Schuurman T, Korte SM (2008) Repeated social defeat in female pigs does not induce neuroendocrine symptoms of depression, but behavioral adaptation. Physiol Behav 93(3):453–460

van Praag HM (2004) Can stress cause depression? Prog Neuropsychopharmacol Biol Psychiatry 28(5):891–907

van Riel E, Meijer OC, Steenbergen PJ, Joels M (2003) Chronic unpredictable stress causes attenuation of serotonin responses in cornu ammonis 1 pyramidal neurons. Neuroscience 120(3):649–658

Venzala E, Garcia-Garcia AL, Elizalde N, Tordera RM (2013) Social vs. environmental stress models of depression from a behavioural and neurochemical approach. Eur Neuropsychopharmacol 23(7):697–708

Ver Hoeve ES, Kelly G, Luz S, Ghanshani S, Bhatnagar S (2013) Short-term and long-term effects of repeated social defeat during adolescence or adulthood in female rats. Neuroscience 249:63–73

Vollmann-Honsdorf GK, Flügge G, Fuchs E (1997) Chronic psychosocial stress does not affect the number of pyramidal neurons in tree shrew hippocampus. Neurosci Lett 233(2–3):121–124

Wang CC, Wang SJ (2009) Modulation of presynaptic glucocorticoid receptors on glutamate release from rat hippocampal nerve terminals. Synapse 63(9):745–751

Watson S, Gallagher P, Smith MS, Ferrier IN, Young AH (2006) The dex/CRH test—is it better than the DST? Psychoneuroendocrinology 31(7):889–894

Weinstock M (2005) The potential influence of maternal stress hormones on development and mental health of the offspring. Brain Behav Immun 19(4):296–308

Weninger SC, Muglia LJ, Jacobson L, Majzoub JA (1999) CRH-deficient mice have a normal anorectic response to chronic stress. Regul Pept 84(1–3):69–74

Wiborg O (2013) Chronic mild stress for modeling anhedonia. Cell Tissue Res 354(1):155–169

Wilkinson MB, Xiao G, Kumar A, LaPlant Q, Renthal W, Sikder D, Kodadek TJ, Nestler EJ (2009) Imipramine treatment and resiliency exhibit similar chromatin regulation in the mouse nucleus accumbens in depression models. J Neurosci 29(24):7820–7832

Willner P (2005) Chronic mild stress (CMS) revisited: consistency and behavioural-neurobiological concordance in the effects of CMS. Neuropsychobiology 52(2):90–110

Willner P, Mitchell PJ (2002) The validity of animal models of predisposition to depression. Behav Pharmacol 13(3):169–188

Willner P, Muscat R, Papp M (1992) Chronic mild stress-induced anhedonia: a realistic animal model of depression. Neurosci Biobehav Rev 16(4):525–534

Willner P, Scheel-Kruger J, Belzung C (2013) The neurobiology of depression and antidepressant action. Neurosci Biobehav Rev 37(10 Pt 1):2331–2371

Wilson CL, Weber ET (2013) Chemotherapy drug thioTEPA exacerbates stress-induced anhedonia and corticosteroid responses but not impairment of hippocampal cell proliferation in adult mice. Behav Brain Res 236(1):180–185

Wood GE, Norris EH, Waters E, Stoldt JT, McEwen BS (2008) Chronic immobilization stress alters aspects of emotionality and associative learning in the rat. Behav Neurosci 122(2):282–292

Wu HH, Wang S (2010) Strain differences in the chronic mild stress animal model of depression. Behav Brain Res 213(1):94–102

Wu T-C, Chen H-T, Chang H-Y, Yang C-Y, Hsiao M-C, Cheng M-L, Chen J-C (2013) Mineralocorticoid receptor antagonist spironolactone prevents chronic corticosterone induced depression-like behavior. Psychoneuroendocrinology 38(6):871–883

Wuttke W, Duker E, Vaupel R, Jarry H (1987) The neuroendocrinology of stress. Stress Med 3(3):217–225

Xu Y, Ku B, Tie L, Yao H, Jiang W, Ma X, Li X (2006) Curcumin reverses the effects of chronic stress on behavior, the HPA axis, BDNF expression and phosphorylation of CREB. Brain Res 1122(1):56–64

Xu Z, Zhang Y, Hou B, Gao Y, Wu Y, Zhang C (2011) Chronic corticosterone administration from adolescence through early adulthood attenuates depression-like behaviors in mice. J Affect Disord 131(1–3):128–135

Yau JL, Noble J, Seckl JR (2001) Acute restraint stress increases 5-HT7 receptor mRNA expression in the rat hippocampus. Neurosci Lett 309(3):141–144

Yehuda R, Boisoneau D, Lowy MT, Giller EL Jr (1995) Dose-response changes in plasma cortisol and lymphocyte glucocorticoid receptors following dexamethasone administration in combat veterans with and without posttraumatic stress disorder. Arch Gen Psychiatry 52(7):583–593

Young EA (1995) Normal glucocorticoid fast feedback following chronic 50 % corticosterone pellet treatment. Psychoneuroendocrinology 20(7):771–784

Young EA, Haskett RF, Grunhaus L, Pande A, Weinberg VM, Watson SJ, Akil H (1994) Increased evening activation of the hypothalamic-pituitary-adrenal axis in depressed patients. Arch Gen Psychiatry 51(9):701–707

Young EA, Lopez JF, Murphy-Weinberg V, Watson SJ, Akil H (2000) Hormonal evidence for altered responsiveness to social stress in major depression. Neuropsychopharmacology 23(4):411–418

Yu S, Holsboer F, Almeida OFX (2008) Neuronal actions of glucocorticoids: focus on depression. J Steroid Biochem Mol Biol 108(3–5):300–309

Zannas AS, McQuoid DR, Payne ME, Steffens DC, MacFall JR, Ashley-Koch A, Taylor WD (2013) Negative life stress and longitudinal hippocampal volume changes in older adults with and without depression. J Psychiatr Res 47(6):829–834

Zelena D, Haller J, Halasz J, Makara GB (1999) Social stress of variable intensity: physiological and behavioral consequences. Brain Res Bull 48(3):297–302

Zhang L, Zhang J, Sun H, Liu H, Yang Y, Yao Z (2011) Exposure to enriched environment restores the mRNA expression of mineralocorticoid and glucocorticoid receptors in the hippocampus and ameliorates depressive-like symptoms in chronically stressed rats. Current Neurovascul Res 8(4):286–293

Zhao Y, Liu L-J, Wang C, Li S-X (2012a) Swimming exercise may not alleviate the depressive-like behaviors and circadian alterations of neuroendocrine induced by chronic unpredictable mild stress in rats. Neurology, Psychiatry Brain Res 18(4):202–207

Zhao Y, Wang Z, Dai J, Chen L, Huang Y, Zhan Z (2012b) Beneficial effects of benzodiazepine diazepam on chronic stress-induced impairment of hippocampal structural plasticity and depression-like behavior in mice. Behav Brain Res 228(2):339–350

Zheng H, Liu Y, Li W, Yang B, Chen D, Wang X, Jiang Z, Wang H, Wang Z, Cornelisson G, Halberg F (2006) Beneficial effects of exercise and its molecular mechanisms on depression in rats. Behav Brain Res 168(1):47–55

Zhou QG, Zhu LJ, Chen C, Wu HY, Luo CX, Chang L, Zhu DY (2011) Hippocampal neuronal nitric oxide synthase mediates the stress-related depressive behaviors of glucocorticoids by downregulating glucocorticoid receptor. J Neurosci 31(21):7579–7590

Zorrilla EP, Valdez GR, Nozulak J, Koob GF, Markou A (2002) Effects of antalarmin, a CRF type 1 receptor antagonist, on anxiety-like behavior and motor activation in the rat. Brain Res 952(2):188–199

Zunszain PA, Anacker C, Cattaneo A, Carvalho LA, Pariante CM (2011) Glucocorticoids, cytokines and brain abnormalities in depression. Prog Neuropsychopharmacol Biol Psychiatry 35(3):722–729

Neurobehavioral Mechanisms of Traumatic Stress in Post-traumatic Stress Disorder

M. Danet Lapiz-Bluhm and Alan L. Peterson

Abstract Post-traumatic stress disorder (PTSD) is a debilitating psychiatric disorder that develops following trauma exposure. It is characterized by four symptom clusters: intrusion, avoidance, negative alteration in cognitions and mood, and alterations in arousal and reactivity. Several risk factors have been associated with PTSD, including trauma type and severity, gender and sexual orientation, race and ethnicity, cognitive reserve, pretrauma psychopathology, familial psychiatric history, and genetics. Great strides have been made in understanding the neurobiology of PTSD through animal models and human imaging studies. Most of the animal models have face validity, but they have limitations in the generalization to the human model of PTSD. Newer animal models, such as the "CBC" model, have better validity for PTSD, which takes into account the different components of its diagnostic criteria. To date, fear conditioning and fear extinction animal models have provided support for the hypothesis that PTSD is a dysregulation of the processes related to fear regulation and, especially, fear extinction. More research is needed to further understand these processes as they relate not only to PTSD but also to resilience. Further, this research could be instrumental in the development of novel effective treatments for PTSD.

Keywords Post-traumatic stress disorder · PTSD · Conditioned learning · Fear extinction · Animal models

---

M. D. Lapiz-Bluhm (✉)
Department of Family and Community Health Systems, School of Nursing,
The University of Texas Health Science Center at San Antonio,
7703 Floyd Curl Drive, San Antonio, TX 78229, USA
e-mail: lapiz@uthscsa.edu

A. L. Peterson
Department of Psychiatry, School of Medicine,
The University of Texas Health Science Center at San Antonio,
7703 Floyd Curl Drive, San Antonio, TX 78229, USA
e-mail: petersona3@uthscsa.edu

Contents

1	Introduction	162
2	A Brief History of PTSD	163
3	Trajectory of Trauma-Related Disorders and PTSD	164
4	Epidemiology of PTSD	165
5	Risk Factors for the Development of PTSD	166
6	Neurobiological Basis for PTSD	168
7	Genetic Studies on PTSD	169
8	Animal Models of PTSD	172
	8.1 Trauma or Stress-Based Models	172
	8.2 Mechanism-Based Models: Enhanced Fear Conditioning and Impaired Fear Extinction	173
	8.3 More Current Animal Models	175
9	Human Imaging Studies on PTSD	178
	9.1 Systems Involved in the Extinction of Fear Responses	178
10	NMDA Receptor and Amygdala-Dependent Learning	179
11	BDNF and Fear Learning	179
12	Conclusion	181
References		181

1 Introduction

Post-traumatic stress disorder (PTSD) is a debilitating psychiatric disorder that develops following exposure to a significantly threatening and/or horrifying event (American Psychiatric Association 2013). These traumatic events may include natural disasters (e.g., tsunamis, earthquakes, and tornadoes), accidents (e.g., vehicle and airplane crashes), military combat, victimization, or abuse such as physical and sexual assault, armed robbery, and torture (Gates et al. 2012; Punamaki et al. 2010; Harrison and Kinner 1998; Hoge et al. 2004). The newly revised fifth edition of the *Diagnostic and Statistical Manual of Mental Disorders* (DSM-5) by the American Psychiatric Association has classified PTSD not as an anxiety disorder as in previous versions but as part of a separate category of trauma and stress-related disorders (American Psychiatric Association 2013).

PTSD is characterized by the presence of a certain number of symptoms—lasting for at least 1 month following a traumatic event—from each of the four designated symptom clusters: intrusion; avoidance; negative alteration in cognitions and mood; and alterations in arousal and reactivity (American Psychiatric Association 2013). Intrusion symptoms are reminiscent of the re-experiencing symptoms from the DSM-IV, which include involuntary intrusive memories, nightmares, flashbacks, distress, or marked physiological reactivity after exposure to trauma-related stimuli. Avoidance symptoms include evasion of external reminders and distressing thoughts or feelings related to the trauma. Symptoms associated with negative alterations in cognition and mood include the inability to recall key features of the trauma, persistent and distorted negative beliefs, persistent negative trauma-related emotions, alienation, and the inability to experience positive emotions. Alterations in arousal and reactivity include symptoms of hypervigilance, exaggerated startle response, problems with sleep and concentration, irritable and aggressive behaviors, and reckless or destructive behaviors. The diagnosis of PTSD requires that these symptoms cause significant impairment of the person's life, disrupting both the functional and social aspects of daily activities. Thus, PTSD is a highly debilitating psychiatric disorder that impacts the individual and others (e.g., the individual's family, workplace, and social circle). Not surprisingly, PTSD is also associated with several adverse outcomes through its course, including breakdown of social and familial relationships, lower quality of life, work-related impairment, and medical illness (Marshall et al. 2001; Resnick and Rosenheck 2008).

With these considerations, it is important to understand the neurobiological mechanisms that contribute to the development of PTSD following exposure to traumatic stress. Although several studies have been performed, the etiopathology of PTSD remains unclear. Various hypotheses have been put forward, including one which posits that PTSD is a failure to recover from a traumatic experience due to the inability to extinguish fear and anxiety associated with conditioned sensory cues (Institute of Medicine 2012). This chapter reviews the traumatic stress-induced neurobehavioral mechanisms associated with the development of PTSD, as evidenced by studies using animal models and neuroimaging in humans. It also reviews briefly the history of PTSD and the risk factors associated with the disorder, with emphasis on more recent genetic studies.

2 A Brief History of PTSD

Compared to other psychiatric disorders, the history of PTSD as a fully conceptualized mental health disorder is relatively young. Although post-traumatic stress symptoms have been identified in writings that date back over 2,500 years (Crocq and Crocq 2000), PTSD was not formally codified by the American Psychiatric Association as a mental health disorder until 1980. In the nineteenth century, soldiers fighting the American Civil War were diagnosed with nostalgia or melancholia, characterized by lethargy, withdrawal, and excessive emotionality (Birmes et al. 2003). Other diagnoses included exhaustion, effort syndrome, and heart-related conditions, such as *irritable heart*, *soldier's heart*, and *cardiac muscular exhaustion* (Birmes et al. 2003). During World War I, combat veterans were diagnosed with shell shock and disordered action of the heart (Jones 2006). Symptoms for shell shock included fatigue, memory loss, difficulty sleeping, nightmares, and poor concentration. Vietnam War veterans suffering from chronic psychological problems that resulted in social and occupational dysfunction were

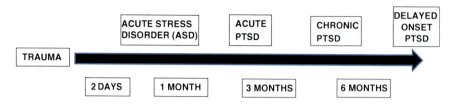

Fig. 1 Classification of stress and trauma-related disorders as characterized by presence of required symptoms as a time-dependent function following exposure to traumatic stress: acute stress disorder (ASD) (symptoms >2 days but <1 month), acute PTSD (symptoms >1 month but <3 months), chronic PTSD (symptoms ≥3 months), and delayed onset PTSD (symptom onset ≥6 months)

diagnosed with combat fatigue (Institute of Medicine 2007). Large-scale studies to examine combat-related issues of Vietnam veterans contributed to the formal recognition of PTSD as a distinct mental health disorder.

3 Trajectory of Trauma-Related Disorders and PTSD

Following exposure to a traumatic stressor—which may include an event which an individual experienced directly, witnessed, or learned about that threatened death or injury to themselves or another—an individual may have a wide range of reactions. The development and progression of symptoms following the traumatic event exposure determine the classification of the trauma response, which may range from normal to acute stress disorder (ASD) and PTSD (see Fig. 1). ASD is characterized by clinically significant dissociative symptoms (such as numbing and detachment or amnesia), trauma re-experiencing, situation avoidance, and increased arousal with significant functional impairment lasting more than 2 days but <1 month after the trauma (American Psychiatric Association 2013). If the symptoms are experienced for more than 1 month and meet full diagnostic criteria as stated above, PTSD can be classified as acute, chronic, or delayed onset (American Psychiatric Association 2000; Institute of Medicine 2012). In acute PTSD, symptoms develop immediately or soon after experiencing a traumatic event and persist longer than 1 month but <3 months. In chronic PTSD, the symptom duration is longer than 3 months. In delayed onset PTSD, a person does not express symptoms for months (≥6 months) or even years after the traumatic event. If a person does not meet the full diagnostic criteria for PTSD, or if the symptoms are not in the correct distribution as per required number of symptoms, the condition is considered partial or subthreshold PTSD (Institute of Medicine 2012).

PTSD may remit with time, with the largest remission reported during the first 12 months after diagnosis (Institute of Medicine 2012). Treatment with cognitive behavioral therapies significantly improves remission rates. A randomized controlled trial of female sexual assault victims with PTSD by Resick and colleagues found that 80 % of participants treated with cognitive processing therapy or prolonged exposure therapy no longer met the criteria for PTSD at post-treatment and at the 9-month follow-up (Resick et al. 2002). Long-term follow-up of this cohort conducted 5–10 years after the completion of treatment indicated that these patients remained in remission (Resick et al. 2012). However, most individuals with PTSD do not receive any treatment, and only a small percentage of those who do receive treatment are given an evidence-based treatment (Foa et al. 2013). Approximately one-third of PTSD cases do not remit, even after many years of treatment. Chronic PTSD has been found to be most strongly associated with men who reported combat as their worst trauma (Prigerson et al. 2001), history of childhood trauma and alcohol abuse or dependence (Zlotnick et al. 1999), and exposure to ongoing stressors and other traumatic events throughout life (Galea et al. 2008).

PTSD has been shown to co-occur with other psychiatric disorders such as depressive disorders, substance dependence, panic disorder, agoraphobia, generalized anxiety disorder, social phobia, bipolar disorder, and somatization (American Psychiatric Association 2013). These disorders can precede or present simultaneously with PTSD (Institute of Medicine 2012). They may also resolve before, after, or simultaneously with PTSD. A New Zealand birth cohort study showed that 93.5 % of those meeting the criteria for lifetime prevalence of PTSD at age 26 had also previously met the criteria for diagnosis of another mental health disorder such as major depression, anxiety disorder, conduct disorder, marijuana dependence, or alcohol dependence (Scherrer et al. 2008). There is also an association between PTSD and suicide ideation, attempts, and completions (Marshall et al. 2001). An increased number of PTSD symptoms were associated with a linear increase in current suicide ideation. Among individuals with chronic PTSD attending a clinic, 38.3 % reported to have suicide ideation, while 9.6 % reported to have had a suicide attempt (Marshall et al. 2001).

4 Epidemiology of PTSD

According to the DSM-5 (American Psychiatric Association 2013), the projected lifetime risk of PTSD at age 75 in the United States is 8.7 %, using the previous DSM-IV-TR criteria. National surveys conducted at different time periods have estimated the overall lifetime prevalence of PTSD among individuals 18 years and older. The 1990–1992 United States National Comorbidity Survey (NCS), conducted to examine the distribution of and factors associated with psychiatric disorders, reported a lifetime overall prevalence of 7.8 % for PTSD (Kessler et al. 1995). The NCS Replication (NCS-R), conducted 10 years after the original survey, estimated the overall prevalence of PTSD to be 6.8 % (Kessler et al. 2005). In 2004–2005, the National Epidemiologic Survey on Alcohol and Related Conditions estimated the lifetime prevalence of PTSD to be 7.3 % overall (Roberts et al. 2011). Estimates of lifetime PTSD prevalence in United States service members deployed to Operation Enduring Freedom (OEF) and Operation Iraqi Freedom (OIF) are two or three times higher compared to the general population (Institute of Medicine 2012). Current PTSD prevalence estimates in OEF and OIF service members range from 13 to 20 %, depending on the assessment measures used (Hoge et al. 2004; Seal et al. 2007; Vasterling et al. 2010). Worldwide, the estimates for lifetime prevalence of PTSD range from a low of 0.3 % in China to 6.1 % in New Zealand (Kessler et al. 2008). However, the statistics reported from various countries may not be directly comparable due to methodological differences in survey administration and sampling strategies.

5 Risk Factors for the Development of PTSD

Studies have found associations of factors in the development of PTSD, including trauma type and severity, gender and sexual orientation, race and ethnicity, cognitive reserve, pretrauma psychopathology, familial psychiatric history, and genetics.

The type and severity of the trauma are primary determinants of the development of PTSD (Institute of Medicine 2012). Most of the risk-factor differences found in various subgroupings in epidemiological studies can be best explained by differences in trauma type and severity. Higher risk of PTSD has been associated with traumas that involve physical injuries (either penetrating or assault), perception of the trauma as a true threat to one's life, and major losses (Holbrook et al. 2001; Ozer et al. 2003). Increased risk of PTSD has also been associated with lack or loss of social support after the traumatic event and ongoing life stress, including loss of employment, financial strain, and disability (Ozer et al. 2003; Brewin et al. 2000a).

Gender and sexual orientation are also considered risk factors for PTSD. In the original NCS, the prevalence of PTSD was twice as high in women as in men (Kessler et al. 1995). The NCS-R estimated that women were 2.7 times more likely to develop PTSD than men (Harvard Medical School 2007). These gender differences are thought to be explained primarily by differences in trauma exposure, such as sexual assault, which is much more likely to occur in females. Although men were more likely to report having experienced traumatic events over their lifetime, women with PTSD were more likely to develop more comorbid psychiatric disorders (Seedat et al. 2005), experience PTSD symptoms longer than men (Chilcoat and Breslau 1998a), and more likely to report poorer quality of life (Holbrook et al. 2001; Seedat et al. 2005). Sexual minorities have been reported to have a higher risk of PTSD compared to a heterosexual reference group (Roberts et al. 2010). However, sexual minorities have also been reported to have earlier and greater exposure to violence and traumatic events (Roberts et al. 2010).

Race and ethnicity may be risk factors for PTSD, although the evidence is inconsistent (Institute of Medicine 2012). The 2004–2005 National Epidemiologic Survey on Alcohol and Related Conditions survey showed that the risk of PTSD is significantly higher in blacks and lower in Asians than in whites (Roberts et al. 2011).

In a sample of survivors of physical trauma, Hispanic whites were more likely to report PTSD and with greater symptom severity compared to non-Hispanic whites (Marshall et al. 2009). They also reported more symptoms related to cognitive and sensory perception, such as hypervigilance and emotional reactivity.

Cognitive reserve is considered to be an important etiologic factor in the development of PTSD (Institute of Medicine 2012; Barnett et al. 2006). Intelligence quotient or IQ is a marker of cognitive reserve. IQ has been shown to be inversely related to the risk of PTSD and other psychiatric disorders (Batty et al. 2005). Breslau and colleagues reported that children who had an IQ >115 at 6-years old had decreased conditional risk of PTSD after trauma exposure (Breslau et al. 2006). It has also been reported that IQ assessed at age 5 was inversely associated with the risk of developing PTSD at age 32 (Koenen et al. 2007).

The evidence to support the association between family psychiatric history and PTSD is inconsistent. Parental mental health disorders were associated with increased risk of PTSD, even after controlling for previous traumatic events (Bromet et al. 1998). Maternal depression was also associated with increased risk of PTSD (Koenen et al. 2007). Statistically significant associations between PTSD and family psychiatric history of depression, anxiety, and psychosis have also been reported (Breslau et al. 1991). However, a meta-analysis of risk factors for PTSD did not find this association to be significant (Brewin et al. 2000a, b).

PTSD may have a genetic component. Recent genetic studies showed that relatives of probands (persons serving as index cases in genetic investigations of families) who had PTSD had higher risk of the disorder than relatives of similarly trauma-exposed controls who did not develop PTSD (Institute of Medicine 2012). Twin studies of male Vietnam veterans established a genetic influence of about 30 % for the vulnerability of PTSD, even after genetic influences on trauma exposure are accounted for (True et al. 1993). Another twin study among young women reported PTSD vulnerability at 7.2 % (Sartor et al. 2011). Twin and family studies also provided evidence that most of the genes that affect the risk of PTSD also influence the risk of other psychiatric disorders, including major depression, generalized anxiety disorder, and substance abuse, and vice versa (Institute of Medicine 2012). Genetic variation in PTSD can be accounted for by the genetic influences of generalized anxiety disorder and panic disorder symptoms, alcohol and drug dependence (Xian et al. 2000), and nicotine dependence (Koenen et al. 2005).

The psychopathology of the individual prior to the trauma has also been implicated as a risk for developing PTSD, with increased risks-associated externalizing and anxiety problems. Children rated as having externalizing problems above the normal range at age 6 were more likely to develop PTSD than children who were rated as normal externalizers (Breslau et al. 2006; Breslau 2006). Children diagnosed with anxiety disorder at age 6 were more likely to develop PTSD than a young adult (Breslau 2006; Breslau et al. 2006). Likewise, children who were categorized as highly anxious or having depressive mood at first grade were also at higher risk of PTSD at age 15 when exposed to traumatic events compared to their peers who did not have these psychologic symptoms (Storr et al. 2007). Children with difficult temperaments, fewer friends, or antisocial behaviors were more likely to develop PTSD than their peers who did not have these characteristics (Koenen et al. 2007).

Exposure to prior trauma has been implicated in PTSD. A meta-analysis of nine studies suggests that childhood abuse is a risk factor for PTSD (Brewin et al. 2000b). Women who experienced physical abuse during childhood had a higher risk of lifetime PTSD (Ozer et al. 2003). Persons who experienced a traumatic event before the target stressor reported higher levels of PTSD symptoms than those who did not, especially among individuals who experienced noncombat interpersonal violence (Ozer et al. 2003). However, more recent studies have shown that it is not the prior trauma experience per se, but the development of PTSD symptoms in response to a prior trauma that increased the risk of PTSD after a later trauma (Breslau et al. 2008; Breslau and Peterson 2010).

PTSD can affect people at any age (Institute of Medicine 2012). However, the 1990–1992 United States NCS reported the lowest prevalence for PTSD was in men 15–24-years old and in women 45–54-years old at 2.8 and 8.7 %, respectively (Kessler et al. 1995). The NCS Replication done 10 years later showed that the lowest lifetime prevalence of PTSD at 2.8 % was with individuals who are 59-years old and over (Harvard Medical School 2007). In addition, the reported highest lifetime and 12-month prevalence rates of PTSD were with 45–59-years old at 9.2 and 5.3 %, respectively. The prevalence estimates of PTSD by age groups may be confounded by historical events, such as wars (Vietnam, Iraq, and Afghanistan) or major natural disasters.

Drug and alcohol use and dependence are associated with PTSD. In the NCS, individuals with PTSD were twice as likely to have substance abuse disorder (Kessler et al. 1995). Likewise, Chilcoat and Breslau reported that individuals with a history of PTSD were four times more likely to have drug use or dependence (Chilcoat and Breslau 1998b). They suggested that the association between substance abuse and PTSD may be related to self-medication. Individuals exposed to trauma initiate the use of drugs and other psychoactive substances to help them cope with PTSD symptoms (Brown and Wolfe 1994; Khantzian 1985). Active PTSD also increased the risk of smoking, independent of one's genetic make-up (Koenen et al. 2006). Further, preexisting nicotine dependence increased the risk of PTSD in male veterans (Koenen et al. 2005).

6 Neurobiological Basis for PTSD

Several neurobiological systems have been implicated in the pathologic and protective responses to stress and development of PTSD, including the sympathetic nervous system, the hypothalamic–pituitary–adrenal (HPA) axis, serotonin system, opiate system, and sex steroidal system (Institute of Medicine 2007, 2012). Other chapters in this book discuss the changes in these systems in response to stress. Here we will look at the role of these neurobiological systems in PTSD as studied using advances in molecular genetics, animal models, and human neuroimaging studies.

7 Genetic Studies on PTSD

Advances in molecular genetics have contributed to the body of knowledge on PTSD. Studies have progressed from genotype–phenotype associations to the identification of epigenetic signatures associated with the disorder (Uddin et al. 2010, 2011). They have also examined how individual differences in epigenetic programming may modify the risk of PTSD in association with trauma exposure (Koenen et al. 2011). One approach examines variation in polymorphisms to identify specific genetic variants that may be associated with increased risk or resilience (Amstadter et al. 2009b). Although this approach could potentially provide useful knowledge on the etiology of PTSD, it offers limited interpretation of the research findings or determination of functional genetic variants (Amstadter et al. 2009b). This section provides an overview, but not a comprehensive review, of candidate genes for PTSD using this approach.

Table 1 shows a summary of some of the candidate genes for PTSD. This is by no means exhaustive but offers information about the diversity of candidate genes that modulate dopaminergic, serotonergic, HPA axis, noradrenergic, and neurotrophic systems. The current evidence for a specific genetic variant that increases vulnerability or resilience to PTSD is not robust (Institute of Medicine 2012). The associations between specific genetic variants and PTSD lack consistency. This may be due to small sample size and differences between studies that are not consistently accounted for, i.e., modification of genetic variants by environmental factors (Institute of Medicine 2012).

Genetic effects may be modified by the environment through molecular mechanisms such as deoxyribonucleic acid (DNA) methylation (Bernstein et al. 2007). DNA methylation alters transcriptional activity of the loci through chemical modifications that regulate DNA accessibility. Increased methylation in specific gene regions (i.e., promoter region) is associated with reduced transcriptional activity and therefore reduced gene expression. For example, individuals with PTSD were distinguished by methylation profiles that suggest upregulation of immune system-related genes and downregulation of genes involved in neurogenesis and the startle response (Uddin et al. 2010). Upregulation of these genes was associated with higher concentrations of biomarkers for immune system reactivity of these patients, i.e., increased cytomegalovirus, interleukin-2, interleukin-4, and tumor necrosis factor-alpha (Uddin et al. 2010). Koenen and colleagues showed that methylation of the gene that encodes the serotonin transporter (SLC6A4) modified the effects of traumatic events on the development of PTSD when the SLC6A4 genotype was controlled for (Koenen et al. 2011). Lower methylation levels were observed in individuals who experienced more traumatic events and were therefore of higher risk of PTSD. Hence, gene-specific methylation patterns may be associated with increased risk of and resilience to PTSD (Koenen et al. 2011). However, these PTSD-associated epigenetic differences were not shown to be associated with downstream differences in gene expression.

Table 1 Some identified candidate genes for PTSD. Modified and updated from Institute of Medicine (2012)

System *Common name(s)*	Gene	Location	Significant findings	Null findings
Dopamine				
Dopamine receptor DR	*RD2 (D2R, D2DR)*	11q23	Comings et al. (1991, 1996); Hemmings (2013); Voisey et al. (2009); Young et al. (2002)	Bailey et al. (2011); Gelertner et al. (1999)
Dopamine receptor D4	*DRD4*	11p15.5	Dragan and Oniszczenko (2009)	
Dopamine transport	*SLC6A3 (DAT1)*	5p15.3	Chang et al. (2012); Hoexter et al. (2012); Drury et al. (2009, 2013); Segman et al. (2002); Valente et al. (2011)	Bailey et al. (2011)
Dopamine beta-hydroxylase	*DBH*	9q34		Mustapic et al. (2007); Tang et al. (2010)
Serotonin (5-HT)				
Serotonin (5-hydroxytryptamine or 5-HT) transporter	*SLC6A4 (HTT, 5HTT, SERT, 5-HTTPLPR*	17q11	Grabe et al. (2009); Kilpatrick et al. (2007); Koenen et al. (2009a, 2009b, 2011); Kolassa et al. (2010); Lee et al. (2005); Morey et al. (2011); Thakur et al. (2009); Wang et al. (2011); Xie et al. (2009, 2012)	Mellman et al. (2009); Sayin et al. (2010); Valente et al. (2011)
5-HT receptor 2A	*5HTR2*	13q14q21	Lee et al. (2007)	
Hypothalamic–pituitary–adrenal axis				
FK506-binding protein 5	*FKBP5*	6p21	Binder et al. (2008); Boscarino et al. (2011); Koenen et al. (2005); Xie et al. (2010)	
Glucocorticoid receptor	*GCCR (NR3C1)*	5q31.3	Amstadter et al. (2011); White et al. (2013)	Bachmann et al. (2005)
Corticotrophin-releasing hormone receptor 1	*CRHR1*	17q12–22		

(continued)

Table 1 (continued)

System Common name(s)	Gene	Location	Significant findings	Null findings
Regulator of G protein signaling 2	RGS2	1q31	Amstadter et al. (2009a)	
Cannabinoid receptor 1 (brain)	CNR1 (CB1, CNR)	6q14–q15		Lu et al. (2008)
Apolipoprotein E	APOE	19q13	Freeman et al. (2005)	
Brain-derived neurotrophic factor	BDNF Val66 Met	11p13	Hemmings et al. (2013); Zhang et al. (2014)	Mustapic et al. (2007)
Neuropeptide Y	NPY	7p15.1	Nelson et al. (2009)	
Gamma aminobutyric acid (GABA)4	GABRA2	4p12		Lappalainen et al. (2002)
Catechol-*O*-methyl transferase	COMT	22q11	Amstadter et al. (2009b); Boscarino et al. (2011); Clark et al. (2013); Kolassa et al. (2010b)	
Adenylate cyclase-activating polypeptide 1 receptor	ADCYAP1R1	7p14	Ressler et al. (2011)	Chang et al. (2012)
Dystrobrevin-binding protein	DTNBP1	6p22	Voisey et al. (2010)	
Cholinergic receptor, neuronal nicotinic alpha polypeptide	CRNA5	15q25.1	Boscarino et al. (2011)	

Microarray-based studies have assessed gene expression changes in ribonucleic acid (RNA) derived from peripheral blood mononuclear cells or whole blood. A study of whole-blood-derived genes of individuals affected and not affected by the September 11, 2001, New York City attack showed a differential expression in 16 genes (Yehuda et al. 2009). Several of these genes are involved in signal transduction, brain and immune cell function, and HPA axis activity. Interestingly, the largest difference in expression was the gene for mannosidase, alpha, class 2C, member 1 (MAN2C1). MAN2C1 distinguished between those who have and those who do not have PTSD on the basis of gene expression (Yehuda et al. 2009; Uddin et al. 2011) and methylation (Uddin et al. 2011).

These studies suggest that genotype, methylation, and gene expression are promising areas of research to help understand the etiology of PTSD. To date, there is a lack of robust definitive findings on any single gene or gene system in the etiology of PTSD. More studies incorporating perhaps all three genetic approaches and perhaps even genome-wide association studies will help further the understanding of PTSD.

8 Animal Models of PTSD

Since it is unethical to subject humans to traumatic events to study the consequences of such exposure, experimental studies on trauma are limited to animal models. Animal models include the earlier trauma/stress-based models, mechanism-based models, and the more recent "chronic plus acute prolonged stress" (CAPS) and "cut-off behavioral criteria" (CBC) model of PTSD.

8.1 Trauma or Stress-Based Models

Trauma or stress-based animal models are based on exposure to a traumatic or stressful event. Extremely stressful experiences aimed at engendering a sense of threat and helplessness in the animal are used, with focus on the intensity and type of experience. Others have combined intensity with an attempt to design an ethologically valid experience, i.e., one that an animal might encounter in its natural environment. For example, rodents are exposed to fear-provoking and stressful predator stimuli (cat, cat odor, fox odor, or trimethylthiazoline, a synthetic compound isolated from fox feces). These stimuli have been shown to produce long-lasting behavioral and physiological responses. These paradigms, in which adult rodents are exposed to feline predators for 5–10 min in a closed environment (i.e., inescapable exposure), have been validated (Adamec 1997; File et al. 1993; Blanchard et al. 1998; Griebel et al. 1995). Predator stress has ecological validity in that it mimics brief intense threatening experiences with lasting affective consequences (Adamec et al. 2006a, b). The predator stress paradigm has proven to be effective in inducing the expected range of behavioral and physiological responses (Adamec et al. 2006a, b, 2007). These responses include freezing, avoidance, increased secretion of stress hormones, and changes in transmission from the hippocampus via the ventral angular bundle to the basolateral amygdala and from the central amygdala to the lateral column of the periaqueductal gray (Adamec et al. 2006b, 2007; Apfelbach et al. 2005; File et al. 1993; Roseboom et al. 2007; Mazor et al. 2009; Sullivan and Gratton 1998; Kozlovsky et al. 2008; Takahashi et al. 2005). The plasticity of these neurobiological pathways is associated with aversive learning. The potency of predator stimuli is comparable with that of a variety of paradigms in which the threat is more tangible and immediate. These include paradigms based on inescapable pain or electric shock, swimming and near-drowning, a small raised platform, and even direct proximity to a kitten or a cat (separated by a mesh divide or a solid divide with an opening large enough for the rodent to slip through).

8.2 Mechanism-Based Models: Enhanced Fear Conditioning and Impaired Fear Extinction

Other animal models of PTSD consider potential neurobiological mechanisms that might underlie post-traumatic stress. These models involve behavioral protocols that mimic the activation of such mechanisms. One key aspect considered is that exposure to stress alone does not sufficiently explain the persistence of psychologic and biologic fear responses long after the trauma exposure. This led some to suggest that fear conditioning may underlie the phenomenon of PTSD (Yehuda and LeDoux 2007).

Fear conditioning is an adaptive and evolutionary advantageous response to traumatic events (Morrison and Ressler 2013). Following stress or trauma exposure, the normal fear responses involve the consolidation and manifestation of fear memories in fearful situations and also the suppression and extinction of fear behaviors in safe situations. The extinction of fear memories involves the gradual decline in fear responses upon repeated presentations of the fearful cue in non-threatening situations. When the processes involved in fear regulation become dysregulated, sensitization and overgeneralizations can take place. Dysregulated fear responses characterize PTSD and most anxiety disorders. Hence, models of fear response have been used to understand the neurobiology of PTSD.

The neural mechanisms involved in the acquisition and extinction of learned fear responses have been studied using the classical Pavlovian fear-conditioning paradigms in animal models (Morrison and Ressler 2013). In this paradigm, a conditioned stimulus (CS; e.g., a light or tone that is initially inoffensive) is paired with an aversive unconditioned stimulus (US; such as a mild foot shock). After several CS–US pairings, the subject exhibits a conditioned response (CR) to presentation of the CS (tone or light). In rodent models, the conditioned fear responses are measured with freezing (complete lack of bodily movements except those involved in respiration) and fear-potentiated startle response (increase in acoustically elicited startle response).

Extinction of learned fear is manipulated through the repeated or prolonged exposure of the previously fear-conditioned organism to the CS (tone or light) in the absence of the aversive US (shock). The repeated or prolonged exposures result in fear extinction and the gradual decline in the CR (Myers and Davis 2007). The diminished CRs following extinction training are often not permanent and are subject to reinstatement, renewal, and spontaneous recovery. Renewal consists of re-emergence of the extinguished CR when animals are exposed to the CS in a novel context. Spontaneous recovery refers to the reappearance of the extinguished CRs after enough time has passed following extinction training. Reinstatement occurs when the extinguished fear response is triggered and reappears upon exposure to the US after the organism has undergone extinction training.

Although several brain regions are involved in fear processing and fear-related behaviors, the key brain regions include the hippocampus, amygdala, and prefrontal cortex or PFC (Morrison and Ressler 2013). Other regions involved are the parahippocampal gyrus, orbitofrontal cortex, sensorimotor cortex, the thalamus, and anterior cingulate cortex. The activation of the amygdala is the hallmark of all fear-related disorders. Increased amygdala activation has been shown during the presentation of fearful faces and fearful cues as well as during fear acquisition and expression (Hamilton et al. 2012). Individuals diagnosed with PTSD or other fear-related disorders exhibit hyperactive amygdala activity compared to normal subjects.

The amygdala consists of the basolateral complex (lateral, basal, and accessory basal nuclei) and the central nuclei (CeA). The basolateral amygdala (BLA) is critical in the acquisition, expression, and extinction of fear (Fanselow and LeDoux 1999). In the context of the classical conditioning paradigm, multimodal sensory information from thalamic and sensory cortical areas (auditory, visual, somatosensory cortex, etc.) specific to the CS project to the lateral nucleus of the amygdala (LA) (LeDoux et al. 1990; Campeau and Davis 1995). Also, information specific to the US is relayed to the LA from somatosensory thalamic and cortical areas and the periaqueductal gray (Lanuza et al. 2004). The LA is thus considered as a critical site for synaptic plasticity and Hebbian learning that occurs during paired presentations of the CS and US during fear learning (Pape and Pare 2010; Sah et al. 2008). The CeA has primarily been regarded as the fear output structure that sends projections to brain regions, which activate a host of downstream behavioral fear responses and symptoms. Long-term potentiation (LTP) and synaptic plasticity at any point along this circuit contribute to alterations in pathways that underlie the fear response (Sah et al. 2008; Pape and Pare 2010).

Several brain regions, including the PFC and hippocampus, modulate the activity of the amygdala. In the context of fear-conditioning models, the PFC is considered to provide inhibitory control to the amygdala, although some prefrontal regions appear to be positively correlated with amygdala activation and others are negative correlated. The infralimbic (IL) cortex (ventromedial PFC in humans) is required for fear extinction but not fear acquisition. The prelimbic (PL) cortex (considered to be homologous to the dorsal cingulate cortex in humans) is required for fear acquisition but not extinction (Sierra-Mercado et al. 2011). The hippocampus also modulates amygdala activity. In the context of fear learning and memory, the hippocampus is crucial for the regulation and discrimination of contextual learning and extinction (Heldt et al. 2007; Knight et al. 2004).

In PTSD, Morrison and Ressler proposed that fear memories of a traumatic event may become overgeneralized and difficult to extinguish, leading to the development of PTSD symptoms (Morrison and Ressler 2013). In terms of classical Pavlovian fear conditioning, intrusion symptoms may result from a trauma-related cue that triggers a painful emotional and/or physiological response, in addition to concomitant nightmares and flashbacks. The avoidance symptom cluster may be thought of as a type of operant conditioning, in which the avoidance of reminders of the trauma in and of itself becomes a reinforcing process (Morrison and Ressler 2013). The negative alterations in cognition and mood may reflect the dysregulation of the prefrontal–amygdala circuitry. Finally, the symptoms associated with arousal and reactivity may result from the activation of central and autonomic nervous system processes that lead to behaviors such as being easily startled or having trouble sleeping (Morrison and Ressler 2013).

8.3 More Current Animal Models

While fear conditioning and extinction could explain some of the symptoms of PTSD, these models have also been shown to be insufficient to produce the PTSD phenotype (Pitman et al. 2012). Newer animal models have now attempted to incorporate construct validity by capitalizing on the increasing understanding of the pathophysiology of PTSD. These models include predator exposure, exposure to single prolonged stress (SPS), and exposure to foot shock with additional stressors. These models have used one or more "PTSD-specific" endpoints, such as abnormal fear learning, exaggerated acoustic startle response and startle habituation, enhanced glucocorticoid signaling and negative feedback inhibition, and an exaggerated autonomic nervous system (Matar et al. 2013). An example of these animal models is the "CAPS" for rats (Green et al. 2011; Roth et al. 2012). The CAPS is used to model some of the stressful events that can lead to PTSD in humans. The paradigm consisted of chronic intermittent cold stress (4 °C, 6 h/day, 14 days) followed on day 15 by a single session of sequential acute stressors such as social defeat, immobilization, and cold swim (Green et al. 2011). The CAPS model has been shown to enhance acute fear responses and impair extinction of conditioned fear, and to reduce expression of glucocorticoid receptors in the medial prefrontal cortex (Green et al. 2011). Behaviorally, the CAPS treatment was shown to decrease active burying behavior and increase immobility in the shock probe defensive burying test. CAPS-treated rats displayed increased latency to feed in the novelty-suppressed feeding test. Further, CAPS treatment reduced HPA response to a subsequent acute immobilization stress. Taken together, this validated CAPS treatment as a rat model for PTSD as its effects resembles many aspects of human PTSD: impaired fear extinction, shifted behavior from active to more passive strategy, increased anxiety, and altered HPA reactivity (Roth et al. 2012).

To model a vulnerability factor that may produce stable changes in central stress response system in PTSD, Green and colleagues utilized the prenatal stress model prior to CAPS treatment (Green et al. 2011). Prenatal stress elevated basal corticosterone decreased GR protein levels in the hippocampus and prefrontal cortex, and decreased tyrosine hydroxylase mRNA expression in noradrenergic neurons in the dorsal pons. Rats exposed to prenatal stress and CAPS showed attenuated extinction of cue-conditioned fear. Thus, prenatal stress may induce vulnerability to subsequent adult stress (Green et al. 2011), thereby increasing risk for the development of PTSD.

Most animal studies tend to report the results of the entire exposed population versus control populations without distinction, even though individual subjects display a variable range of response to stress paradigms. This has often been considered problematic for behavioral models, where significant response to stress exposure is expected. However, Matar and colleagues highlighted that the heterogeneity in animal responses may actually confirm the validity of animal studies, since humans do not clearly respond homogenously to potentially traumatic experience (Matar et al. 2013). Subsequently, they validated a "cut-off behavioral criteria" or CBC model in the rat, which uses the predator scent stressor (15-min exposure to cat urine) as the threatening stimuli. For a trauma reminder, the rat was exposed to unused cat litter, no less than 8 days after the initial stress exposure. They used the elevated plus maze and acoustic startle response paradigms to measure anxiety-like, fearful, avoidant, and hypervigilant/hyperalert behaviors, all of which parallel aspects of traumatic stress-induced behaviour in humans (Adamec 1997; File et al. 1993). According to the behavioral response in the elevated plus maze and acoustic startle, rats were subsequently classified as having extreme behavioral response (EBR), partial behavioral response (PBR), or minimal behavioral response (MBR). Data were then analyzed according to the CBC response classification. The prevalence rates of rats with EBR and MBR were comparable with the PTSD data in the human population, where 15–35 % meets the full criteria for PTSD, while 20–30 % displays a partial or sub-symptomatic clinical picture (Breslau et al. 1999). The prevalence of the EBR tapered with time to about 25 % at day 7, remained stable until day 30, and then tapered down to 15 % at day 90, comparable to the trajectory of trauma response and PTSD (see Fig. 1).

The CBC model found associations between EBR, MBR, and PBR behavior patterns and biomolecular, physiological, and morphological consequences of traumatic stress. The detailed description of these effects is available elsewhere (Matar et al. 2013). Briefly, the behavioral classification correlated with HPA axis (corticosterone, dehydroepiandrosterone, and its derivative dehydroepiandrosterone sulfate levels), autonomic nervous system (heart and heart rate variability), and immune system activation (Cohen 2003; Kozlovsky et al. 2009). The EBR was characterized by significantly more disturbances on all measures, whereas MBR

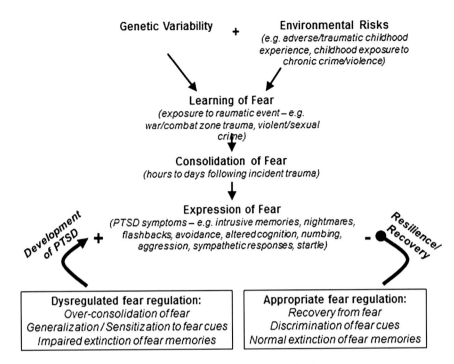

Fig. 2 Genetic and environmental factors contribute to the individual's response to exposure to traumatic stress. Following stress or trauma exposure, the normal fear responses involve the consolidation and manifestation of fear memories in fearful situations and also the suppression and extinction of fear behaviors in safe situations. The extinction of fear memories involves the gradual decline in fear responses upon repeated presentations of the fearful cue in the nonthreatening situation. Dysregulated fear regulation may result in the overconsolidation of fear and impaired extinction of fear memories, resulting in sensitization to fear cues and overgeneralizations. This mechanism has been hypothesized to contribute to the development of PTSD. Figure adapted from Morrison and Ressler (2013), with permission and with modifications

rats displayed almost none. In addition, EBR was associated with a distinct pattern of long-term and persistent downregulation of brain-derived neurotrophic factor (BDNF; mRNA and protein levels) and synaptophysin; it was also associated with upregulation of glucocorticoid receptor (GR) protein levels and tyrosine kinase receptor mRNA in the CA1 region of the hippocampus (Kozlovsky et al. 2009). EBR was also associated with significant downregulation of growth-associated protein 43, signal-regulated kinase–mitogen-activated protein kinase (ERK1/2) and phosphor-ERK1/2, p-38 and phosphor-38 in the hippocampus. There was association with an upregulation of post-synaptic density-95 in the same region. These studies suggest a relationship between the type of behavioral response and the expression of key intracellular and intercellular biomolecules associated with neuromodulation, synaptic plasticity, and receptor systems. These data support the validity of the CBC as an animal model for PTSD (Fig. 2).

9 Human Imaging Studies on PTSD

Studies on PTSD involving human subjects seem to support the hypothesis that PTSD develops from a dysfunction in fear learning and extinction. As discussed, the key brain region involved in fear learning and extinction in humans is the amygdala. The amygdala, in turn, receives input from other brain regions, including the lateral and medial prefrontal cortex, hippocampus, and insula. Neuroimaging studies have contributed greatly to the understanding of the involvement of these regions in PTSD. These studies included resting activity using positron emission tomography, structural magnetic resonance imaging (MRI) studies, and functional MRI studies of patients performing a variety of emotional tasks or viewing emotional stimuli. This section reports an overview, but not a comprehensive review, of some of the results from imaging studies.

Some imaging studies on PTSD used symptom provocation, where patients are reminded of traumatic events while their brains are being scanned. The brain scans are then analyzed for decreases and increases in blood flow in particular brain regions. Patients with PTSD exposed to reminders of traumatic events had decreased blood flow in the medial frontal gyrus (Bremner et al. 1999; Shin et al. 2004). The blood flow to the medial prefrontal gyrus was inversely correlated with changes in amygdala blood flow. There was a positive correlation between changes in amygdala blood flow and symptom severity and a negative correlation between changes in medial frontal gyrus blood flow and symptom severity (Shin et al. 2004). Subjects with PTSD who viewed fearful faces during functional MRI had heightened amygdala activity (Shin et al. 2004) and diminished ventromedial prefrontal cortex activity (Shin et al. 2005). PTSD patients were reported to have impaired function of the ventromedial prefrontal cortex and amygdala in response to presentation of nontrauma-related stressful cues (Gold et al. 2011; Phan et al. 2006). In addition, PTSD patients who underwent cognitive tasks were reported to have abnormal resting state and functional reactivation in the rostral and more dorsal areas of the anterior cingulate cortex (Shin and Handwerger 2009; Shin et al. 2005). Other studies have reported the involvement of the insula, a brain region involved in interoception, and monitoring of internal states that predict autonomic responses to fear. Patients with PTSD were found to have exaggerated insula activation in a number of different paradigms, including responses to presentation of fearful faces, painful stimuli, and traumatic memories (Simmons et al. 2008; Strigo et al. 2010). These results have provided relevant information about the neural circuitry associated with PTSD.

9.1 Systems Involved in the Extinction of Fear Responses

Several biologic systems play a role in the extinction of fear responses. These systems will not be discussed here, as a review is available elsewhere (Morrison and Ressler 2013). This section will discuss advances in novel systems currently explored as potential target systems for the development of newer treatments for PTSD, i.e., the glutamatergic N-methyl-D-aspartate (NMDA) receptor and the BDNF–tyrosine kinase B (TrkB)-induced signaling pathways.

10 NMDA Receptor and Amygdala-Dependent Learning

NMDA receptors are highly expressed in the amygdala and are important for fear learning. Fear learning requires NMDA receptor activation. In turn, learning is blocked by NMDA antagonists. For example, studies have shown that systemic and direct administrations of NMDA antagonists to the BLA before fear extinction training block the extinction of fear and fear-potentiated started (Falls et al. 1992).

The NMDA partial agonist D-cycloserine (DCS) activates the NMDA receptor glycine/serine modulatory site, resulting in increased calcium influx upon glutamate binding. Systemic or direct infusion of DCS into the amygdala prior to or immediately following extinction training facilitates the extinction of fear-potentiated (Walker et al. 2002). DCS-facilitated extinction is more resistant to reinstatement and also generalizes the inhibition of fear. Since DCS was previously approved by the Food and Drug Administration (FDA) for tuberculosis, it has been tested in exposure psychotherapy in human subjects. DCS has been shown to enhance exposure-based psychotherapy for a number of phobia and fear-related disorders (Norberg et al. 2012) and may also be useful for PTSD. In fact, a recent randomized controlled trial showed that DCS-facilitated virtual reality exposure (VRE) resulted in significantly earlier and greater improvement in PTSD symptoms as well as greater improvements in depression, anger, expression, and sleep (Difede et al. 2014). However, positive results were not found in a recent randomized, double-blind, placebo-controlled trial to determine whether DCS augments exposure therapy for PTSD in veterans returning from Iraq and Afghanistan (Litz et al. 2012). The results indicated that veterans in the exposure therapy plus DCS condition experienced significantly less symptom reduction than those in the exposure therapy plus placebo condition over the course of the treatment. These results suggest additional research is needed.

11 BDNF and Fear Learning

Neurotrophins are a class of proteins that serve as survival factors for central nervous system neurons. BDNF is a neurotrophin that plays a role in the limbic system by regulating synaptic plasticity, memory processes, and behavior (Angelucci et al. 2014). BDNF and its intracellular kinase-activating receptor TrkB have been implicated in the neurobiological mechanisms underlying the clinical manifestations of PTSD, especially those related to synaptic efficacy and neural plasticity. BDNF and its action at the TrkB receptor plays a significant role in both the acquisition and extinction of fear learning in both human and animal models (Andero et al. 2012; Andero and Ressler 2012).

There are few studies of BDNF levels in patients with PTSD, with contrasting results. Bonne and colleagues reported that the BDNF levels of the cerebrospinal fluid of female civilians with PTSD are comparable with healthy controls (Bonne et al. 2011), although the sample size was small and patients had moderate PTSD severity. Another study reported patients with PTSD or ASD have significantly higher serum BDNF compared to healthy controls (Hauck et al. 2010). On the other hand, drug-naïve PTSD patients (without comorbid psychiatric symptoms) have significantly lower BDNF levels compared to healthy controls (Dell'Osso et al. 2009). Lower serum BDNF levels were also found in individuals who were exposed to trauma and developed PTSD, compared to individuals who were exposed to trauma but did not develop PTSD (Angelucci et al. 2014).

A single nucleotide polymorphism (SNP) of the BDNF gene has also been implicated in PTSD. The Val66Met SNP is an SNP in the proregion of BDNF that consists of a Met substitution for Val at position 66. It has been implicated in several psychiatric disorders including depression, schizophrenia and PTSD (Frielingsdorf et al. 2010). There is evidence to suggest that this polymorphism may disrupt BDNF signaling and, in turn, affect emotional learning and memory. Individuals with the Val66Met SNP release less BDNF, have decreased hippocampal volume, and exhibit deficits in declarative memory and fear extinction (Bueller et al. 2006; Soliman et al. 2010).

Animal models support these findings. Mice with the knock-in allele of the human BDNF Val66Met allele display reduced hippocampal dendritic arborization, decreased hippocampal volume, and impaired long-term potentiation as well as deficits in declarative memory and decreased fear extinction (Bueller et al. 2006; Frielingsdorf et al. 2010; Soliman et al. 2010; Ninan et al. 2010). Data from rodent models have also demonstrated that BDNF-TrkB signaling is necessary for the acquisition of fear conditioning and consolidation of fear extinction in the amygdala, hippocampus, and PFC (Chhatwal et al. 2006; Rattiner et al. 2004). BDNF plays distinct roles in different regions of PFC. Direct BDNF infusion in the IL region enhances fear extinction (Peters et al. 2010). In contrast, BDNF deletion, by injecting Cre recombinase expressing lentivirus into the brain of floxed BDNF transgenic mice, in the PL region of the PFC results in deficits in fear acquisition but does not affect fear extinction. The BDNF deletion-induced fear acquisition deficits in the PL may be rescued by the administration of 7,8-dihydroxyflavone (7,8-DHF), a small molecule compound that activated the TrkB receptor, thus mimicking the actions of endogenous BDNF. Systemic administration of a single dose of 7,8-DHF has been found to activate TrkB receptors in the amygdala and also enhance the acquisition and extinction of fear in mice (Andero et al. 2012). 7,8-DHF also rescues the extinction deficit present in a mouse model of stress (Andero et al. 2012; Andero and Ressler 2012). These data suggest that small molecule agonists and antagonist that target the BDNF-TrkB signaling pathway may provide novel approaches in the enhancement of extinction and possible treatment of fear-related disorders, including PTSD.

12 Conclusion

PTSD is a debilitating psychiatric disorder that is prevalent in the general population, with adult lifetime prevalence rate estimated at 8.7 % (American Psychiatric Association 2013). This rate is much increased in military and veteran populations that have been exposed to many more traumatic and severe events. Some factors have been identified to be associated with the risk of developing PTSD, such as trauma type and severity, gender and sexual orientation, race and ethnicity, lower IQ, prior trauma exposure and pretrauma psychopathology, familial psychiatric history, and genetics. The recently published DSM-5 has classified PTSD as a trauma and stress-related disorder which is characterized by intrusion symptoms, avoidance, negative alteration in cognitions and mood, and alterations in arousal and reactivity for at least 1 month following exposure to trauma (American Psychiatric Association 2013). Studies using animal models and human imaging have contributed much to the understanding of the neurobiology of PTSD. However, there is still much to be learned and understood in terms of the etiopathology of the disorder. Through animal and human studies, correlations have been found between a stressor or a risk factor and PTSD symptoms. While the knowledge gained so far has been important, causal studies are needed that would firmly implicate the identified neurobiological mechanisms in PTSD or resilience. These studies are unfortunately lacking because they will require invasive and risky procedures that cannot be ethically undertaken in humans. Therefore, animal models on learning and stress systems that consider the diagnostic criteria of PTSD need to be developed.

Current evidence indicates that the development of PTSD may be influenced by genetic variation as well as environmental factors. The nature of the genes–environment interactions needs to be studied in order to understand how these factors contribute to the onset and severity of PTSD. Data from these studies would have important implications for PTSD prevention and treatment. Another important area of science is the identification of biomarkers and a brain imaging model for the diagnosis of PTSD to help reduce the dependence on self-reported symptoms. The identification of biomarkers also has the potential to inform the development of treatment strategies as well as tools to assess treatment success.

References

Adamec R (1997) Transmitter systems involved in neural plasticity underlying increased anxiety and defense–implications for understanding anxiety following traumatic stress. Neurosci Biobehav Rev 21(6):755–765

Adamec R, Burton P, Blundell J, Murphy DL, Holmes A (2006a) Vulnerability to mild predator stress in serotonin transporter knockout mice. Behav Brain Res 170(1):126–140

Adamec R, Head D, Blundell J, Burton P, Berton O (2006b) Lasting anxiogenic effects of feline predator stress in mice: sex differences in vulnerability to stress and predicting severity of anxiogenic response from the stress experience. Physiol Behav 88(1–2):12–29

Adamec R, Muir C, Grimes M, Pearcey K (2007) Involvement of noradrenergic and corticoid receptors in the consolidation of the lasting anxiogenic effects of predator stress. Behav Brain Res 179(2):192–207

American Psychiatric Association (2000) Diagnostic and statistical manual of mental disorders, fourth edition, text revision (DSM-IV-TR). American Psychiatric Association, Washington, DC

American Psychiatric Association (2013) Diagnostic Statistical Manual of Mental Disorders, 5th edn, DSM-5. American Psychiatric Association, Washington, DC

Amstadter AB, Koenen KC, Ruggiero KJ, Acierno R, Galea S, Kilpatrick DG, Gelernter J (2009a) Variant in RGS2 moderates posttraumatic stress symptoms following potentially traumatic event exposure. J Anxiety Disord 23(3):369–373

Amstadter AB, Nugent NR, Koenen KC, Ruggiero KJ, Acierno R, Galea S, Kilpatrick DG, Gelernter J (2009b) Association between COMT, PTSD, and increased smoking following hurricane exposure in an epidemiologic sample. Psychiatry 72(4):360–369

Amstadter AB, Richardson L, Meyer A, Sawyer G, Kilpatrick DG, Tran TL, Trung LT, Tam NT, Tuan T, Buoi LT, Ha TT, Thach TD, Gaboury M, Acierno R (2011) Prevalence and correlates of probable adolescent mental health problems reported by parents in Vietnam. Soc Psychiatry Psychiatr Epidemiol 46(2):95–100

Andero R, Ressler KJ (2012) Fear extinction and BDNF: translating animal models of PTSD to the clinic. Genes Brain Behav 11(5):503–512

Andero R, Daviu N, Escorihuela RM, Nadal R, Armario A (2012) 7,8-dihydroxyflavone, a TrkB receptor agonist, blocks long-term spatial memory impairment caused by immobilization stress in rats. Hippocampus 22(3):399–408

Angelucci F, Ricci V, Gelfo F, Martinotti G, Brunetti M, Sepede G, Signorelli M, Aguglia E, Pettorruso M, Vellante F, Di Giannantonio M, Caltagirone C (2014) BDNF serum levels in subjects developing or not post-traumatic stress disorder after trauma exposure. Brain Cogn 84(1):118–122

Apfelbach R, Blanchard CD, Blanchard RJ, Hayes RA, McGregor IS (2005) The effects of predator odors in mammalian prey species: a review of field and laboratory studies. Neurosci Biobehav Rev 29(8):1123–1144

Bachmann AW, Sedgley TL, Jackson RV, Gibson JN, Young RM, Torpy DJ (2005) Glucocorticoid receptor polymorphisms and post-traumatic stress disorder. Psychoneuroendocrinology 30(3):297–306

Bailey JN, Goenjian AK, Noble EP, Walling DP, Ritchie T, Goenjian HA (2011) PTSD and dopaminergic genes, DRD2 and DAT, in multigenerational families exposed to the Spitak earthquake. Psychiatry Res 178(3):507–510

Barnett JH, Salmond CH, Jones PB, Sahakian BJ (2006) Cognitive reserve in neuropsychiatry. Psychol Med 36(8):1053–1064

Batty GD, Mortensen EL, Osler M (2005) Childhood IQ in relation to later psychiatric disorder: evidence from a Danish birth cohort study. British J Psychiatry J Mental Sci 187:180–181

Bernstein BE, Meissner A, Lander ES (2007) The mammalian epigenome. Cell 128(4):669–681

Binder EB, Bradley RG, Liu W, Epstein MP, Deveau TC, Mercer KB, Tang Y, Gillespie CF, Heim CM, Nemeroff CB, Schwartz AC, Cubells JF, Ressler KJ (2008) Association of FKBP5 polymorphisms and childhood abuse with risk of posttraumatic stress disorder symptoms in adults. JAMA: J Am Med Assoc 299(11):1291–1305

Birmes P, Brunet A, Carreras D, Ducasse JL, Charlet JP, Lauque D, Sztulman H, Schmitt L (2003) The predictive power of peritraumatic dissociation and acute stress symptoms for posttraumatic stress symptoms: a three-month prospective study. Am J Psychiatry 160(7):1337–1339

Blanchard RJ, Hebert MA, Ferrari PF, Palanza P, Figueira R, Blanchard DC, Parmigiani S (1998) Defensive behaviors in wild and laboratory (Swiss) mice: the mouse defense test battery. Physiol Behav 65(2):201–209

Bonne O, Gill JM, Luckenbaugh DA, Collins C, Owens MJ, Alesci S, Neumeister A, Yuan P, Kinkead B, Manji HK, Charney DS, Vythilingam M (2011) Corticotropin-releasing factor, interleukin-6, brain-derived neurotrophic factor, insulin-like growth factor-1, and substance P in the cerebrospinal fluid of civilians with posttraumatic stress disorder before and after treatment with paroxetine. J Clin Psychiatry 72(8):1124–1128

Boscarino JA, Erlich PM, Hoffman SN, Rukstalis M, Stewart WF (2011) Association of FKBP5, COMT and CHRNA5 polymorphisms with PTSD among outpatients at risk for PTSD. Psychiatry Res 188(1):173–174

Bremner JD, Narayan M, Staib LH, Southwick SM, McGlashan T, Charney DS (1999) Neural correlates of memories of childhood sexual abuse in women with and without posttraumatic stress disorder. Am J Psychiatry 156(11):1787–1795

Breslau N (2006) Neurobiological research on sleep and stress hormones in epidemiological samples. Ann N Y Acad Sci 1071:221–230

Breslau N, Peterson EL (2010) Assaultive violence and the risk of posttraumatic stress disorder following a subsequent trauma. Behav Res Ther 48(10):1063–1066

Breslau N, Davis GC, Andreski P, Peterson E (1991) Traumatic events and posttraumatic stress disorder in an urban population of young adults. Arch Gen Psychiatry 48(3):216–222

Breslau N, Chilcoat HD, Kessler RC, Davis GC (1999) Previous exposure to trauma and PTSD effects of subsequent trauma: results from the Detroit Area Survey of Trauma. Am J Psychiatry 156(6):902–907

Breslau N, Lucia VC, Alvarado GF (2006) Intelligence and other predisposing factors in exposure to trauma and posttraumatic stress disorder: a follow-up study at age 17 years. Arch Gen Psychiatry 63(11):1238–1245

Breslau N, Peterson EL, Schultz LR (2008) A second look at prior trauma and the posttraumatic stress disorder effects of subsequent trauma: a prospective epidemiological study. Arch Gen Psychiatry 65(4):431–437

Brewin CR, Andrews B, Rose S (2000a) Fear, helplessness, and horror in posttraumatic stress disorder: investigating DSM-IV criterion A2 in victims of violent crime. J Trauma Stress 13(3):499–509

Brewin CR, Andrews B, Valentine JD (2000b) Meta-analysis of risk factors for posttraumatic stress disorder in trauma-exposed adults. J Consult Clin Psychol 68(5):748–766

Bromet E, Sonnega A, Kessler RC (1998) Risk factors for DSM-III-R posttraumatic stress disorder: findings from the National Comorbidity Survey. Am J Epidemiol 147(4):353–361

Brown PJ, Wolfe J (1994) Substance abuse and post-traumatic stress disorder comorbidity. Drug Alcohol Depend 35(1):51–59

Bueller JA, Aftab M, Sen S, Gomez-Hassan D, Burmeister M, Zubieta JK (2006) BDNF Val66Met allele is associated with reduced hippocampal volume in healthy subjects. Biol Psychiatry 59(9):812–815

Campeau S, Davis M (1995) Involvement of subcortical and cortical afferents to the lateral nucleus of the amygdala in fear conditioning measured with fear-potentiated startle in rats trained concurrently with auditory and visual conditioned stimuli. J Neurosci Official J Soc Neurosci 15(3 Pt 2):2312–2327

Chang SC, Xie P, Anton RF, De Vivo I, Farrer LA, Kranzler HR, Oslin D, Purcell SM, Roberts AL, Smoller JW, Uddin M, Gelernter J, Koenen KC (2012) No association between ADCYAP1R1 and post-traumatic stress disorder in two independent samples. Mol Psychiatry 17(3):239–241

Chhatwal JP, Stanek-Rattiner L, Davis M, Ressler KJ (2006) Amygdala BDNF signaling is required for consolidation but not encoding of extinction. Nat Neurosci 9(7):870–872

Chilcoat HD, Breslau N (1998a) Investigations of causal pathways between PTSD and drug use disorders. Addict Behav 23(6):827–840

Chilcoat HD, Breslau N (1998b) Posttraumatic stress disorder and drug disorders: testing causal pathways. Arch Gen Psychiatry 55(10):913–917

Clark R, DeYoung CG, Sponheim SR, Bender TL, Polusny MA, Erbes CR, Arbisi PA (2013) Predicting post-traumatic stress disorder in veterans: interactions of traumatic load with COMT gene variation. J Psychiatr Res 47(12):1849–1856

Cohen PL (2003) Studies of murine systemic autoimmunity. Immunol Res 27(2–3):179–184

Comings DE, Comings BG, Muhleman D, Dietz G, Shahbahrami B, Tast D, Knell E, Kocsis P, Baumgarten R, Kovacs BW (1991) The dopamine D2 receptor locus as a modifying gene in neuropsychiatric disorders. JAMA: J Am Med Assoc 266(13):1793–1800

Comings DE, Muhleman D, Gysin R (1996) Dopamine D2 receptor (DRD2) gene and susceptibility to posttraumatic stress disorder: a study and replication. Biol Psychiatry 40(5): 368–372

Crocq MA, Crocq L (2000) From shell shock and war neurosis to posttraumatic stress disorder: a history of psychotraumatology. Dialogues Clin Neurosci 1:47–55

Dell'Osso L, Carmassi C, Del Debbio A, Catena Dell'Osso M, Bianchi C, da Pozzo E, Origlia N, Domenici L, Massimetti G, Marazziti D, Piccinni A (2009) Brain-derived neurotrophic factor plasma levels in patients suffering from post-traumatic stress disorder. Prog Neuropsychopharmacol Biol Psychiatry 33(5):899–902

Difede J, Cukor J, Wyka K, Olden M, Hoffman H, Lee FS, Altemus M (2014) D-cycloserine augmentation of exposure therapy for post-traumatic stress disorder: a pilot randomized clinical trial. Neuropsychopharmacol 39(5):1052–1058

Dragan WL, Oniszczenko W (2009) The association between dopamine D4 receptor exon III polymorphism and intensity of PTSD symptoms among flood survivors. Anxiety Stress Coping 22(5):483–495

Drury SS, Theall KP, Keats BJ, Scheeringa M (2009) The role of the dopamine transporter (DAT) in the development of PTSD in preschool children. J Trauma Stress 22(6):534–539

Drury SS, Brett ZH, Henry C, Scheeringa M (2013) The association of a novel haplotype in the dopamine transporter with preschool age posttraumatic stress disorder. J Child Adolesc Psychopharmacol 23(4):236–243

Falls WA, Miserendino MJ, Davis M (1992) Extinction of fear-potentiated startle: blockade by infusion of an NMDA antagonist into the amygdala. J Neurosci Official J Soc Neurosci 12(3):854–863

Fanselow MS, LeDoux JE (1999) Why we think plasticity underlying Pavlovian fear conditioning occurs in the basolateral amygdala. Neuron 23(2):229–232

File SE, Andrews N, al-Farhan M (1993) Anxiogenic responses of rats on withdrawal from chronic ethanol treatment: effects of tianeptine. Alcohol Alcohol (Oxford, Oxfordshire) 28(3):281–286

Foa EB, Gillihan SJ, Bryant RA (2013) Challenges and successes in dissemination of evidence-based treatments for posttraumatic stress: Lessons learned from prolonged exposure therapy for PTSD. Psychol Sci Public Interest Suppl 14(2):65–111

Freeman T, Roca V, Guggenheim F, Kimbrell T, Griffin WS (2005) Neuropsychiatric associations of apolipoprotein E alleles in subjects with combat-related posttraumatic stress disorder. J Neuropsychiatry Clin Neurosci 17(4):541–543

Frielingsdorf H, Bath KG, Soliman F, Difede J, Casey BJ, Lee FS (2010) Variant brain-derived neurotrophic factor Val66Met endophenotypes: implications for posttraumatic stress disorder. Ann N Y Acad Sci 1208:150–157

Galea MP, Levinger P, Lythgo N, Cimoli C, Weller R, Tully E, McMeeken J, Westh R (2008) A targeted home- and center-based exercise program for people after total hip replacement: a randomized clinical trial. Arch Phys Med Rehabil 89(8):1442–1447

Gates MA, Holowka DW, Vasterling JJ, Keane TM, Marx BP, Rosen RC (2012) Posttraumatic stress disorder in veterans and military personnel: epidemiology, screening, and case recognition. Psychol Serv 9(4):361–382

Gelertner J, Southwick S, Goodson S, Morgan A, Nagy L, Charney DS (1999) No association between D2 dopamine receptor (DRD2) "A" system alleles, or DRD2 haplotypes, and posttraumatic stress disorder. Biol Psychiatry 45(5):620–625

Gold AL, Shin LM, Orr SP, Carson MA, Rauch SL, Macklin ML, Lasko NB, Metzger LJ, Dougherty DD, Alpert NM, Fischman AJ, Pitman RK (2011) Decreased regional cerebral blood flow in medial prefrontal cortex during trauma-unrelated stressful imagery in Vietnam veterans with post-traumatic stress disorder. Psychol Med 41(12):2563–2572

Grabe HJ, Spitzer C, Schwahn C, Marcinek A, Frahnow A, Barnow S, Lucht M, Freyberger HJ, John U, Wallaschofski H, Volzke H, Rosskopf D (2009) Serotonin transporter gene (SLC6A4) promoter polymorphisms and the susceptibility to posttraumatic stress disorder in the general population. Am J Psychiatry 166(8):926–933

Green MK, Rani CS, Joshi A, Soto-Pina AE, Martinez PA, Frazer A, Strong R, Morilak DA (2011) Prenatal stress induces long term stress vulnerability, compromising stress response systems in the brain and impairing extinction of conditioned fear after adult stress. Neuroscience 192:438–451

Griebel G, Blanchard DC, Jung A, Blanchard RJ (1995) A model of 'antipredator' defense in Swiss-Webster mice: effects of benzodiazepine receptor ligands with different intrinsic activities. Behav Pharmacol 6(7):732–745

Hamilton JP, Etkin A, Furman DJ, Lemus MG, Johnson RF, Gotlib IH (2012) Functional neuroimaging of major depressive disorder: a meta-analysis and new integration of base line activation and neural response data. Am J Psychiatry 169(7):693–703

Harrison CA, Kinner SA (1998) Correlates of psychological distress following armed robbery. J Trauma Stress 11(4):787–798

Harvard Medical School (2007) Lifetime prevalence DSM-IV/WMH-CIDI disorders by sex and cohort (n = 9282). http://www.hcp.med.harvard.edu/ncs/ftpdir/NCS-R_Lifetime_Prevalence_Estimates.pdf. Accessed 10 Jan 2014

Hauck S, Kapczinski F, Roesler R, de Moura Silveira E Jr, Magalhaes PV, Kruel LR, Schestatsky SS, Ceitlin LH (2010) Serum brain-derived neurotrophic factor in patients with trauma psychopathology. Prog Neuro-psychopharmacol Biol Psychiatry 34(3):459–462

Heldt SA, Stanek L, Chhatwal JP, Ressler KJ (2007) Hippocampus-specific deletion of BDNF in adult mice impairs spatial memory and extinction of aversive memories. Mol Psychiatry 12(7):656–670

Hemmings SM, Martin LI, Klopper M, van der Merwe L, Aitken L, de Wit E, Black GF, Hoal EG, Walzl G, Seedat S (2013) BDNF Val66Met and DRD2 Taq1A polymorphisms interact to influence PTSD symptom severity: a preliminary investigation in a South African population. Prog Neuro-psychopharmacol Biol Psychiatry 40:273–280

Hoexter MQ, Fadel G, Felicio AC, Calzavara MB, Batista IR, Reis MA, Shih MC, Pitman RK, Andreoli SB, Mello MF, Mari JJ, Bressan RA (2012) Higher striatal dopamine transporter density in PTSD: an in vivo SPECT study with [(99 m)Tc]TRODAT-1. Psychopharmacology 224(2):337–345

Hoge CW, Castro CA, Messer SC, McGurk D, Cotting DI, Koffman RL (2004) Combat duty in Iraq and Afghanistan, mental health problems, and barriers to care. N Engl J Med 351(1):13–22

Holbrook TL, Hoyt DB, Stein MB, Sieber WJ (2001) Perceived threat to life predicts posttraumatic stress disorder after major trauma: risk factors and functional outcome. J Trauma 51(2):287–292 (discussion 292–293)

Institute of Medicine (2007) PTSD compensation and military service. The National Academies Press, Washington

Institute of Medicine (2012) Treatment of posttraumatic stress disorder in military and veteran populations: Initial assessment. The National Academies Press, Washington

Jones E (2006) Historical approaches to post-combat disorders. Philos Trans R Soc Lond Ser B Biol Sci 361(1468):533–542

Kessler RC, Sonnega A, Bromet E, Hughes M, Nelson CB (1995) Posttraumatic stress disorder in the National Comorbidity Survey. Arch Gen Psychiatry 52(12):1048–1060

Kessler RC, Demler O, Frank RG, Olfson M, Pincus HA, Walters EE, Wang P, Wells KB, Zaslavsky AM (2005) Prevalence and treatment of mental disorders, 1990 to 2003. N Engl J Med 352(24):2515–2523

Kessler RC, Keane TM, Ursano RJ, Mokdad A, Zaslavsky AM (2008) Sample and design considerations in post-disaster mental health needs assessment tracking surveys. Int J Methods Psychiatr Res 17(Suppl 2):S6–S20

Khantzian EJ (1985) The self-medication hypothesis of addictive disorders: focus on heroin and cocaine dependence. Am J Psychiatry 142(11):1259–1264

Kilpatrick DG, Koenen KC, Ruggiero KJ, Acierno R, Galea S, Resnick HS, Roitzsch J, Boyle J, Gelernter J (2007) The serotonin transporter genotype and social support and moderation of posttraumatic stress disorder and depression in hurricane-exposed adults. Am J Psychiatry 164(11):1693–1699

Knight DC, Smith CN, Cheng DT, Stein EA, Helmstetter FJ (2004) Amygdala and hippocampal activity during acquisition and extinction of human fear conditioning. Cogn Affect Behav Neurosci 4(3):317–325

Koenen KC, Hitsman B, Lyons MJ, Niaura R, McCaffery J, Goldberg J, Eisen SA, True W, Tsuang M (2005) A twin registry study of the relationship between posttraumatic stress disorder and nicotine dependence in men. Arch Gen Psychiatry 62(11):1258–1265

Koenen KC, Caspi A, Moffitt TE, Rijsdijk F, Taylor A (2006) Genetic influences on the overlap between low IQ and antisocial behavior in young children. J Abnorm Psychol 115(4):787–797

Koenen KC, Moffitt TE, Poulton R, Martin J, Caspi A (2007) Early childhood factors associated with the development of post-traumatic stress disorder: results from a longitudinal birth cohort. Psychol Med 37(2):181–192

Koenen KC, Aiello AE, Bakshis E, Amstadter AB, Ruggiero KJ, Acierno R, Kilpatrick DG, Gelernter J, Galea S (2009a) Modification of the association between serotonin transporter genotype and risk of posttraumatic stress disorder in adults by county-level social environment. Am J Epidemiol 169(6):704–711

Koenen KC, Amstadter AB, Nugent NR (2009b) Gene-environment interaction in posttraumatic stress disorder: an update. J Trauma Stress 22(5):416–426

Koenen KC, Uddin M, Chang SC, Aiello AE, Wildman DE, Goldmann E, Galea S (2011) SLC6A4 methylation modifies the effect of the number of traumatic events on risk for posttraumatic stress disorder. Depression Anxiety 28(8):639–647

Kolassa IT, Ertl V, Eckart C, Glockner F, Kolassa S, Papassotiropoulos A, de Quervain DJ, Elbert T (2010a) Association study of trauma load and SLC6A4 promoter polymorphism in posttraumatic stress disorder: evidence from survivors of the Rwandan genocide. J Clin Psychiatry 71(5):543–547

Kolassa IT, Kolassa S, Ertl V, Papassotiropoulos A, De Quervain DJ (2010b) The risk of posttraumatic stress disorder after trauma depends on traumatic load and the catechol-o-methyltransferase Val(158)Met polymorphism. Biol Psychiatry 67(4):304–308

Kozlovsky N, Kaplan Z, Zohar J, Matar MA, Shimon H, Cohen H (2008) Protein synthesis inhibition before or after stress exposure results in divergent endocrine and BDNF responses disassociated from behavioral responses. Depression Anxiety 25(5):E24–E34

Kozlovsky N, Matar MA, Kaplan Z, Zohar J, Cohen H (2009) A distinct pattern of intracellular glucocorticoid-related responses is associated with extreme behavioral response to stress in an animal model of post-traumatic stress disorder. Eur Neuropsychopharmacol J Eur College Neuropsychopharmacol 19(11):759–771

Lanuza GM, Gosgnach S, Pierani A, Jessell TM, Goulding M (2004) Genetic identification of spinal interneurons that coordinate left-right locomotor activity necessary for walking movements. Neuron 42(3):375–386

Lappalainen J, Kranzler HR, Malison R, Price LH, Van Dyck C, Rosenheck RA, Cramer J, Southwick S, Charney D, Krystal J, Gelernter J (2002) A functional neuropeptide Y Leu7Pro polymorphism associated with alcohol dependence in a large population sample from the United States. Arch Gen Psychiatry 59(9):825–831

LeDoux JE, Cicchetti P, Xagoraris A, Romanski LM (1990) The lateral amygdaloid nucleus: sensory interface of the amygdala in fear conditioning. J Neurosci Official J Soc Neurosci 10(4):1062–1069

Lee HJ, Lee MS, Kang RH, Kim H, Kim SD, Kee BS, Kim YH, Kim YK, Kim JB, Yeon BK, Oh KS, Oh BH, Yoon JS, Lee C, Jung HY, Chee IS, Paik IH (2005) Influence of the serotonin transporter promoter gene polymorphism on susceptibility to posttraumatic stress disorder. Depression Anxiety 21(3):135–139

Lee BH, Kim H, Park SH, Kim YK (2007) Decreased plasma BDNF level in depressive patients. J Affect Disord 101(1–3):239–244

Litz BT, Salters-Pedneault K, Steenkamp MM, Hermos JA, Bryant RA, Otto MW, Hofmann SG (2012) A randomized placebo-controlled trial of D-cycloserine and exposure therapy for posttraumatic stress disorder. J Psychiatr Res 46(9):1184–1190

Lu AT, Ogdie MN, Jarvelin MR, Moilanen IK, Loo SK, McCracken JT, McGough JJ, Yang MH, Peltonen L, Nelson SF, Cantor RM, Smalley SL (2008) Association of the cannabinoid receptor gene (CNR1) with ADHD and post-traumatic stress disorder. Am J Med Gen Part B Neuropsychiatr Gen Official Publ Int Soc Psychiatr Gen 147B(8):1488–1494

Marshall RD, Olfson M, Hellman F, Blanco C, Guardino M, Struening EL (2001) Comorbidity, impairment, and suicidality in subthreshold PTSD. Am J Psychiatry 158(9):1467–1473

Marshall GN, Schell TL, Miles JN (2009) Ethnic differences in posttraumatic distress: Hispanics' symptoms differ in kind and degree. J Consult Clin Psychol 77(6):1169–1178

Matar MA, Zohar J, Cohen H (2013) Translationally relevant modeling of PTSD in rodents. Cell Tissue Res 354(1):127–139

Mazor A, Matar MA, Kaplan Z, Kozlovsky N, Zohar J, Cohen H (2009) Gender-related qualitative differences in baseline and post-stress anxiety responses are not reflected in the incidence of criterion-based PTSD-like behaviour patterns. World J Biol Psychiatry Official J World Fed Soc Biol Psychiatry 10(4 Pt 3):856–869

Mellman TA, Alim T, Brown DD, Gorodetsky E, Buzas B, Lawson WB, Goldman D, Charney DS (2009) Serotonin polymorphisms and posttraumatic stress disorder in a trauma exposed African American population. Depression Anxiety 26(11):993–997

Morey RA, Hariri AR, Gold AL, Hauser MA, Munger HJ, Dolcos F, McCarthy G (2011) Serotonin transporter gene polymorphisms and brain function during emotional distraction from cognitive processing in posttraumatic stress disorder. BMC Psychiatry 11:76-244X-11-76

Morrison FG, Ressler KJ (2013) From the neurobiology of extinction to improved clinical treatments. Depression Anxiety

Mustapic M, Pivac N, Kozaric-Kovacic D, Dezeljin M, Cubells JF, Muck-Seler D (2007) Dopamine beta-hydroxylase (DBH) activity and -1021C/T polymorphism of DBH gene in combat-related post-traumatic stress disorder. Am J Med Gen Part B Neuropsychiatr Gen Official Publ Int Soc Psychiatr Gen 144B(8):1087–1089

Myers KM, Davis M (2007) Mechanisms of fear extinction. Mol Psychiatry 12(2):120–150

Nelson EC, Agrawal A, Pergadia ML, Lynskey MT, Todorov AA, Wang JC, Todd RD, Martin NG, Heath AC, Goate AM, Montgomery GW, Madden PA (2009) Association of childhood trauma exposure and GABRA2 polymorphisms with risk of posttraumatic stress disorder in adults. Mol Psychiatry 14(3):234–235

Ninan I, Bath KG, Dagar K, Perez-Castro R, Plummer MR, Lee FS, Chao MV (2010) The BDNF Val66Met polymorphism impairs NMDA receptor-dependent synaptic plasticity in the hippocampus. J Neurosci Official J Soc Neurosci 30(26):8866–8870

Norberg MM, Battisti RA, Copeland J, Hermens DF, Hickie IB (2012) Two sides of the same coin: Cannabis dependence and mental health problems in help-seeking adolescent and young adult outpatients. Int J Mental Health Addict 10(6):818–828

Ozer EJ, Best SR, Lipsey TL, Weiss DS (2003) Predictors of posttraumatic stress disorder and symptoms in adults: a meta-analysis. Psychol Bull 129(1):52–73

Pape HC, Pare D (2010) Plastic synaptic networks of the amygdala for the acquisition, expression, and extinction of conditioned fear. Physiol Rev 90(2):419–463

Peters J, Dieppa-Perea LM, Melendez LM, Quirk GJ (2010) Induction of fear extinction with hippocampal-infralimbic BDNF. Sci (N Y) 328(5983):1288–1290

Phan KL, Britton JC, Taylor SF, Fig LM, Liberzon I (2006) Corticolimbic blood flow during nontraumatic emotional processing in posttraumatic stress disorder. Arch Gen Psychiatry 63(2):184–192

Pitman RK, Rasmusson AM, Koenen KC, Shin LM, Orr SP, Gilbertson MW, Milad MR, Liberzon I (2012) Biological studies of post-traumatic stress disorder. Nat Rev Neurosci 13(11):769–787

Prigerson HG, Maciejewski PK, Rosenheck RA (2001) Combat trauma: trauma with highest risk of delayed onset and unresolved posttraumatic stress disorder symptoms, unemployment, and abuse among men. J Nerv Ment Dis 189(2):99–108

Punamaki RL, Qouta SR, El Sarraj E (2010) Nature of torture, PTSD, and somatic symptoms among political ex-prisoners. J Trauma Stress 23(4):532–536

Rattiner LM, Davis M, French CT, Ressler KJ (2004) Brain-derived neurotrophic factor and tyrosine kinase receptor B involvement in amygdala-dependent fear conditioning. J Neurosci Official J Soc Neurosci 24(20):4796–4806

Resick PA, Nishith P, Weaver TL, Astin MC, Feuer CA (2002) A comparison of cognitive-processing therapy with prolonged exposure and a waiting condition for the treatment of chronic posttraumatic stress disorder in female rape victims. J Consult Clin Psychol 70(4):867–879

Resick PA, Williams LF, Suvak MK, Monson CM, Gradus JL (2012) Long-term outcomes of cognitive-behavioral treatments for posttraumatic stress disorder among female rape survivors. J Consult Clin Psychol 80(2):201–210

Resnick SG, Rosenheck RA (2008) Posttraumatic stress disorder and employment in veterans participating in veterans health administration compensated work therapy. J Rehabil Res Dev 45(3):427–435

Ressler KJ, Mercer KB, Bradley B, Jovanovic T, Mahan A, Kerley K, Norrholm SD, Kilaru V, Smith AK, Myers AJ, Ramirez M, Engel A, Hammack SE, Toufexis D, Braas KM, Binder EB, May V (2011) Post-traumatic stress disorder is associated with PACAP and the PAC1 receptor. Nature 470(7335):492–497

Roberts NP, Kitchiner NJ, Kenardy J, Bisson JI (2010) Early psychological interventions to treat acute traumatic stress symptoms. Cochrane Database Syst Rev 3:CD007944

Roberts AL, Gilman SE, Breslau J, Breslau N, Koenen KC (2011) Race/ethnic differences in exposure to traumatic events, development of post-traumatic stress disorder, and treatment-seeking for post-traumatic stress disorder in the United States. Psychol Med 41(1):71–83

Roseboom PH, Nanda SA, Bakshi VP, Trentani A, Newman SM, Kalin NH (2007) Predator threat induces behavioral inhibition, pituitary-adrenal activation and changes in amygdala CRF-binding protein gene expression. Psychoneuroendocrinology 32(1):44–55

Roth MK, Bingham B, Shah A, Joshi A, Frazer A, Strong R, Morilak DA (2012) Effects of chronic plus acute prolonged stress on measures of coping style, anxiety, and evoked HPA-axis reactivity. Neuropharmacology 63(6):1118–1126

Sah P, Westbrook RF, Luthi A (2008) Fear conditioning and long-term potentiation in the amygdala: what really is the connection? Ann N Y Acad Sci 1129:88–95

Sartor CE, McCutcheon VV, Pommer NE, Nelson EC, Grant JD, Duncan AE, Waldron M, Bucholz KK, Madden PA, Heath AC (2011) Common genetic and environmental contributions to post-traumatic stress disorder and alcohol dependence in young women. Psychol Med 41(7):1497–1505

Sayin A, Kucukyildirim S, Akar T, Bakkaloglu Z, Demircan A, Kurtoglu G, Demirel B, Candansayar S, Mergen H (2010) A prospective study of serotonin transporter gene promoter (5-HTT gene linked polymorphic region) and intron 2 (variable number of tandem repeats) polymorphisms as predictors of trauma response to mild physical injury. DNA Cell Biol 29(2):71–77

Scherrer JF, Xian H, Lyons MJ, Goldberg J, Eisen SA, True WR, Tsuang M, Bucholz KK, Koenen KC (2008) Posttraumatic stress disorder; combat exposure; and nicotine dependence, alcohol dependence, and major depression in male twins. Compr Psychiatry 49(3):297–304

Seal KH, Bertenthal D, Miner CR, Sen S, Marmar C (2007) Bringing the war back home: mental health disorders among 103,788 US veterans returning from Iraq and Afghanistan seen at Department of Veterans Affairs facilities. Arch Intern Med 167(5):476–482

Seedat S, Stein DJ, Carey PD (2005a) Post-traumatic stress disorder in women: epidemiological and treatment issues. CNS Drugs 19(5):411–427

Seedat S, Stein MB, Forde DR (2005b) Association between physical partner violence, posttraumatic stress, childhood trauma, and suicide attempts in a community sample of women. Violence Vict 20(1):87–98

Segman RH, Cooper-Kazaz R, Macciardi F, Goltser T, Halfon Y, Dobroborski T, Shalev AY (2002) Association between the dopamine transporter gene and posttraumatic stress disorder. Mol Psychiatry 7(8):903–907

Shin LM, Handwerger K (2009) Is posttraumatic stress disorder a stress-induced fear circuitry disorder? J Trauma Stress 22(5):409–415

Shin LM, Orr SP, Carson MA, Rauch SL, Macklin ML, Lasko NB, Peters PM, Metzger LJ, Dougherty DD, Cannistraro PA, Alpert NM, Fischman AJ, Pitman RK (2004) Regional cerebral blood flow in the amygdala and medial prefrontal cortex during traumatic imagery in male and female Vietnam veterans with PTSD. Arch Gen Psychiatry 61(2):168–176

Shin LM, Wright CI, Cannistraro PA, Wedig MM, McMullin K, Martis B, Macklin ML, Lasko NB, Cavanagh SR, Krangel TS, Orr SP, Pitman RK, Whalen PJ, Rauch SL (2005) A functional magnetic resonance imaging study of amygdala and medial prefrontal cortex responses to overtly presented fearful faces in posttraumatic stress disorder. Arch Gen Psychiatry 62(3):273–281

Sierra-Mercado D, Padilla-Coreano N, Quirk GJ (2011) Dissociable roles of prelimbic and infralimbic cortices, ventral hippocampus, and basolateral amygdala in the expression and extinction of conditioned fear. Neuropsychopharmacol Official Publ Am Coll Neuropsychopharmacol 36(2):529–538

Simmons AN, Paulus MP, Thorp SR, Matthews SC, Norman SB, Stein MB (2008) Functional activation and neural networks in women with posttraumatic stress disorder related to intimate partner violence. Biol Psychiatry 64(8):681–690

Soliman F, Glatt CE, Bath KG, Levita L, Jones RM, Pattwell SS, Jing D, Tottenham N, Amso D, Somerville LH, Voss HU, Glover G, Ballon DJ, Liston C, Teslovich T, Van Kempen T, Lee FS, Casey BJ (2010) A genetic variant BDNF polymorphism alters extinction learning in both mouse and human. Sci (N Y) 327(5967):863–866

Storr CL, Ialongo NS, Anthony JC, Breslau N (2007) Childhood antecedents of exposure to traumatic events and posttraumatic stress disorder. Am J Psychiatry 164(1):119–125

Strigo IA, Simmons AN, Matthews SC, Craig AD (2010) The relationship between amygdala activation and passive exposure time to an aversive cue during a continuous performance task. PloS One 5(11):e15093

Sullivan RM, Gratton A (1998) Relationships between stress-induced increases in medial prefrontal cortical dopamine and plasma corticosterone levels in rats: role of cerebral laterality. Neuroscience 83(1):81–91

Takahashi LK, Nakashima BR, Hong H, Watanabe K (2005) The smell of danger: a behavioral and neural analysis of predator odor-induced fear. Neurosci Biobehav Rev 29(8):1157–1167

Tang YL, Li W, Mercer K, Bradley B, Gillespie CF, Bonsall R, Ressler KJ, Cubells JF (2010) Genotype-controlled analysis of serum dopamine β-hydroxylase activity in civilian posttraumatic stress disorder. Prog Neuropsychopharmacol Biol Psychiatry 34(8):1396–401

Thakur GA, Joober R, Brunet A (2009) Development and persistence of posttraumatic stress disorder and the 5-HTTLPR polymorphism. J Trauma Stress 22(3):240–243

True WR, Rice J, Eisen SA, Heath AC, Goldberg J, Lyons MJ, Nowak J (1993) A twin study of genetic and environmental contributions to liability for posttraumatic stress symptoms. Arch Gen Psychiatry 50(4):257–264

Uddin M, Aiello AE, Wildman DE, Koenen KC, Pawelec G, de Los Santos R, Goldmann E, Galea S (2010) Epigenetic and immune function profiles associated with posttraumatic stress disorder. Proc Nat Acad Sci USA 107(20):9470–9475

Uddin M, Koenen KC, Aiello AE, Wildman DE, de los Santos R, Galea S (2011) Epigenetic and inflammatory marker profiles associated with depression in a community-based epidemiologic sample. Psychol Med 41(5):997–1007

Valente NL, Vallada H, Cordeiro Q, Miguita K, Bressan RA, Andreoli SB, Mari JJ, Mello MF (2011) Candidate-gene approach in posttraumatic stress disorder after urban violence: association analysis of the genes encoding serotonin transporter, dopamine transporter, and BDNF. J Mol Neurosci: MN 44(1):59–67

Vasterling JJ, Proctor SP, Friedman MJ, Hoge CW, Heeren T, King LA, King DW (2010) PTSD symptom increases in Iraq-deployed soldiers: comparison with nondeployed soldiers and associations with baseline symptoms, deployment experiences, and postdeployment stress. J Trauma Stress 23(1):41–51

Voisey J, Swagell CD, Hughes IP, Morris CP, van Daal A, Noble EP, Kann B, Heslop KA, Young RM, Lawford BR (2009) The DRD2 gene 957C > T polymorphism is associated with posttraumatic stress disorder in war veterans. Depression Anxiety 26(1):28–33

Voisey J, Swagell CD, Hughes IP, Connor JP, Lawford BR, Young RM, Morris CP (2010) A polymorphism in the dysbindin gene (DTNBP1) associated with multiple psychiatric disorders including schizophrenia. Behav Brain Funct: BBF 6:41-9081-6-41

Walker J, Curtis V, Shaw P, Murray RM (2002) Schizophrenia and bipolar disorder are distinguished mainly by differences in neurodevelopment. Neurotox Res 4(5–6):427–436

Wang Z, Baker DG, Harrer J, Hamner M, Price M, Amstadter A (2011) The relationship between combat-related posttraumatic stress disorder and the 5-HTTLPR/rs25531 polymorphism. Depression Anxiety 28(12):1067–1073

White S, Acierno R, Ruggiero KJ, Koenen KC, Kilpatrick DG, Galea S, Gelernter J, Williamson V, McMichael O, Vladimirov VI, Amstadter AB (2013) Association of CRHR1 variants and posttraumatic stress symptoms in hurricane exposed adults. J Anxiety Disord 27(7):678–683

Xian H, Chantarujikapong SI, Scherrer JF, Eisen SA, Lyons MJ, Goldberg J, Tsuang M, True WR (2000) Genetic and environmental influences on posttraumatic stress disorder, alcohol and drug dependence in twin pairs. Drug Alcohol Depend 61(1):95–102

Xie P, Kranzler HR, Poling J, Stein MB, Anton RF, Brady K, Weiss RD, Farrer L, Gelernter J (2009) Interactive effect of stressful life events and the serotonin transporter 5-HTTLPR genotype on posttraumatic stress disorder diagnosis in 2 independent populations. Arch Gen Psychiatry 66(11):1201–1209

Xie P, Kranzler HR, Poling J, Stein MB, Anton RF, Farrer LA, Gelernter J (2010) Interaction of FKBP5 with childhood adversity on risk for post-traumatic stress disorder. Neuropsychopharmacol Official Publ Am Coll Neuropsychopharmacol 35(8):1684–1692

Xie P, Kranzler HR, Farrer L, Gelernter J (2012) Serotonin transporter 5-HTTLPR genotype moderates the effects of childhood adversity on posttraumatic stress disorder risk: a replication study. Am J Med Gen Part B Neuropsychiatr Gen Official Publ Int Soc Psychiatr Gen 159B(6):644–652

Yehuda R, LeDoux J (2007) Response variation following trauma: a translational neuroscience approach to understanding PTSD. Neuron 56(1):19–32

Yehuda R, Cai G, Golier JA, Sarapas C, Galea S, Ising M, Rein T, Schmeidler J, Muller-Myhsok B, Holsboer F, Buxbaum JD (2009) Gene expression patterns associated with posttraumatic stress disorder following exposure to the World Trade Center attacks. Biol Psychiatry 66(7):708–711

Young RM, Lawford BR, Noble EP, Kann B, Wilkie A, Ritchie T, Arnold L, Shadforth S (2002) Harmful drinking in military veterans with post-traumatic stress disorder: association with the D2 dopamine receptor A1 allele. Alcohol Alcohol (Oxford, Oxfordshire) 37(5):451–456

Zhang L, Benedek DM, Fullerton CS, Forsten RD, Naifeh JA, Li XX, Hu XZ, Li H, Jia M, Xing GQ, Benevides KN, Ursano RJ (2014) PTSD risk is associated with BDNF Val66Met and BDNF overexpression. Mol Psychiatry 19(1):8–10

Zlotnick C, Warshaw M, Shea MT, Allsworth J, Pearlstein T, Keller MB (1999) Chronicity in posttraumatic stress disorder (PTSD) and predictors of course of comorbid PTSD in patients with anxiety disorders. J Trauma Stress 12(1):89–100

Investigation of Cortisol Levels in Patients with Anxiety Disorders: A Structured Review

Hesham Yousry Elnazer and David S. Baldwin

Abstract Anxiety disorders are common and distressing medical conditions, which typically arise in adolescence or early adult life. They can persist for many years, reducing quality of life, limiting academic and occupational achievement, and being responsible for considerable economic pressures. Although a range of psychological and pharmacological treatments are available, their success is often limited, and many patients remain troubled by significant symptom-related disability for long periods. The detailed pathophysiology of each anxiety disorder is not established, and novel treatments that are based solely on current understanding of conventional neurotransmitter function are unlikely to be substantially more effective or better tolerated than current treatments. Investigations of hypothalamo-pituitary axis function across panic disorder, generalized anxiety disorder, specific phobias and social anxiety disorder have produced intriguing findings but not revealed a consistent pattern of endocrine disturbance, perhaps reflecting differences in methodology and the nature and size of the clinical samples. There is a persistent need for large, prospective studies using standardized methods for investigation and data analysis (164 *words*).

Keywords Anxiety disorder · Panic · Phobia · Cortisol

Submitted for inclusion within Bluhm and Pariante (eds) *Behavioral Neurobiology of Stress-related Disorders* February 2014

H. Y. Elnazer · D. S. Baldwin
Clinical and Experimental Sciences Academic Unit (CNS and Psychiatry),
Faculty of Medicine, University of Southampton, Southampton, UK

H. Y. Elnazer · D. S. Baldwin (✉)
University Department of Psychiatry, College Keep, 4-12 Terminus Terrace,
Southampton SO14 3DT, UK
e-mail: dsb1@soton.ac.uk

Contents

1	Background	192
2	Method for the Literature Review	192
3	Panic Disorder and Agoraphobia	193
4	Generalized Anxiety Disorder	201
5	Specific (Simple) Phobia	205
6	Social Anxiety Disorder (Social Phobia)	206
7	Summary	209
	References	210

1 Background

Anxiety disorders are common, usually have an early onset, typically run a chronic or relapsing course, cause substantial personal distress, impair social and occupational function, reduce quality of life and impose a substantial personal and societal burden. Unfortunately the effectiveness of pharmacological and psychological treatment interventions for patients with anxiety disorders in real-world clinical practice is often disappointing. Advances in genetics, imaging genetics, psychoneuroimmunology and psychophysiology have all deepened understanding of the causes of anxiety disorders, but it remains hard to attribute particular psychopathological states to specific neuropsychobiological substrates. Despite advances in investigation of the biological, environmental and temperamental mediators of resilience to traumatic adversity, on an individual level it is difficult to predict who will become troubled by anxiety symptoms (Baldwin et al. 2010). Steadily growing awareness of the importance of disturbances of the hypothalamo-pituitary adrenal (HPA) axis in the pathophysiology and potentially in the treatment of mood disorders and psychosis (Pariante 2009) has naturally encouraged parallel investigations into the role of neuroendocrine disturbances in the origin, investigation and treatment of anxiety disorders, and this area of enquiry is the focus of this chapter.

2 Method for the Literature Review

We searched all titles listed in 'Pub Med' up to March 2013, for all anxiety disorders, excluding obsessive-compulsive disorder (OCD) and post-traumatic stress disorder (PTSD), being mindful of changes in the categorization of the latter two conditions within DSM-5 (American Psychiatric Association 2013); however, papers which examined cortisol and in which OCD or PTSD were co-morbid with an anxiety disorder were included. The terms glucocorticoid, cortisol and hypercortisolism were used, combined with each disorder to compile separate lists for each condition. For generalized anxiety disorder (GAD) we included the terms GAD, generalized and GAD; for phobic disorders, we used the terms phobia, social phobia, simple phobia and specific phobia. A combined list was generated and duplications were eliminated; all letters, and papers that were not available in English, were eliminated. The search terms were as follows: agoraphobia (with or without a history of panic disorder) with glucocorticoids, cortisol, hypercortisolism; generalized anxiety disorder or GAD with glucocorticoids, cortisol, hypercortisolism; panic disorder with glucocorticoids, cortisol, hypercortisolism; phobic disorders, phobia, social phobia, simple phobia, specific phobia, with glucocorticoids cortisol and hypercortisolism.

3 Panic Disorder and Agoraphobia

Evaluation of cortisol levels has encompassed investigations of urinary, salivary and plasma cortisol levels, non-suppression following dexamethasone administration, and cortisol response to psychological and pharmacological challenges. In an early investigation, no significant differences in mean urinary free cortisol or plasma 3-methoxy-4-hydroxyphenethyleneglycol (MHPG) were found between patients with panic disorder ($n = 12$) or healthy controls (Uhde et al. 1988). Significant group differences in urinary free cortisol between patients with panic disorder ($n = 65$) and healthy controls were reported in a larger investigation, but only in the sub-group of patients with more marked depressive symptoms or agoraphobic avoidance (Kathol et al. 1988); an extension of this study demonstrating that elevated urinary cortisol was less marked than that in depressed patients (Kathol et al. 1989). In patients with panic disorder ($n = 66$), coexisting depression and agoraphobia were found to be associated with significantly elevated urinary free cortisol levels when compared to healthy controls; these levels decrease during treatment with the benzodiazepines alprazolam or diazepam (Lopez et al. 1990). Nocturnal levels of urinary cortisol, epinephrine and norepinephrine were found to be persistently elevated in patients with panic disorder ($n = 16$) (Bandelow et al. 1997), though not to decline following successful treatment with the selective serotonin reuptake inhibitor (SSRI) paroxetine, exercise or relaxation (Wedekind et al. 2008) (Table 1).

Plasma cortisol (and growth hormone) levels were found to be elevated in patients with 'panic anxiety' compared to those in healthy controls (Nesse et al. 1984). Another comparative study found that plasma cortisol levels were not significantly different between patients with panic disorder ($n = 10$) and healthy controls (Villacres et al. 1987). Both total and free plasma levels, and salivary levels of cortisol were found to be elevated in patients with panic disorder ($n = 47$) compared to healthy controls (Wedekind et al. 2000). In further investigations, salivary cortisol levels taken during a panic attack were found to decrease 24 h later in patients with panic disorder ($n = 25$) (Bandelow et al.

Table 1 Investigations of cortisol in panic disorder

Urinary cortisol levels	
Uhde et al. 1988	No significant differences in mean urinary cortisol or plasma MHPG
Kathol et al. 1988	Elevation of urinary free cortisol in patients with more marked depression or agoraphobia
Kathol et al. 1989	Elevation of urinary cortisol less marked than in depressed patients
Lopez et al. 1990	Elevated urinary free cortisol levels decline during benzodiazepine treatment
Bandelow et al. 1997	Elevation of nocturnal cortisol levels, persisting SSRI treatment, exercise or relaxation
Plasma, salivary and hair cortisol levels	
Nesse et al. 1984	Elevation of cortisol and growth hormone levels
Villacres et al. 1987	No difference in levels
Wedekind et al. 2000	Elevation of total and free plasma (and salivary) cortisol levels
Bandelow et al. 2000a	Decline in salivary cortisol levels following a panic attack
Bandelow et al. 2000b	Daytime salivary cortisol levels correlated to symptom severity
Staufenbiel et al. 2013	Evidence of low levels of cortisol in panic disorder and GAD
ACTH and cortisol levels	
Roy-Byrne et al. 1986	Elevation of cortisol and ACTH, diminished response to CRH
Brambilla et al. 1992	Elevation of ACTH levels, ACTH response to CRH disturbed
Curtis et al. 1997	No significant difference in ACTH and cortisol response to CRH challenge
Erhardt et al. 2006	No significant difference in CRH challenge following dexamethasone administration
Petrowski et al. 2012	Diminished cortisol (but not ACTH) response to CRH challenge
Kellner et al. 2004	No significant difference in reduction of cortisol levels following metyrapone administration
Fava et al. 1989	Higher ratio of DHEA to cortisol, declining during benzodiazepine and placebo treatment
Den Boer and Westenberg 1990	No reduction in plasma cortisol levels with successful SSRI treatment
Abelson and Curtis 1996a	Increased nocturnal cortisol levels
Abelson et al. 1996	Reduction in hypercortisolaemia with successful alprazolam treatment
Abelson and Curtis 1996b	Baseline cortisol levels predictive of symptom-related disability 2 years later
Herran et al. 2005	Decline in cortisol levels with successful SSRI treatment
Dexamethasone non-suppression	
Sheehan et al. 1983	Non-suppression in 11.8 % of patients
Lieberman et al. 1983	Non-suppression in none of 10 patients
Whiteford and Evans 1984	Less non-suppression in panic patients (29 %) than depressed patients (64 %)
Judd et al. 1987	Non-suppression in 29 % of patients (not significantly greater than in controls)

(continued)

Table 1 (continued)

Cottraux and Claustrat 1984	Greater non-suppression with co-morbid depression or family history of depressive disorders
Peterson et al. 1985	Non-suppression independent of depressive symptom severity
Faludi et al. 1986	Non-suppression significantly less frequent (16.7 %) than in depressed patients (56.7 %)
Goldstein et al. 1987	Non-suppression significantly less frequent than in depressed patients
Grunhaus et al. 1987	Co-morbid panic disorder does not affect chance of non-suppression in depressed patients
Westberg et al. 1991	Non-suppression more common in patients with co-morbid agoraphobia
Carson et al. 1988	Dexamethasone distribution may be altered in panic disorder
Coryell et al. 1985	Reduction in non-suppression with successful treatment
Bridges et al. 1986	Persistent non-suppression predictive of relapse following withdrawal of treatment
Coryell and Noyes 1988	Baseline non-suppression not predictive of response to alprazolam
Coryell et al. 1989, 1991	Baseline non-suppression predictive of relapse following response to benzodiazepines
Vreeburg et al. 2010	No significant difference in non-suppression compared to controls
Psychological challenge	
Woods et al. 1987	No decrease in cortisol or MHPG levels with exposure to feared situations
Stones et al. 1999	No decline in salivary cortisol levels following detailed personal assessment
Siegmund et al. 2011	No increase in plasma cortisol or ACTH during in vivo exposure
Garcia-Leal et al. 2005	No increase in salivary cortisol levels during simulated public speaking task
Petrowski et al. 2013	Reduced cortisol response following Trier Social Stress Test (TSST)
Petrowski et al. 2010	Cortisol response in TSST not influenced by depression or cortisol awakening response
Relationship between cortisol and 5-HT	
Westenberg et al. 1989	No significant difference in response to 5-HTP
Van Vliet et al. 1996	Enhanced cortisol response following 5-HTP administration is only transient
Schruers et al. 2002	No evidence of enhanced cortisol response following 5-HTP administration
Targum and Marshall 1989	Significantly greater increase in cortisol and prolactin levels following fenfluramine challenge
Targum 1990	Significantly greater increase in cortisol and prolactin levels following fenfluramine challenge
Judd et al. 1994	No significant difference in cortisol response to fenfluramine challenge
Charney et al. 1987b	No significant difference in cortisol, growth hormone or prolactin response to mCPP challenge
Germine et al. 1994	No significant difference in cortisol response to mCPP challenge
Kahn et al. 1988	Exaggerated response to mCPP challenge, compared to controls and depressed patients

(continued)

Table 1 (continued)

Wetzler et al. 1996	No significant difference in cortisol response to mCPP challenge
Van der Wee et al. 2004	No difference in cortisol response to mCPP challenge
Lesch et al. 1992	Reduction in hypothermic and ACTH/cortisol to ipsapirone challenge
Broocks et al. 2000	Reduction in hypothermic and ACTH/cortisol to ipsapirone challenge
Broocks et al. 2002	Enhanced response to ipsapirone challenge, particularly in smokers
Broocks et al. 2003	Hypothermic response to ipsapirone challenge corrected with clomipramine treatment
Cortisol and adrenergic function	
Stein et al. 1988	No difference in decline in cortisol level after clonidine (alpha-2 adrenergic agonist) challenge
Brambilla et al. 1992	Altered cortisol response following clonidine challenge
Coplan et al. 1997	Increased MHPG volatility following clonidine challenge, lessening with SSRI treatment
Charney et al. 1987a	Increased cortisol and MHGP following yohimbine (alpha-2 adrenergic antagonist) challenge
Lactate infusion	
Liebowitz et al. 1985	No consistent increase in cortisol or epinephrine levels
Levin et al. 1987	No increase in ACTH or cortisol in association with panic attacks
Hollander et al. 1989	Elevated cortisol at baseline predictive of panic attacks following lactate infusion
Coplan et al. 1998	Elevated cortisol at baseline predictive of panic attacks following infusion of saline
Targum 1990	Infusion accompanied by increase in cortisol in panic disorder and depression plus panic
Seier et al. 1997	No increase in cortisol levels
Ströhle et al. 1998	No increase in cortisol levels with either lactate or flumazenil
Peskind et al. 1998	Panic attacks induced by hypertonic solutions
Kellner et al. 1998	Enhanced release of ANP exerts inhibitory role on ACTH and cortisol release
Cholecystokinin challenge	
Abelson et al. 1991, 1994	No difference in neuroendocrine response to pentagastrin challenge
Kellner et al. 1997	Prior administration of clonidine enhances response to CCK challenge
Shlik et al. 1997	Successful SSRI treatment reduces anxiety but not endocrine response to CCK-4 challenge
Ströhle et al. 2000	ACTH levels following challenge greater in those experiencing panic attacks
Wiedemann et al. 2001	ANP reduces likelihood of panic following administration of CCK-4
Abelson et al. 2005	Cognitive intervention diminishes ACTH and cortisol response to pentagastrin challenge
Carbon dioxide challenge	
Sinha et al. 1999	Reduction in cortisol levels following CO_2 challenge
Van Duinen et al. 2004	No difference in cortisol response to CO_2 challenge
Belgorodsky et al. 2005	Pre-challenge metyrapone reduces cortisol levels but does not affect anxiety response

2000a) and daytime salivary cortisol levels were correlated with symptom severity (Bandelow et al. 2000b).

In an early investigation, levels of cortisol and adrenocorticotrophin (ACTH) were found to be elevated in patients with panic disorder ($n = 30$), but the ACTH and cortisol response following challenge with corticotrophin releasing hormone (CRH) was diminished compared to healthy controls (Roy-Byrne et al. 1986). Another study found that baseline plasma ACTH levels were elevated, and the ACTH response to stimulation with CRH disturbed in patients with panic disorder ($n = 17$) compared to healthy controls (Brambilla et al. 1992). However, another investigation found no significant difference between patients with panic disorder and healthy controls in the ACTH and cortisol response to CRH challenge (Curtis et al. 1997). A third investigation, involving CRH challenge with dexamethasone administration found no difference between patients with panic disorder ($n = 30$) and depressed patients (Erhardt et al. 2006); and a fourth suggested that the plasma cortisol response, but not the ACTH response, to CRH stimulation was decreased (Petrowski et al. 2012). Administration of the cortisol synthesis inhibitor metyrapone reduced ACTH and cortisol both in patients with panic disorder ($n = 14$) and healthy controls, but with no difference between groups (Kellner et al. 2004). The ratio of dihydroepiandrosterone (DHEA) to cortisol, which is considered a measure of adrenal cortical activity, was found to be significantly higher in patients with panic disorder ($n = 24$) than in depressed patients or healthy controls and, in female patients, to decrease during treatment with alprazolam, clonazepam or placebo (Fava et al. 1989). An early investigation found that successful treatment of patients with panic disorder with the SSRI fluvoxamine was not associated with a reduction in plasma levels of cortisol (Den Boer and Westenberg 1990); however, a reduction in cortisol levels has been reported to be associated with successful treatment with other SSRIs (Herran et al. 2005).

In an early study, non-suppression following dexamethasone administration was found in 11.8 % of patients experiencing panic attacks (Sheehan et al. 1983); in a second, none of ten patients with panic disorder showed non-suppression compared to 9 of 22 patients with depressive disorders (Lieberman et al. 1983). In a third investigation, non-suppression was found in 29 % of patients with agoraphobia, compared to 64 % of depressed patients, and 12 % of healthy controls (Whiteford and Evans 1984). A similar proportion of patients with panic disorder showing non-suppression (29 %) was seen in another early study, but non-suppression was not significantly more prevalent than in healthy controls (9.5 %) (Judd et al. 1987). Another investigation found that 20 % of patients with agoraphobia and panic attacks were 'non-suppressors', the likelihood being greater in patients with co-morbid depression or a family history of depressive disorders (Cottraux and Claustrat 1984). In another study, non-suppression following dexamethasone administration was found in 12.4 % of patients with agoraphobia ($n = 97$), independent of depressive symptom severity (Peterson et al. 1985). In a comparative study, non-suppression was significantly less common (16.7 %) among patients with panic disorder ($n = 30$) than in patients with major depressive episodes (56.7 %) (Faludi et al. 1986). Similar findings were reported in another comparative study involving patients with panic disorder ($n = 24$), depressed patients and healthy controls (Goldstein et al. 1987). Dexamethasone non-suppression was similarly prevalent in depressed patients, with or without co-morbid panic disorder (Grunhaus et al. 1987). Non-suppression was more common in panic disorder patients with agoraphobia (28 %) than in those without agoraphobia (Westberg et al. 1991). It should be noted that the distribution of dexamethasone itself may be altered in patients with panic disorder, with significantly lower levels being achieved following administration of a standard dose (Carson et al. 1988).

Successful treatment of patients with agoraphobia with panic attacks was associated with 'escape' from dexamethasone non-suppression (Coryell et al. 1985). Another study found the response to treatment with benzodiazepine anxiolytics or placebo was not associated with change in response to dexamethasone challenge: but persistent non-suppression despite successful treatment was predictive of relapse in symptoms, following the withdrawal of treatment (Bridges et al. 1986). A further investigation found that dexamethasone non-suppression at baseline was not predictive of response to treatment with alprazolam (Coryell and Noyes 1988), but in another investigation baseline dexamethasone non-suppression was predictive of relapse following successful treatment of patients with panic disorder ($n = 82$) with alprazolam, diazepam or placebo (Coryell et al. 1989; Coryell et al. 1991). In a series of studies, in which patients with panic disorder ($n = 20$) were found to have evidence of increased nocturnal cortisol levels compared to healthy controls (Abelson and Curtis 1996a), subsequent successful treatment with alprazolam was associated with a reduction in hypercortisolaemia (Abelson et al. 1996), and at 2-year follow-up, mean 24-h cortisol levels at baseline, prior to alprazolam treatment, were predictive of greater symptom-related disability (Abelson and Curtis, 1996b).

There have been many investigations of the cortisol response to psychological and pharmacological challenge. An early investigation found that exposure to feared situations was associated with increased reported fear, but not with increases in plasma cortisol or MHPG in patients with agoraphobia ($n = 18$) (Woods et al. 1987). Detailed psychometric and physiological assessment of patients with panic disorder ($n = 24$) was associated with a failure of salivary cortisol levels to decline, in contrast to the diminution seen in healthy controls (Stones et al. 1999). A more recent study involving repeated in vivo exposure to phobic situations ('flooding') found this therapeutic challenge was not associated with increases in plasma cortisol or ACTH levels; patients with lower responses having the least benefit from treatment (Siegmund et al. 2011). A simulated public speaking task engendered anxiety but did not increase salivary cortisol levels in either remitted ($n = 16$) or symptomatic ($n = 18$) patients with panic disorder (Garcia-Leal et al. 2005). Challenge through the Trier Social Stress Test (TSST) was associated with significantly lower increases in plasma and salivary cortisol in patients with panic disorder ($n = 27$) than in healthy controls (Petrowski et al. 2013), regardless of the presence of depressive symptoms and a normal cortisol awakening response (Petrowski et al. 2010).

There have been many investigations on the effects of cortisol on experimental alterations of 5-hdroxytryptamine (5-HT, serotonin) levels and receptor function. Patients with panic disorder ($n = 7$) did not differ markedly from healthy controls in either the cortisol or beta-endorphin response, following administration of the 5-HT precursor 5-hydroxytryptophan (5-HTP) (Westenberg et al. 1989); an enhanced cortisol response being seen only transiently at the highest 5-HTP dosage (van Vliet et al. 1996). The lack of an enhanced cortisol response (assessed with salivary cortisol levels) to 5-HTP challenge was confirmed in a subsequent investigation ($n = 24$) (Schruers et al. 2002). Challenge with the 5-HT releasing agent fenfluramine elicited both a significantly greater panic response, and greater increases in prolactin and cortisol levels in patients with panic disorder than in depressed patients or healthy controls (Targum and Marshall 1989 [$n = 9$]; Targum 1990 [$n = 17$]). However, in another investigation involving fenfluramine challenge, there were no significant differences in the cortisol response between patients with panic disorder ($n = 16$) and healthy controls (Judd et al. 1994).

Administration of the non-selective 5-HT$_{2B}$ and 5-HT$_{2C}$ agonist meta-chlorophenylpiperazine (mCPP) appeared no more likely to cause panic attacks in patients with panic disorder than in healthy controls, with no significant differences between groups in cortisol, growth hormone or prolactin responses (Charney et al. 1987a). No significant differences between patients with panic disorder ($n = 27$) and controls in the cortisol response were seen in another study involving intravenous mCPP infusion (Germine et al. 1994). Another study found evidence of an exaggerated response to mCPP challenge, when patients with panic disorder ($n = 15$) were compared to depressed patients and healthy controls, with a positive correlation between cortisol response and anxiety level (Kahn et al. 1988); but a further investigation found no differences between patients and controls in the cortisol response to mCPP challenge (Wetzler et al. 1996).

Administration of the 5-HT$_{1A}$ agonist ipsapirone was found to result in a diminution of the hypothermic and ACTH/cortisol response in patients with panic disorder ($n = 14$), compared to healthy controls, suggesting that 5-HT$_{1A}$ receptor-related serotonergic dysfunction may be a factor in the pathophysiology of panic disorder (Lesch et al. 1992). A diminution of the cortisol response and hypothermic response was also seem following ipsapirone challenge in patients with panic disorder ($n = 40$) compared to healthy controls, in a study in which administration of mCPP was associated with a trend towards a greater increase in cortisol levels, together suggesting opposite changes in the responsiveness of 5-HT$_{1A}$ and 5-HT$_{2C}$ receptors (Broocks et al. 2000). However, in another investigation in patients with panic disorder ($n = 39$), plasma cortisol levels rose significantly in response to challenge with ipsapirone, this increase being particularly marked in smokers (Broocks et al. 2002), the hypothermic response to ipsapirone challenge being reduced with successful treatment with clomipramine (Broocks et al. 2003). However, although rapid intravenous administration of mCPP induced panic attacks significantly more frequently in patients with panic disorder ($n = 10$) than healthy controls, there were no differences between groups in the neuroendocrine response (van der Wee et al. 2004).

Plasma cortisol levels have been found to increase during challenge of healthy volunteers with air 'enriched' with 35 % carbon dioxide (CO_2), a common challenge test for inducing panic attacks in patients with panic disorder (Sasaki et al. 1996; van Duinen et al. 2005; Hood et al. 2006). However patients may not differ from controls in the cortisol response to CO_2 challenge, despite marked differences between groups in the induction of anxiety symptoms (van Duinen et al. 2004, 2007); indeed in one investigation, cortisol levels decreased significantly following CO_2 challenge (Sinha et al. 1999). Furthermore, administration of the cortisol synthesis inhibitor metyrapone prior to CO_2 challenge reduces cortisol levels prior to challenge, but does not affect the anxiety response (Belgorodsky et al. 2005).

Caffeine administration was associated with a significantly greater increase in anxiety and nervousness in patients with panic disorder and agoraphobia ($n = 21$) compared to healthy controls, but with no difference between groups in change in plasma levels of cortisol or MHPG (Charney et al. 1985). Challenge with the alpha 2-adrenergic receptor agonist clonidine has been used as an investigational tool for evaluating norepinephrine function in panic disorder. The cortisol response to challenge with clonidine differed between patients with panic disorder ($n = 12$) and controls (Brambilla et al. 1995). However in another study, the decline in cortisol levels following challenge with clonidine did not differ significantly between patients with panic disorder ($n = 10$), patients with mood disorders or healthy controls (Stein et al. 1988). In another investigation involving clonidine challenge before and after treatment with the SSRI fluoxetine, which demonstrated significantly increased 'volatility' (i.e. within-subject oscillatory activity) of plasma MHPG levels in patients with panic disorder ($n = 17$) compared to healthy controls at baseline, successful treatment was accompanied by a reduction in volatility to levels seen in controls (Coplan et al. 1997). By contrast, induction of panic attacks through challenge with yohimbine, an alpha 2-adrenergic receptor antagonist, was characterized by increases in both cortisol and MHPG (Charney et al. 1987b).

Hyperventilation, a common feature of panic attacks, can lead to an increase in serum lactate levels (Maddock et al. 1991) and infusion of sodium lactate is often used as an anxiogenic challenge, both in healthy volunteers and in patients with anxiety disorders. Induction of panic through intravenous sodium lactate infusion in a mixed sample of patients with panic disorder or agoraphobia with panic ($n = 43$) was not consistently associated with increases in plasma cortisol or epinephrine levels (Liebowitz et al. 1985). In addition, neither ACTH nor cortisol increased with lactate-induced panic attacks in patients with panic disorder and agoraphobia (Levin et al. 1987). However, in another study the presence of elevated plasma cortisol level at baseline was found to be predictive of late panic attacks following lactate infusion, in a mixed sample of patients with panic disorder or agoraphobia with panic attacks ($n = 103$) (Hollander et al. 1989). Elevated cortisol levels (along with higher reported fear and evidence of hyperventilation) at baseline were predictive of a greater likelihood of experiencing panic during placebo with placebo-controlled lactate infusion studies (Coplan et al. 1998). In another study, lactate infusion was accompanied by an increase in cortisol levels in patients with panic disorder ($n = 17$) and in patients with major depression and panic attacks ($n = 12$), but not in depressed patients without panic attacks ($n = 27$) or healthy controls (Targum 1990). A further investigation found no evidence that lactate infusion enhanced the cortisol response in either patients or healthy controls (Seier et al. 1997), nor did infusion of lactate or the GABA$_A$ receptor antagonist flumazenil in patients with panic disorder ($n = 10$) (Ströhle et al. 1998). The lack of consistent evidence for an enhanced cortisol response following lactate infusion has led to speculations that panic attacks may result from infusion of hypertonic solutions (being seen with hypertonic saline as well as sodium lactate) (Peskind et al. 1998), with enhanced release of atrial natiuretic peptide (ANP) exerting an inhibitory role on ACTH and cortisol release (Kellner et al. 1998). Furthermore, ANP may have anxiolytic effects, as prior administration of ANP reduces the likelihood of experiencing panic following administration of CCK-4 (Wiedemann et al. 2001).

Intravenous infusion with cholecystokinin tetrapeptide (CCK-4) can induce panic attacks in patients with panic disorder in a dose-dependent manner (van Megen et al. 1996). Early investigations suggested that neuroendocrine responses to induction of panic through intravenous infusion of pentagastrin (a cholecystokinin-B receptor agonist) did not differ between patients with panic disorder ($n = 10$) and healthy controls (Abelson et al. 1991, 1994). However, in another investigation of patients with panic disorder ($n = 24$), ACTH levels were significantly higher in patients experiencing panic attacks than in those without attacks; and even patients without attacks had brief but mild increases in ACTH levels (Ströhle et al. 2000). The ACTH and cortisol response to pentagastrin challenge can be reduced by prior cognitive intervention, both in patients with panic disorder and healthy controls (Abelson et al. 2005); and can be enhanced through prior administration of clonidine (Kellner et al. 1997). Successful treatment of patients ($n = 8$) with the SSRI citalopram has been found to reduce the panic response, but not the cortisol, prolactin or growth response, to challenge with CCK-4 delivered as bolus injection (Shlik et al. 1997). Other intravenous challenge tests include infusion of physostigmine and insulin. A small study of intravenous infusion of physostigmine found that patients with panic disorder ($n = 9$) did not differ from controls in anxiety symptoms or cortisol response (Rapaport et al. 1991), whereas administration of an intravenous insulin bolus was associated with an attenuated cortisol (and growth hormone and prolactin) response in patients with panic disorder compared to healthy controls (Jezova et al. 2010).

4 Generalized Anxiety Disorder

Similar to endeavours in panic disorder, evaluation of the influence of cortisol in GAD has included investigations of plasma and salivary cortisol, dexamethasone non-suppression, and cortisol response to psychological and pharmacological challenge. As with major depression, GAD appears common among patients with a primary diagnosis of Cushing's disease, defined as Cushing's syndrome associated with an ACTH-secreting pituitary microadenoma, and characterized by hypercortisolism (Loosen et al. 1992). Successful treatment of Cushing's syndrome by correction of hypercortisolism is associated with a gradual reduction in the presence of mood and anxiety disorders (Dorn et al. 1997). However, investigations of hypothalamo-pituitary adrenal axis function among patients with GAD have produced variable findings, with no consistent evidence of hypercortisolism. In an early study, significant diurnal changes in plasma cortisol levels were reported in a small sample ($n = 13$) of patients with GAD (Hoehn-Saric et al. 1991). A current or lifetime history of GAD or phobic disorder (but not post-traumatic syndromes) was found to be associated with a pattern of up-regulated diurnal cortisol secretion in a large population study in elderly individuals (Chaudieu et al. 2008). Both GAD ($n = 12$) and major depression ($n = 8$) were characterized by a failure of the pattern of cortisol-induced serotonin uptake in lymphocytes, seen in matched healthy controls ($n = 8$) (Tafet et al. 2001). However, a comparison of morning plasma cortisol and DHEA sulphate levels in the Vietnam-era US army veterans with GAD, major depression or co-morbid depression and GAD, found that depressed and co-morbid patients, but not patients with GAD alone, had evidence of hypocortisolism (Phillips et al. 2011) (Table 2).

Recent investigations have been focused on salivary cortisol levels and concentrations of cortisol in hair. An investigation involving serial saliva sampling found the cortisol awakening response to be less elevated in patients with GAD than in patients with panic disorder, with neither group showing more dexamethasone non-suppression than in matched controls (Vreeburg et al. 2010). A lower cortisol awakening response was seen in individuals with GAD, drawn from a large population-based study of older people (aged 65 years and above) (Hek et al. 2013). A small case–control study in pre-pubescent children found no difference in bedtime salivary cortisol levels between 'anxious patients' (with a primary diagnosis of GAD) and healthy controls (Alfano et al. 2013). In a case–control study of cortisol concentrations in hair, which may provide a reflection of cortisol levels over time, there were significantly lower levels among patients with GAD ($n = 15$) than in age- and gender-matched controls (Steudte et al. 2011). This observation is supported by the findings of a recent meta-analysis of hair cortisol and stress exposure, which found evidence of hypocortisolism in both GAD and panic disorder (Staufenbiel et al. 2013).

An early investigation involving the dexamethasone suppression test (using a minimum cortisol value of 5 mcg/dl to indicate non-suppression) found no significant group differences between medication-free patients with GAD ($n = 26$), panic disorder ($n = 22$), agoraphobia with panic attacks ($n = 13$), or 'primary affective disorder' ($n = 60$) (Avery et al. 1985). A subsequent investigation in 79 patients with GAD found a non-suppression rate of 27 %, similar to that reported in patients with major depression, but greater than the previously reported rate in panic disorder, the presence of non-suppression being independent of depressive symptom severity (Schweizer et al. 1986). Non-suppression in the dexamethasone

Table 2 Investigations of cortisol in generalized anxiety disorder

Plasma, salivary and hair cortisol levels	
Hoehn-Saric et al. 1991	Significant diurnal changes in plasma cortisol levels
Chaudieu et al. 2008	Up-regulated diurnal cortisol secretion in elderly individuals with current or past GAD
Tafet et al. 2001	Abnormal cortisol-induced lymphocyte serotonin uptake similar to that seen in depression
Phillips et al. 2011	Co-morbid depression and GAD associated with hypocortisolism
Vreeburg et al. 2010	Lesser elevation of cortisol awakening response than in depressed patients
Hek et al. 2013	Lower cortisol awakening response in older individuals with GAD
Alfano et al. 2013	No significant difference in bedtime salivary cortisol levels in pre-pubescent children
Steudte et al. 2011	Lower cortisol levels in hair than in controls
Staufenbiel et al. 2013	Evidence of low levels of cortisol in GAD and panic disorder
Dexamethasone non-suppression	
Avery et al. 1985	No evidence of increased non-suppression compared to controls
Schweizer et al. 1986	Non-suppression rate of 27 %, presence independent of depressive symptom severity
Schittecatte et al. 1995	Non-suppression rate similar in GAD and major depression
Vreeburg et al. 2010	No significant difference in non-suppression compared to controls
Psychological and pharmacological challenge	
Roy-Byrne et al. 1991	Reduction in cortisol levels following IV diazepam administration
Gerra et al. 2000	No differences in cortisol or ACTH levels in adolescents in GAD after psychological challenge
Rosnick et al. 2013	Lowering of salivary cortisol levels following detailed psychological assessment
Seddon et al. 2011	No change in cortisol levels after inhalation of 7.5 % CO_2
Cohn et al. 1986	Buspirone administration lowers anxiety but has no effect on cortisol levels
Klein et al. 1995	No reduction in plasma cortisol levels with alprazolam treatment
Pomara et al. 2005	Diazepam administration reduces cortisol levels in GAD and healthy controls
Lenze et al. 2011	Reduction in elevated cortisol levels associated with greater reduction in symptom severity
Lenze et al. 2012	Reduction in elevated cortisol levels associated with greater improvement in memory
Tiller et al. 1988	Successful psychological treatment converts previous dexamethasone non-suppression
Tafet et al. 2005	Successful cognitive therapy accompanied by decline in plasma cortisol levels

test was found to have little value in distinguishing between patients with GAD ($n = 15$) or major depression ($n = 15$), in an investigation of the suppression of rapid eye movement sleep by clonidine administration in depressed patients and healthy controls (Schittecatte et al. 1995).

It is uncertain whether change in cortisol levels, or dexamethasone non-suppression, is predictive of the response to experimental pharmacological or psychological challenge. In a mixed group of patients with GAD ($n = 8$) or panic disorder ($n = 13$), challenge by intravenous diazepam administration was associated with a reduction in cortisol levels (but increase in levels of growth hormone and ACTH) (Roy-Byrne et al. 1991). A case–control study involving psychological challenge in adolescents with GAD ($n = 20$) found no significant changes in cortisol or ACTH during challenge in either cases or controls, in contrast to increases seen in norepinephrine, growth hormone and testosterone (Gerra et al. 2000). A case–control study in older individuals with GAD ($n = 69$) found that their participation in detailed neuropsychological assessment was associated with a lowering of salivary cortisol levels (Rosnick et al. 2013). In a small study ($n = 12$) inhalation of air enriched with 7.5 % carbon dioxide was associated with increased subjective anxiety and with autonomic responses seen in heightened anxiety, but not with a change in cortisol levels (Seddon et al. 2011).

Despite early contrary findings, it seems possible that change in cortisol levels and in other indices of HPA function are altered during the response to pharmacological or psychological treatment. A placebo-controlled study of the 5-HT$_{1A}$ partial agonist buspirone in 23 patients with GAD, which found that buspirone was effective in reducing anxiety symptom severity, found no association with change in plasma levels of cortisol (or prolactin or growth hormone) (Cohn et al. 1986). In a mixed group of patients with GAD ($n = 35$) or panic disorder ($n = 36$), treatment with alprazolam was associated with a reduction in plasma cortisol levels in only the panic disorder group, the group with GAD having a reduction in plasma epinephrine levels (Klein et al. 1995). However, a case–control study of the effects of acute challenge and subsequent prolonged administration of diazepam in individuals with GAD found it was associated with a reduction in plasma cortisol levels, particularly in elderly patients, in both cases and controls (Pomara et al. 2005). In addition, in a placebo-controlled study of treatment with the SSRI escitalopram in elderly patients with GAD, reduction in previously elevated cortisol levels was associated with a more marked reduction in symptom severity (Lenze et al. 2011), and with improvements in measures of immediate and delayed memory (Lenze et al. 2012). Furthermore, in an investigation which found a 27 % rate of non-suppression among 30 patients with GAD, successful psychological treatment was associated with 'conversion' to suppression, though post-treatment concentrations remained significantly lower in the initial non-suppressors (Tiller et al. 1988). Finally, a controlled investigation of cognitive therapy in 24 patients with GAD found that successful treatment was accompanied by a significant decline in plasma cortisol levels (Tafet et al. 2005).

5 Specific (Simple) Phobia

There have been comparatively few investigations of cortisol levels in individuals with specific phobia 'at rest'. In children and adolescents with major depressive disorder, co-morbidity with phobic disorders or panic disorder was associated with an absence of the elevated cortisol levels that were seen in depressed patients without co-morbidity (Herbert et al. 1996). Another investigation in children and adolescents with varying anxiety disorders ($n = 99$) found no difference between the disorders in salivary cortisol or diurnal cortisol rhythm (Kallen et al. 2008). Compared to pregnant but healthy controls, pregnant women with blood-injection phobia ($n = 110$) showed evidence of higher cortisol output, but no difference in diurnal cortisol rhythm (Lilliecruz et al. 2011).

A series of investigations have suggested that experimental exposure of individuals with specific phobia to a feared object or situation is associated with an enhanced cortisol response, but again not all evidence is consistent. An early investigation of cortisol, electrodermal activity and subjective distress in a mixed sample ($n = 12$) of individuals with blood-injection phobia or animal phobia found that experimental exposure to pictorial images of feared objects (but not exposure to neutral objects) elicited cortisol excretion (Fredrikson et al. 1985). Cortisol levels (and levels of epinephrine, norepinephrine, growth hormone and insulin) were also found to rise during therapeutic in vivo exposure to feared animals in a small sample ($n = 10$) of women with various animal phobias (Nesse et al. 1985). Cortisol levels were found to increase during exposure therapy in two patients with height phobia (Abelson and Curtis 1989). Although baseline cortisol levels did not differ between groups, a significantly greater increase in cortisol levels was reported immediately before, during and immediately after a driving task in individuals with driving phobia, when compared to healthy controls (Alpers et al. 2003). But not all evidence is supportive, few differences being seen in an investigation in women with spider phobia ($n = 46$) and healthy control women, when challenged with neutral or feared images (Knopf and Possel 2009), nor in an investigation in individuals with spider phobia ($n = 16$), in whom cortisol levels did not increase following presentation of the feared stimulus despite increases in self-reported fear (Van Duinen et al. 2010) (Table 3).

Investigations of the interaction between cortisol exposure and the effectiveness of exposure therapy have produced intriguing findings. In a placebo-controlled study, cortisol administration an hour prior to experimental exposure to feared social situations or animals was found to significantly reduce stimulus-induced self-reported fear (but not to reduce more general non-phobic anxiety) (Soravia et al. 2006). In patients with specific phobia, social phobia and PTSD, prior cortisol administration was also associated with reduced stimulus-induced fear, both immediately after exposure and 2 days later (de Quervain and Margraf 2008). In a placebo-controlled study in individuals ($n = 40$) with height phobia, cortisol administration prior to virtual reality therapeutic exposure was found to significantly reduce both reported fear and the degree of exposure-induced increased skin

Table 3 Investigations of cortisol in specific (simple) phobia

Herbert et al. 1996	Co-morbid phobic disorders associated with lower likelihood of hypercortisolism
Kellner et al. 1998	No difference from other anxiety disorders in salivary cortisol or diurnal cortisol rhythm
Lilliecreutz et al. 2011	Blood-injection phobia in pregnant women associated with hypercortisolism
Fredrikson et al. 1985	Exposure to pictorial images of feared objects elicits rise in cortisol levels
Nesse et al. 1985	Cortisol levels rise during in vivo exposure to feared animals
Abelson and Curtis 1989	Cortisol levels rise during exposure therapy for height phobia
Alpers et al. 2003	Greater increase in cortisol levels before, during and after driving task in driving phobia
Knopf and Possel 2009	No differences in cortisol levels following exposure to feared or neutral images
Van Duinen et al. 2010	Increased fear following presentation of feared stimuli not accompanied by increase in cortisol
Soravia et al. 2006	Cortisol administration prior to exposure to feared situations reduces stimulus-induced fear
De Quervain et al. 2008	Cortisol administration associated with reduction in stimulus-induced fear
De Quervain et al. 2011	Cortisol administration reduces fear and exposure-induced increased skin conductance
Brand et al. 2011	Reduction in cortisol levels with in vivo exposure in protective mask phobia

conductance (de Quervain et al. 2011). Intensive therapeutic in vivo exposure in military personnel ($n = 46$) with protective mask phobia (a form of simple phobia) was associated with a reduction in salivary cortisol levels, this reduction not being seen in controls from emergency responder services (Brand et al. 2011).

6 Social Anxiety Disorder (Social Phobia)

Compared to panic disorder and GAD, there have been relatively few investigations of cortisol in patients with social phobia. Although a prospective study in a community sample ($n = 238$) found that elevated afternoon salivary cortisol levels in early childhood were predictive of subsequent social phobia in adolescence (Essex et al. 2010), a series of investigations have suggested that cortisol levels are not elevated in patients with social phobia in the 'resting' or unchallenged state. An early investigation found no significant differences in urinary cortisol levels between patients with social phobia ($n = 10$) and healthy controls (Potts et al. 1991): this finding being replicated in an analysis of 24-h urinary cortisol levels, which also found no evidence of dexamethasone non-suppression in a larger group ($n = 64$) of patients with social phobia (Uhde et al. 1994). No significant differences were found between patients with social phobia ($n = 26$) and healthy controls, in plasma levels of cortisol, pregnenalone or DHEA (Laufer et al. 2005). Although patients with social phobia ($n = 43$) differed from healthy controls in salivary alpha-amylase (a marker of sympathetic autonomic nervous system activity), there were no differences in salivary cortisol levels (van Veen et al. 2008). A case–control study of men with social phobia ($n = 12$) found that salivary cortisol levels were significantly lower than in healthy controls, with strong negative correlations between cortisol levels and 5-HT$_{1A}$ binding in the amygdala, hippocampus and retrosplenial cortex (Lanzenberger et al. 2010).

Investigations of the response to psychological challenge have produced reasonably consistent findings. An early study found that challenge through the Trier Social Stress Test (TSST) was associated with a significant elevation in salivary cortisol levels in adolescent girls with social phobia ($n = 27$), but no more so than in healthy controls (Martel et al. 1999). By contrast, a public speaking task (but not physical exercise) was associated with a significantly greater increase in salivary cortisol in patients with social phobia ($n = 18$) compared to healthy controls (Furlan et al. 2001). Performance of a public speaking task was associated with a significantly greater increase in salivary cortisol in children with social phobia ($n = 25$) compared to healthy controls (van West et al. 2008). In addition, performance in a social approach-avoidance task and challenge through the TSST was associated with a significantly greater increase in salivary cortisol in patients with social phobia ($n = 18$) compared to healthy controls and patients with PTSD (Roelofs et al. 2009). However, in children with social phobia ($n = 41$) undergoing challenge with the TSST, the increase in salivary cortisol was not significantly greater than in healthy controls (Krämer et al. 2012). In a functional imaging study involving a public speaking task in patients with social phobia ($n = 12$), the increase in salivary cortisol levels was associated with increased regional cerebral blood flow in the hypothalamus (especially the mamillary bodies) but decreased flow in the medial prefrontal cortex (Ahs et al. 2006) (Table 4).

Pharmacological challenge tests and pharmacological treatment studies suggest a complex interaction between cortisol and serotonin in patients with social phobia. An early investigation found no correlation between plasma cortisol level and the degree of improvement during treatment with the SSRI fluvoxamine in patients with social phobia (DeVane et al. 1999). In a placebo-controlled case–control study in generalized social phobia ($n = 21$), single dose pharmacological challenge with fenfluramine was associated with a significantly augmented cortisol response, compared to that seen in healthy volunteers (Tancer et al. 1994). Single dose pharmacological challenge with mCPP was associated with a trend towards a greater cortisol response in patients with social phobia ($n = 18$) than in controls or in patients with OCD (Hollander et al. 1998), but in a further investigation involving mCPP challenge, the cortisol response was not significantly different between patients with social phobia ($n = 7$) and healthy controls (van Veen et al. 2007). An investigation involving placebo-controlled single dose intravenous administration of the SSRI citalopram found no difference in plasma cortisol or prolactin responses, between patients with social phobia ($n = 18$) and healthy

Table 4 Investigations of cortisol in social anxiety disorder (social phobia)

Essex et al. 2010	Elevated afternoon salivary cortisol level in childhood predicts social phobia in adolescence
Potts et al. 1991	No difference in urinary cortisol levels
Uhde et al. 1994	No difference in 24-h cortisol levels and no evidence of dexamethasone non-suppression
Laufer et al. 2005	No difference in levels of cortisol, pregnenalone or DHEA
Van Veen et al. 2008	No difference in salivary cortisol levels despite difference in salivary alpha-amylase
Lanzenberger et al. 2010	Significantly lower cortisol levels and negative correlations with 5-HT$_{1A}$ binding
Martel et al. 1999	No significant difference in elevation of cortisol levels following TSST
Furlan et al. 2001	Enhancement of increase in salivary cortisol levels following public speaking task
Van West et al. 2008	Enhancement of increase in salivary cortisol levels following public speaking task
Roelofs et al. 2009	Greater increase in salivary cortisol levels with psychological task following TSST challenge
Ahs et al. 2006	Increase in salivary cortisol levels associated with increased blood flow in hypothalamus
Krämer et al. 2012	No significant difference in rise in salivary cortisol level following TSST
DeVane et al. 1994	No correlation between cortisol level and degree of improvement with SSRI treatment
Tancer et al. 1994	Augmentation of cortisol response following fenfluramine challenge
Hollander et al. 1998	Trend towards enhancement of cortisol response following mCPP challenge
Van Veen et al. 2007	No significant difference in cortisol response following mCPP challenge
Shlik et al. 2004	No significant difference in cortisol response to citalopram challenge
Van Veen et al. 2009	No increase in cortisol level following acute tryptophan depletion and public speaking task
Soravia et al. 2006	Cortisol administration reduces fear before, during and after social-evaluative stress task
Van Peer et al. 2009, 2010	Cortisol administration enhances processing of social stimuli and event-related amplitudes
Katzman et al. 2004	No difference in cortisol response following CCK-4 challenge

controls (Shlik et al. 2004). Following successful SSRI treatment of patients ($n = 18$) with social phobia, dual pharmacological and psychological challenge—through placebo-controlled transient tryptophan depletion and performance in a public speaking task—was accompanied by significantly increased salivary amylase activity, but not by an increase in cortisol, suggesting hyper-responsivity of the autonomic nervous system but not the HPA (van Veen et al. 2009).

As in patients with simple phobia, cortisol administration significantly reduced self-reported fear prior to, during and after a social-evaluative stress task (Soravia et al. 2006). Cortisol administration prior to a reaction time task enhanced the processing of social stimuli and enhanced event-related potential amplitudes (particularly angry faces, when compared to neutral and happy faces) in patients with social phobia ($n = 20$), in a manner influenced by symptom severity and motivational context (van Peer et al. 2009, 2010). Pharmacological challenge with intravenous CCK-4, found no differences in the ACTH, cortisol, growth hormone or prolactin response between patients with social anxiety disorder ($n = 12$) or obsessive-compulsive disorder or healthy controls (Katzman et al. 2004).

7 Summary

Although findings are inconsistent, panic disorder appears to be characterized by an elevation of urinary cortisol levels, by a decline in cortisol levels with successful pharmacological treatment, and by non-suppression following dexamethasone administration in a proportion greater than in healthy controls but less than that in depressed patients. There is much uncertainty about whether it is characterized by elevated plasma cortisol levels, whether cortisol levels fall following psychological challenge, and whether the anxiety response to panicogenic challenges is accompanied by changes in endocrine function. GAD appears to be characterized by a decline in cortisol levels with successful psychological or pharmacological treatment; it is uncertain whether it is also characterized by elevated cortisol levels prior to treatment, and by whether dexamethasone non-suppression is more common than in healthy controls. Specific phobia appears characterized by cortisol levels which rise during experimental exposure to feared objects or situations, but which decline with successful exposure therapy; social anxiety disorder is possibly characterized by cortisol levels that rise during psychological challenge.

Compared to the extensive literature on HPA axis function in patients with depressive illness, the evidence base relating to HPA axis function in patients with the principal anxiety disorders is limited; the number of investigations in panic disorder and GAD is reasonably extensive, but there have been rather few studies in specific (simple) phobia and social anxiety disorder (social phobia). However, it is clear that there is no unifying disturbance of HPA axis function across these anxiety disorders; furthermore, within each disorder, the findings of investigations using similar methodology have often produced inconsistent findings. These apparent disparities in results from studies of similar design are possibly influenced by the small sample size that is typical of most investigations, and by variations in the nature of the clinical sample. Achieving greater consensus on study objectives, the detailed characterization of patient groups, the methodological protocols for investigation and the preferred mode of statistical analysis would be an important step forward in further evaluations of HPA function in anxiety disorders (Baldwin et al. 2013).

Acknowledgments Hesham Yousry Elnazer receives a bursary from KSS Deanery to support his postgraduate doctorate studies. David Baldwin chairs the Anxiety Disorders Research Network (ADRN) within the European College of Neuropsychopharmacology Network Initiative (ECNP-NI), and the University Department of Psychiatry in Southampton receives funding from the ECNP-NI for administrative support designed to foster the further development of the ADRN.

References

Abelson JL, Curtis GC (1989) Cardiac and neuroendocrine responses to exposure therapy in height phobics: desynchrony within the physiological response system. Behav Res Ther 27:561–567

Abelson JL, Nesse RM, Vinik A (1991) Stimulation of corticotropin release by pentagastrin in normal subjects and patients with panic disorder. Biol Psychiatry 29:1220–1223

Abelson JL, Nesse RM, Vinik AI (1994) Pentagastrin infusions in patients with panic disorder. II. Neuroendocrinology. Biol Psychiatry 36:84–96

Abelson JL, Curtis GC (1996a) Hypothalamic-pituitary-adrenal axis activity in panic disorder. 24-hour secretion of corticotropin and cortisol. Arch Gen Psychiatry 53:323–331

Abelson JL, Curtis GC (1996b) Hypothalamic-pituitary-adrenal axis activity in panic disorder: prediction of long-term outcome by pretreatment cortisol levels. Am J Psychiatry 153:69–73

Abelson JL, Curtis GC, Cameron OG (1996) Hypothalamic-pituitary-adrenal axis activity in panic disorder: effects of alprazolam on 24 h secretion of adrenocorticotropin and cortisol. J Psychiatr Res 30:79–93

Abelson JL, Liberzon I, Young EA et al (2005) Cognitive modulation of the endocrine stress response to a pharmacological challenge in normal and panic disorder subjects. Arch Gen Psychiatry 62:668–675

Ahs F, Furmark T, Michelgård A et al (2006) Hypothalamic blood flow correlates positively with stress-induced cortisol levels in subjects with social anxiety disorder. Psychosom Med 68:859–862

Alfano CA, Reynolds K, Scott N et al (2013) Polysomnographic sleep patterns of non-depressed, non-medicated children with generalized anxiety disorder. J Affect Disord 147(1–3):379–384

Alpers GW, Abelson JL, Wilhelm FH et al (2003) Salivary cortisol response during exposure treatment in driving phobics. Psychosom Med 65:679–687

American Psychiatric Association (2013) Diagnostic and statistical manual of mental disorders (DSM-5). American Psychiatric Publishing, Washington

Avery DH, Osgood TB, Ishiki DM et al (1985) The DST in psychiatric outpatients with generalized anxiety disorder, panic disorder, or primary affective disorder. Am J Psychiatry 142:844–848

Baldwin DS, Allgulander C, Altamura AC et al (2010) Manifesto for a european anxiety disorders research network. Eur Neuropsychopharmacol 20:426–432

Baldwin DS, Pallanti S, Zwanzger P (2013) Developing a european research network to address unmet needs in anxiety disorders. Neurosci Biobehav Rev 37:2312–2317

Bandelow B, Sengos G, Wedekind D et al (1997) Urinary excretion of cortisol, norepinephrine, testosterone, and melatonin in panic disorder. Pharmacopsychiatry 30:113–117. Erratum in: Pharmacopsychiatry 1997 30, 278

Bandelow B, Wedekind D, Pauls J et al (2000a) Salivary cortisol in panic attacks. Am J Psychiatry 157:454–456

Bandelow B, Wedekind D, Sandvoss V et al (2000b) Diurnal variation of cortisol in panic disorder. Psychiatry Res 95:245–250

Belgorodsky A, Knyazhansky L, Loewenthal U et al (2005) Effects of the cortisol synthesis inhibitor metyrapone on the response to carbon dioxide challenge in panic disorder. Depress Anxiety 21:143–148. Erratum in: Depress Anxiety 2005 21:203

Brambilla F, Bellodi L, Perna G et al (1992) Psychoimmunoendocrine aspects of panic disorder. Neuropsychobiol 26:12–22

Brambilla F, Bellodi L, Arancio C et al (1995) Alpha 2-adrenergic receptor sensitivity in panic disorder: II. Cortisol response to clonidine stimulation in panic disorder. Psychoneuroendocrinol 20:11–19

Brand S, Annen H, Holsboer-Trachsler E et al (2011) Intensive 2-day cognitive-behavioral intervention decreases cortisol secretion in soldiers suffering from specific phobia to wear protective mask. J Psychiatr Res 45:1337–1345

Bridges M, Yeragani VK, Raincy JM et al (1986) Dexamethasone suppression test in patients with panic attacks. Biol Psychiatry 21(8–9):853–855

Broocks A, Bandelow B, George A et al (2000) Increased psychological responses and divergent neuroendocrine responses to m-CPP and ipsapirone in patients with panic disorder. Int Clin Psychopharmacol 15:153–161

Broocks A, Bandelow B, Koch K et al (2002) Smoking modulates neuroendocrine responses to ipsapirone in patients with panic disorder. Neuropsychopharmacol 27:270–278

Broocks A, Meyer T, Opitz M et al (2003) 5-HT1A responsivity in patients with panic disorder before and after treatment with aerobic exercise, clomipramine or placebo. Eur Neuropsychopharmacol 13:153–164

Carson SW, Halbreich U, Yeh CM et al (1988) Altered plasma dexamethasone and cortisol suppressibility in patients with panic disorders. Biol Psychiatry 24:56–62

Charney DS, Heninger GR, Jatlow PI (1985) Increased anxiogenic effects of caffeine in panic disorders. Arch Gen Psychiatry 42:233–243

Charney DS, Woods SW, Goodman WK et al (1987a) Neurobiological mechanisms of panic anxiety: biochemical and behavioral correlates of yohimbine-induced panic attacks. Am J Psychiatry 144:1030–1036

Charncy DS, Woods SW, Goodman WK et al (1987b) Serotonin function in anxiety. II. Effects of the serotonin agonist MCPP in panic disorder patients and healthy subjects. Psychopharmacol 92:14–24

Chaudieu I, Beluche I, Norton J et al (2008) Abnormal reactions to environmental stress in elderly persons with anxiety disorders: evidence from a population study of diurnal cortisol changes. J Affect Disord 106:307–313

Cohn JB, Wilcox CS, Meltzer HY (1986) Neuroendocrine effects of buspirone in patients with generalized anxiety disorder. Am J Med 80(3B):36–40

Coplan JD, Papp LA, Pine D et al (1997) Clinical improvement with fluoxetine therapy and noradrenergic function in patients with panic disorder. Arch Gen Psychiatry 54:643–648

Coplan JD, Goetz R, Klein DF et al (1998) Plasma cortisol concentrations preceding lactate-induced panic. Psychological, biochemical, and physiological correlates. Arch Gen Psychiatry 55:130–136

Coryell W, Noyes R Jr, Clancy J et al (1985) Abnormal escape from dexamethasone suppression in agoraphobia with panic attacks. Psychiatry Res 15:301–311

Coryell W, Noyes R (1988) HPA axis disturbance and treatment outcome in panic disorder. Biol Psychiatry 24:762–766

Coryell W, Noyes R Jr, Schlechte J (1989) The significance of HPA axis disturbance in panic disorder. Biol Psychiatry 25:989–1002

Coryell W, Noyes R Jr, Reich J (1991) The prognostic significance of HPA-axis disturbance in panic disorder: a 3-year follow-up. Biol Psychiatry 29:96–102

Cottraux J, Claustrat B 91984) The dexamethasone suppression test in agoraphobia with panic attacks. 30 cases. Encephale 10:267–272 (French)

Curtis GC, Abelson JL, Gold PW (1997) Adrenocorticotropic hormone and cortisol responses to corticotropin-releasing hormone: changes in panic disorder and effects of alprazolam treatment. Biol Psychiatry 41:76–85

Den Boer JA, Westenberg HG (1990) Serotonin function in panic disorder: a double blind placebo controlled study with fluvoxamine and ritanserin. Psychopharmacology (Berl) 102:85–94

de Quervain DJ, Margraf J (2008) Glucocorticoids for the treatment of post-traumatic stress disorder and phobias: a novel therapeutic approach. Eur J Pharmacol 583:365–371

de Quervain DJ, Bentz D, Michael T et al (2011) Glucocorticoids enhance extinction-based psychotherapy. Proc Natl Acad Sci USA 108:6621–6625

DeVane CL, Ware MR, Emmanuel NP et al (1999) Evaluation of the efficacy, safety and physiological effects of fluvoxamine in social phobia. Int Clin Psychopharmacol 14:345–351

Dorn LD, Burgess ES, Friedman TC et al (1997) The longitudinal course of psychopathology in Cushing's syndrome after correction of hypercortisolism. J Clin Endocrinol Metab 82:912–919

Erhardt A, Ising M, Unschuld PG et al (2006) Regulation of the hypothalamic-pituitary-adrenocortical system in patients with panic disorder. Neuropsychopharmacol 31:2515–2522

Essex MJ, Klein MH, Slattery MJ et al (2010) Early risk factors and developmental pathways to chronic high inhibition and social anxiety disorder in adolescence. Am J Psychiatry 167:40–46

Faludi G, Kaskó M, Perényi A et al (1986) The dexamethasone suppression test in panic disorder and major depressive episodes. Biol Psychiatry 21:1008–1014

Fava M, Rosenbaum JF, MacLaughlin RA et al (1989) Dehydroepiandrosterone-sulfate/cortisol ratio in panic disorder. Psychiatry Res 28:345–350

Fredrikson M, Sundin O, Frankenhaeuser M (1985) Cortisol excretion during the defense reaction in humans. Psychosom Med 47:313–319

Furlan PM, DeMartinis N, Schweizer E et al (2001) Abnormal salivary cortisol levels in social phobic patients in response to acute psychological but not physical stress. Biol Psychiatry 50:254–259

Garcia-Leal C, Parente AC, Del-Ben CM et al (2005) Anxiety and salivary cortisol in symptomatic and nonsymptomatic panic patients and healthy volunteers performing simulated public speaking. Psychiatry Res 133:239–252

Germine M, Goddard AW, Sholomskas DE et al (1994) Response to meta-chlorophenylpiperazine in panic disorder patients and healthy subjects: influence of reduction in intravenous dosage. Psychiatry Res 54:115–133

Gerra G, Zaimovic A, Zambelli U et al (2000) Neuroendocrine responses to psychological stress in adolescents with anxiety disorder. Neuropsychobiol 42:82–92

Goldstein S, Halbreich U, Asnis G et al (1987) The hypothalamic-pituitary-adrenal system in panic disorder. Am J Psychiatry 144:1320–1323

Grunhaus L, Tiongco D, Haskett RF et al (1987) The dexamethasone suppression test in inpatients with panic disorder or agoraphobia with panic attacks. Biol Psychiatry 22:517–521

Hek K, Direk N, Newson RS et al (2013) Anxiety disorders and salivary cortisol levels in older adults: a population-based study. Psychoneuroendocrinol 38:300–305

Herbert J, Goodyer IM, Altham PM et al (1996) Adrenal secretion and major depression in 8–16-year-olds, II. Influence of co-morbidity at presentation. Psychol Med 26:257–263

Herrán A, Sierra-Biddle D, García-Unzueta MT et al (2005) The acute phase response in panic disorder. Int J Neuropsychopharmacol 8:529–535

Hoehn-Saric R, McLeod DR, Lee YB et al (1991) Cortisol levels in generalized anxiety disorder. Psychiatry Res 38:313–315

Hollander E, Liebowitz MR, Gorman JM et al (1989) Cortisol and sodium lactate-induced panic. Arch Gen Psychiatry 46:135–140

Hollander E, Kwon J, Weiller F et al (1998) Serotonergic function in social phobia: comparison to normal control and obsessive-compulsive disorder subjects. Psychiatry Res 79:213–217

Hood SD, Hince DA, Robinson H et al (2006) Serotonin regulation of the human stress response. Psychoneuroendocrinol 31:1087–1097

Jezova D, Vigas M, Hlavacova N et al (2010) Attenuated neuroendocrine response to hypoglycemic stress in patients with panic disorder. Neuroendocrinol 92:112–119

Judd FK, Norman TR, Burrows GD et al (1987) The dexamethasone suppression test in panic disorder. Pharmacopsychiatry 20:99–101

Judd FK, Apostolopoulos M, Burrows GD (1994) Serotonergic function in panic disorder: endocrine responses to D-fenfluramine. Prog Neuropsychopharmacol Biol Psychiatry 18:329–337

Kahn RS, Asnis GM, Wetzler S et al (1988) Neuroendocrine evidence for serotonin receptor hypersensitivity in panic disorder. Psychopharmacology 96:360–364

Kallen VL, Tulen JH, Utens EM et al (2008) Associations between HPA axis functioning and level of anxiety in children and adolescents with an anxiety disorder. Depress Anxiety 25:131–141

Kathol RG, Noyes R Jr, Lopez AL et al (1988) Relationship of urinary free cortisol levels in patients with panic disorder to symptoms of depression and agoraphobia. Psychiatry Res 24:211–221

Kathol RG, Anton R, Noyes R et al (1989) Direct comparison of urinary free cortisol excretion in patients with depression and panic disorder. Biol Psychiatry 25:873–878

Katzman MA, Koszycki D, Bradwejn J (2004) Effects of CCK-tetrapeptide in patients with social phobia and obsessive-compulsive disorder. Depress Anxiety 20:51–58

Kellner M, Yassouridis A, Jahn H et al (1997) Influence of clonidine on psychopathological, endocrine and respiratory effects of cholecystokinin tetrapeptide in patients with panic disorder. Psychopharmacology 133:55–61

Kellner M, Knaudt K, Jahn H et al (1998) Atrial natriuretic hormone in lactate-induced panic attacks: mode of release and endocrine and pathophysiological consequences. J Psychiatr Res 32:37–48

Kellner M, Schick M, Yassouridis A et al (2004) Metyrapone tests in patients with panic disorder. Biol Psychiatry 56:898–900

Klein E, Zinder O, Colin V et al (1995) Clinical similarity and biological diversity in the response to alprazolam in patients with panic disorder and generalized anxiety disorder. Acta Psychiatr Scand 92:399–408

Knopf K, Pössel P (2009) Individual response differences in spider phobia: comparing phobic and non-phobic women of different reactivity levels. Anxiety Stress Coping 22:39–55

Krämer M, Seefeldt WL, Heinrichs N et al (2012) Subjective, autonomic, and endocrine reactivity during social stress in children with social phobia. J Abnorm Child Psychol 40:95–104

Lanzenberger R, Wadsak W, Spindelegger C et al (2010) Cortisol plasma levels in social anxiety disorder patients correlate with serotonin-1A receptor binding in limbic brain regions. Int J Neuropsychopharmacol 13:1129–1143

Laufer N, Maayan R, Hermesh H et al (2005) Involvement of GABAA receptor modulating neuroactive steroids in patients with social phobia. Psychiatry Res 137:131–136. Erratum in: Psychiatry Res 144:95

Lenze EJ, Mantella RC, Shi P et al (2011) Elevated cortisol in older adults with generalized anxiety disorder is reduced by treatment: a placebo-controlled evaluation of escitalopram. Am J Geriatr Psychiatry 19:482–490

Lenze EJ, Dixon D, Mantella RC et al (2012) Treatment-related alteration of cortisol predicts change in neuropsychological function during acute treatment of late-life anxiety disorder. Int J Geriatr Psychiatry 27:454–462

Lesch KP, Wiesmann M, Hoh A et al (1992) 5-HT1A receptor-effector system responsivity in panic disorder. Psychopharmacology 106:111–117

Levin AP, Doran AR, Liebowitz MR et al (1987) Pituitary adrenocortical unresponsiveness in lactate-induced panic. Psychiatry Res 21:23–32

Lieberman JA, Brenner R, Lesser M et al (1983) Dexamethasone suppression tests in patients with panic disorder. Am J Psychiatry 140:917–919

Liebowitz MR, Gorman JM, Fyer AJ et al (1985) Lactate provocation of panic attacks. II. Biochemical and physiological findings. Arch Gen Psychiatry 42:709–719

Lilliecreutz C, Theodorsson E, Sydsjö G et al (2011) Salivary cortisol in pregnant women suffering from blood and injection phobia. Arch Womens Ment Health 14:405–411

Lopez AL, Kathol RG, Noyes R Jr (1990) Reduction in urinary free cortisol during benzodiazepine treatment of panic disorder. Psychoneuroendocrinol 15:23–28

Loosen PT, Chambliss B, DeBold CR et al (1992) Psychiatric phenomenology in Cushing's disease. Pharmacopsychiatry 25:192–198

Maddock RJ, Carter CS, Gietzen DW (1991) Elevated serum lactate associated with panic attacks induced by hyperventilation. Psychiatry Res 38:301–311

Martel FL, Hayward C, Lyons DM et al (1999) Salivary cortisol levels in socially phobic adolescent girls. Depress Anxiety 10:25–27

Nesse RM, Cameron OG, Curtis GC et al (1984) Adrenergic function in patients with panic anxiety. Arch Gen Psychiatry 41:771–776

Nesse RM, Curtis GC, Thyer BA et al (1985) Endocrine and cardiovascular responses during phobic anxiety. Psychosom Med 47:320–332

Pariante CM (2009) Risk factors for development of depression and psychosis. Glucocorticoid receptors and pituitary implications for treatment with antidepressants and glucocorticoids. Ann NY Acad Sci 1179:144–152

Peskind ER, Jensen CF, Pascualy M et al (1998) Sodium lactate and hypertonic sodium chloride induce equivalent panic incidence, panic symptoms, and hypernatremia in panic disorder. Biol Psychiatry 44:1007–1016

Peterson GA, Ballenger JC, Cox DP et al (1985) The dexamethasone suppression test in agoraphobia. J Clin Psychopharmacol 5:100–102

Petrowski K, Herold U, Joraschky P et al (2010) A striking pattern of cortisol non-responsiveness to psychosocial stress in patients with panic disorder with concurrent normal cortisol awakening responses. Psychoneuroendocrinol 35:414–421

Petrowski K, Wintermann GB, Kirschbaum C et al (2012) Dissociation between ACTH and cortisol response in DEX-CRH test in patients with panic disorder. Psychoneuroendocrinol 37:1199–1208

Petrowski K, Wintermann GB, Schaarschmidt M et al (2013) Blunted salivary and plasma cortisol response in patients with panic disorder under psychosocial stress. Int J Psychophysiol doi:pii: S0167-8760(13):00017–2. 10.1016/j.ijpsycho.2013.01.002. (Epub ahead of print)

Phillips AC, Batty GD, Gale CR et al (2011) Major depressive disorder, generalised anxiety disorder, and their comorbidity: associations with cortisol in the vietnam experience study. Psychoneuroendocrinol 36:682–690

Pomara N, Willoughby LM, Sidtis JJ et al (2005) Cortisol response to diazepam: its relationship to age, dose, duration of treatment, and presence of generalized anxiety disorder. Psychopharmacol (Berl) 178:1–8

Potts NL, Davidson JR, Krishnan KR et al (1991) Levels of urinary free cortisol in social phobia. J Clin Psychiatry 52(Suppl):41–42

Rapaport MH, Risch SC, Gillin JC et al (1991) The effects of physostigmine infusion on patients with panic disorder. Biol Psychiatry 29:658–664

Roelofs K, van Peer J, Berretty E et al (2009) Hypothalamus-pituitary-adrenal axis hyperresponsiveness is associated with increased social avoidance behavior in social phobia. Biol Psychiatry 65:336–343

Rosnick CB, Rawson KS, Butters MA et al (2013) Association of cortisol with neuropsychological assessment in older adults with generalized anxiety disorder 17:432–440

Roy-Byrne PP, Uhde TW, Post RM et al (1986) The corticotropin-releasing hormone stimulation test in patients with panic disorder. Am J Psychiatry 143:896–899

Roy-Byrne PP, Cowley DS, Hommer D et al (1991) Neuroendocrine effects of diazepam in panic and generalized anxiety disorders. Biol Psychiatry 30:73–80

Sasaki I, Akiyoshi J, Sakurai R, et al (1996) Carbon dioxide induced panic attack in panic disorder in Japan. Prog Neuropsychopharmacol Biol Psychiatry 20:1145–1157

Seier FE, Kellner M, Yassouridis A et al (1997) Autonomic reactivity and hormonal secretion in lactate-induced panic attacks. Am J Physiol 272(6 Pt 2):H2630–H2638

Schittecatte M, Garcia-Valentin J, Charles G et al (1995) Efficacy of the clonidine REM suppression test (CREST) to separate patients with major depression from controls; a comparison with three currently proposed biological markers of depression. J Affect Disord 33:151–157

Schruers K, van Diest R, Nicolson N et al (2002) L-5-hydroxytryptophan induced increase in salivary cortisol in panic disorder patients and healthy volunteers. Psychopharmacology 161:365–369

Schweizer EE, Swenson CM, Winokur A et al (1986) The dexamethasone suppression test in generalised anxiety disorder. Br J Psychiatry 149:320–322

Seddon K, Morris K, Bailey J et al (2011) Effects of 7.5 % CO_2 challenge in generalized anxiety disorder. J Psychopharmacol 25:43–51

Sheehan DV, Claycomb JB, Surman OS et al (1983) Panic attacks and the dexamethasone suppression test. Am J Psychiatry 140:1063–1064

Shlik J, Aluoja A, Vasar V et al (1997) Effects of citalopram treatment on behavioural, cardiovascular and neuroendocrine response to cholecystokinin tetrapeptide challenge in patients with panic disorder. J Psychiatry Neurosci 22:332–340

Shlik J, Maron E, Tru I et al (2004) Citalopram challenge in social anxiety disorder. Int J Neuropsychopharmacol 7:177–182

Sinha SS, Coplan JD, Pine DS et al (1999) Panic induced by carbon dioxide inhalation and lack of hypothalamic-pituitary-adrenal axis activation. Psychiatry Res 86:93–98

Siegmund A, Köster L, Meves AM et al (2011) Stress hormones during flooding therapy and their relationship to therapy outcome in patients with panic disorder and agoraphobia. J Psychiatr Res 45:339–346

Soravia LM, Heinrichs M, Aerni A et al (2006) Glucocorticoids reduce phobic fear in humans. Proc Natl Acad Sci U S A 103:5585–5590

Staufenbiel SM, Penninx BW, Spijker AT et al (2013) Hair cortisol, stress exposure, and mental health in humans: a systematic review. Psychoneuroendocrinol 38:1220–1235

Stein MB, Uhde TW (1988) Cortisol response to clonidine in panic disorder: comparison with depressed patients and normal controls. Biol Psychiatry 24:322–330

Steudte S, Stalder T, Dettenborn L et al (2011) Decreased hair cortisol concentrations in generalised anxiety disorder. Psychiatry Res 186(2–3):310–314

Stones A, Groome D, Perry D et al (1999) The effect of stress on salivary cortisol in panic disorder patients. J Affect Disord 52(1–3):197–201

Ströhle A, Kellner M, Yassouridis A et al (1998) Effect of flumazenil in lactate-sensitive patients with panic disorder. Am J Psychiatry 155:610–612

Ströhle A, Holsboer F, Rupprecht R (2000) Increased ACTH concentrations associated with cholecystokinin tetrapeptide-induced panic attacks in patients with panic disorder. Neuropsychopharmacol 22:251–256

Tafet GE, Idoyaga-Vargas VP, Abulafia DP et al (2001) Correlation between cortisol level and serotonin uptake in patients with chronic stress and depression. Cogn Affect Behav Neurosci 1:388–393

Tafet GE, Feder DJ, Abulafia DP et al (2005) Regulation of hypothalamic-pituitary-adrenal activity in response to cognitive therapy in patients with generalized anxiety disorder. Cogn Affect Behav Neurosci 5:37–40

Tancer ME, Mailman RB, Stein MB et al (1994–1995) Neuroendocrine responsivity to monoaminergic system probes in generalized social phobia. Anxiety 1:216–223

Targum SD, Marshall LE (1989) Fenfluramine provocation of anxiety in patients with panic disorder. Psychiatry Res 28:295–306

Targum SD (1990) Differential responses to anxiogenic challenge studies in patients with major depressive disorder and panic disorder. Biol Psychiatry 28:21–34

Tiller JW, Biddle N, Maguire KP et al (1988) The dexamethasone suppression test and plasma dexamethasone in generalized anxiety disorder. Biol Psychiatry 23:261–270

Uhde TW, Joffe RT, Jimerson DC et al (1988) Normal urinary free cortisol and plasma MHPG in panic disorder: clinical and theoretical implications. Biol Psychiatry 23:575–585

Uhde TW, Tancer ME, Gelernter CS et al (1994) Normal urinary free cortisol and postdexamethasone cortisol in social phobia: comparison to normal volunteers. J Affect Disord 30:155–161

van der Wee NJ, Fiselier J, van Megen HJ et al (2004) Behavioural effects of rapid intravenous administration of meta-chlorophenylpiperazine in patients with panic disorder and controls. Eur Neuropsychopharmacol 14:413–417 van Duinen MA, Schruers KR, Jaegers E et al (2004) Salivary cortisol in panic: are males more vulnerable? Neuro Endocrinol Lett 25:386–390

van Duinen MA, Schruers KR, Maes M et al (2005) CO_2 challenge results in hypothalamic-pituitary-adrenal activation in healthy volunteers. J Psychopharmacol 19:243–247

van Duinen MA, Schruers KR, Maes M et al (2007) CO_2 challenge induced HPA axis activation in panic. Int J Neuropsychopharmacol 10:797–804

van Duinen MA, Schruers KR, Griez EJ (2010) Desynchrony of fear in phobic exposure. J Psychopharmacol 24:695–699

van Megen HJ, Westenberg HG, Den Boer JA et al (1996) The panic-inducing properties of the cholecystokinin tetrapeptide CCK4 in patients with panic disorder. Eur Neuropsychopharmacol 6:187–194

van Peer JM, Spinhoven P, van Dijk JG et al (2009) Cortisol-induced enhancement of emotional face processing in social phobia depends on symptom severity and motivational context. Biol Psychol 81:123–130

Van Veen JF, Van der Wee NJ, Fiselier J et al (2007) Behavioural effects of rapid intravenous administration of meta-chlorophenylpiperazine (m-CPP) in patients with generalized social anxiety disorder, panic disorder and healthy controls. Eur Neuropsychopharmacol 17:637–642

van Peer JM, Spinhoven P, Roelofs K (2010) Psychophysiological evidence for cortisol-induced reduction in early bias for implicit social threat in social phobia. Psychoneuroendocrinol 35:21–32

van Veen JF, van Vliet IM, Derijk RH et al (2008) Elevated alpha-amylase but not cortisol in generalized social anxiety disorder. Psychoneuroendocrinol 33:1313–1321

van Veen JF, van Vliet IM, de Rijk RH et al (2009) Tryptophan depletion affects the autonomic stress response in generalized social anxiety disorder. Psychoneuroendocrinol 34:1590–1594

van Vliet IM, Slaap BR, Westenberg HG et al (1996) Behavioral, neuroendocrine and biochemical effects of different doses of 5-HTP in panic disorder. Eur Neuropsychopharmacol 6:103–110

van West D, Claes S, Sulon J et al (2008) Hypothalamic-pituitary-adrenal reactivity in prepubertal children with social phobia. J Affect Disord 111:281–290

Villacres EC, Hollifield M, Katon WJ et al (1987) Sympathetic nervous system activity in panic disorder. Psychiatry Res 21:313–321

Vreeburg SA, Zitman FG, van Pelt J et al (2010) Salivary cortisol levels in persons with and without different anxiety disorders. Psychosom Med 72:340–347

Wedekind D, Bandelow B, Broocks A et al (2000) Salivary, total plasma and plasma free cortisol in panic disorder. J Neural Trans 107:831–837

Wedekind D, Sprute A, Broocks A et al (2008) Nocturnal urinary cortisol excretion over a randomized controlled trial with paroxetine vs. placebo combined with relaxation training or aerobic exercise in panic disorder. Curr Pharm Des 14:3518–3524

Westberg P, Modigh K, Lisjö P et al (1991) Higher postdexamethasone serum cortisol levels in agoraphobic than in nonagoraphobic panic disorder patients. Biol Psychiatry 30:247–256

Wetzler S, Asnis GM, DeLecuona JM et al (1996) Serotonin function in panic disorder: intravenous administration of meta-chlorophenylpiperazine. Psychiatry Res 64:77–82

Westenberg HG, den Boer JA (1989) Serotonin function in panic disorder: effect of 1-5-hydroxytryptophan in patients and controls. Psychopharmacology 98:283–285

Whiteford HA, Evans L (1984) Agoraphobia and the dexamethasone suppression test: atypical depression? Aust N Z J Psychiatry 18:374–377

Wiedemann K, Jahn H, Yassouridis A et al (2001) Anxiolytic-like effects of atrial natriuretic peptide on cholecystokinin tetrapeptide-induced panic attacks: preliminary findings. Arch Gen Psychiatry 58:371–377

Woods SW, Charney DS, McPherson CA et al (1987) Situational panic attacks. Behavioral, physiologic, and biochemical characterization. Arch Gen Psychiatry 44:365–375

Stress, Schizophrenia and Bipolar Disorder

Melissa J. Green, Leah Girshkin, Nina Teroganova and Yann Quidé

Abstract The role of stress in precipitating psychotic episodes in schizophrenia and bipolar disorder has long been acknowledged. However, the neurobiological mechanism/s of this association have remained elusive. Current neurodevelopmental models of psychosis implicate early dysfunction in biological systems regulating hypothalamic–pituitary–adrenal axis and immune function, with long-term effects on the development of the brain networks responsible for higher order cognitive processes and stress reactivity in later life. There is also increasing evidence of childhood trauma in psychosis, and its impact on the development of brain systems regulating stress. These findings are emerging in the context of a new era of epigenetic methods facilitating the study of environmental effects on gene expression. The evidence is thus converging: exposure to stress at critical periods in life may be an important factor in the development of the brain dysfunction that *represents* psychosis vulnerability, rather than merely interacting with an independent 'biological vulnerability' to manifest in psychosis.

Keywords Psychosis · Stress-vulnerability · Hypothalamic–pituitary–adrenal (HPA) axis · Epigenetics · Trauma

---

M. J. Green · L. Girshkin · N. Teroganova
School of Psychiatry, University of New South Wales, Kensington, NSW, Australia

M. J. Green · L. Girshkin · Y. Quidé
Schizophrenia Research Institute, Liverpool Street, Darlinghurst, NSW, Australia

M. J. Green
Black Dog Institute, Prince of Wales Hospital, Hospital Road,
Randwick, NSW, Australia

M. J. Green (✉)
Research Unit for Schizophrenia Epidemiology, St. Vincent's Hospital,
O'Brien Centre, Level 4, Darlinghurst, NSW 2031, Australia
e-mail: melissa.green@unsw.edu.au

Contents

1	The 'Stress-Vulnerability' Model of Psychosis	218
	1.1 Stress and the Development of Psychosis	219
2	Neurobiology of Stress	220
3	Levels of Cortisol in Schizophrenia and Bipolar Disorder	221
4	Inflammation and Stress	223
	4.1 Pro-Inflammatory Markers in Schizophrenia and Bipolar Disorder	224
5	Early Life Stress and Psychosis	225
6	Genetic Interactions with Early Traumatic Experiences	226
7	Summary and Conclusions	227
References		228

1 The 'Stress-Vulnerability' Model of Psychosis

Schizophrenia and bipolar disorder are severe neuropsychiatric disorders that share some symptoms and cognitive deficits, and are likely caused by the interaction of multiple biological and environmental factors. Heritability estimates for both disorders approximate 80 % (van Winkel et al. 2008b; Sullivan et al. 2003; McGuffin et al. 2003), suggesting a strong genetic component that is not necessarily expressed with complete penetrance. In the largest genetic epidemiology study of heritability patterns to date, it was shown that the biological relatives of both schizophrenia and bipolar disorder had increased risk for both disorders, with an estimated 30–40 % shared genetic risk factors, and 3–6 % shared environmental risk factors (Lichtenstein et al. 2009). Notably, this evidence has emerged within an era of unparalleled genomic advances implicating common genetic loci in the development of the traditionally distinct 'non-affective' and 'affective' psychoses (Moskvina et al. 2009; O'Donovan et al. 2008; Craddock et al. 2005).

However, there remains a high level of interest in elucidating the undoubtedly complex effects of environmental stressors acting in concert with biological vulnerability for psychosis. Understanding how stress impacts the brain—its development and daily functioning—thus remains key to understanding the aetiology of psychosis. The treatment of stress-related features of illness will likely prove vital in preventing relapse in established cases, and may also assist in averting the onset of frank psychosis in vulnerable individuals.

Clinically, schizophrenia manifests in an episodic, and often deteriorating, course of illness. The overt expression illness includes phases in which the hallmark 'positive' symptoms of psychosis predominate (e.g. delusions, hallucinations and disorganisaton of thought and behaviour), while persistent 'negative' symptoms (impaired motivation and affect, social withdrawal, poverty of speech and impaired cognition) commonly increase over the course of illness, culminating in long-term disabling interpersonal and functional impairments (Liddle 1987). The diagnosis of schizophrenia is confined to patients who experience remittent symptoms for a minimum of 6 months, with at least 1 month of persistent psychosis (APA 2013). Somewhat artificially, schizophrenia has been distinguished from schizoaffective disorder by the presence of significant mood symptoms, interspersed with periods of psychotic symptoms that account for a significant proportion of illness, and which occur also during periods without mood symptoms (APA 2013).

In contrast, bipolar disorder has been primarily understood as a mood disorder characterised by a (similarly) episodic course of illness that includes chronic and recurring episodes of mania (elated mood) and depression (low mood), interspersed by 'euthymic' periods in which the individual is neither affected by extremely high or low moods (Manji et al. 2003). Mania includes feelings of elation, irritability, impulsive behaviour, decreased need for sleep and can mimic a schizophrenia-like psychosis in severe cases (Manji et al. 2003; Berk et al. 2007). While mania and psychosis technically do not equate, psychotic symptoms (such as delusions) frequently occurs during mood episodes, with 20–50 % of patients with acute bipolar mania displaying symptoms of psychosis (Pope and Lipinski 1978). Despite many clinical and neurocognitive features in common (Tamminga et al. 2013; Hill et al. 2013), these mood and psychotic disorders remain as distinct categories of illness in the current DSM-V, with growing acknowledgement of the unequivocal evidence for shared genetic vulnerability and environmental factors contributing to their development.

1.1 Stress and the Development of Psychosis

Typically, overt psychotic symptoms emerge in late adolescence or early adulthood (Kessler et al. 2007) and often the onset of the first (and subsequent) psychotic episode/s can be linked to a significant life stressor (Canton and Fraccon 1985). Retrospective studies show that patients with schizophrenia tend to experience increased numbers of life events, especially before the onset of an acute episode (Bebbington et al. 1993; Canton and Fraccon 1985). Exposure to 'life events' may increase response to stress and predispose to subsequent reactions on later exposures (van Winkel et al. 2008b). Prospective studies of life events and psychosis reveal that, while the number of life events experienced by individual at high risk of psychosis are not particularly elevated, these individuals feel they do not cope with the stressors as well and report greater distress than controls (Mason et al. 2004; Phillips et al. 2006). Stress and inadequate social support have also been shown to predict recurrence in bipolar disorder (Cohen et al. 2004).

This accumulation of evidence might suggest that a core feature of psychosis may involve an aberrant neurobiological response to stress. At the same time, neurodevelopmental models assert the likely impact of early life stress on the developing brain, with accumulating evidence implicating the effect of early life exposures to numerous kinds of environmental insult during pre- or post-natal periods, and/or during childhood or adolescence (Demjaha et al. 2012; Murray et al. 2004). These models highlight both similarities and distinctions in the neurodevelopmental trajectories of schizophrenia and bipolar disorder, that nevertheless point toward aberrant development of stress-related brain circuitry in the aetiology of both 'affective' and 'non-affective' forms of psychosis.

2 Neurobiology of Stress

The primary stress response system of the brain involves components of the hypothalamic–pituitary–adrenal (HPA) axis, which function together with multiple bodily systems to enable individual adaptation to the stressful/environmental situation (McEwen and Seeman 1999). After exposure to a stressor, release of serotonin from the amygdala stimulates the secretion of corticotropin-releasing hormone (CRH) and vasopressin through the medial parvocellular portion of the paraventricular nucleus (PVN) of the hypothalamus, which in turn stimulates the production of adrenocorticotropic hormone (ACTH) from the anterior pituitary gland (Jacobson and Sapolsky 1991; Munck et al. 1984; Sapolsky et al. 1986). ACTH acts on the adrenal cortices, stimulating the production of glucocorticoids, such as cortisol. Glucocorticoids in turn regulate their own release directly by action on both the hypothalamus (CRH) and pituitary gland (ACTH) via negative feedback cycles.

Within this system, cortisol is primarily involved in driving the stress response. It binds to glucocorticoid receptors (GR) to both induce and restrain the stress response. These receptors are expressed throughout the brain with concentrated expression in the PVN, hippocampal area and dentate gyrus, amygdala and lateral septum and prefrontal cortex (PFC) (Pillai et al. 2012). Mineralocorticoid receptors (MR), expressed only in the hippocampus and the lateral septum, have a tenfold greater affinity for cortisol and are primarily active at basal level and important in modulating the circadian pulsatile rhythm of cortisol (Deuschle et al. 1998; Heuser et al. 2000).

Chronic exposure to glucocorticoids results in significant changes in neurophysiology (Belanoff et al. 2001), through genomic and non-genomic pathways, [see (Joels et al. 2012) for review] in brain regions which plays an important role in memory and cognition (Eichenbaum et al. 1992; Herman et al. 2005). Elevated levels of glucocorticoids or cumulative exposure can lead to hippocampal degeneration (McEwen and Sapolsky 1995; Sapolsky et al. 1986). The hippocampus plays an important role in explicit memory, and the interaction between glucocorticoids and the hippocampus may explain its effect on cognition (Belanoff et al. 2001). Indeed, longitudinal study of associations between childhood cognitive performance and later cortisol levels in adulthood have shown that lower cognitive ability in childhood predicted lower morning levels of cortisol and a blunted cortisol awakening response later in life (Power et al. 2008). Prefrontal cortex and amygdala housing of substantial corticosteroid receptors may similarly account for variability in cognitive domains; executive function, working memory and emotion regulation (McCormick et al. 2007; Wingenfeld et al. 2011).

Both abnormally high and low cortisol levels, reflecting aberrant HPA function, have been linked with early life stressors including severe childhood abuse, and post-institutionalisation (Doom et al. 2013; Quevedo et al. 2012), with recent evidence of heightened cortisol levels in adult schizophrenia participants with a history of childhood trauma (Braehler et al. 2005). In accordance with the diathesis stress model, evidence also suggests there may be an ability in healthy individuals to 'bounce back' from childhood trauma, confirmed by an attenuated Dexamethasone/CRH response in individuals exposed to early trauma (Klaassens 2010).

3 Levels of Cortisol in Schizophrenia and Bipolar Disorder

Cortisol levels have been found to be higher in chronic sufferers of schizophrenia compared to healthy controls, irrespective of age (Muck-Seler et al. 2004). Similarly, higher cortisol levels are evident in bipolar disorder in the morning and afternoon, relative to healthy controls (Gallagher et al. 2007). Possible explanations of these results come from evidence incorporating symptom severity and disease progression. One study found that first-episode psychosis participants' reduction in cortisol over a 12-week period was directly related to improvements in depressive, negative and positive psychotic symptoms (Garner et al. 2011). Similarly, a longitudinal study found that adolescents followed for 4 years from the onset of prodrome to psychosis demonstrated a pronounced increase in cortisol secretion over the course of the study period (Walker et al. 2010).

Several studies have also shown that repeated administration of corticosteroids to participants initially free of psychiatric illness results in approximately 25 % of individuals meeting diagnostic criteria for mania (Brown et al. 2002; Bolanos et al. 2004; Naber et al. 1996); in addition, those with a greater number of episodes demonstrated more dysfunctional cortisol patterns than those with relatively less severe bipolar disorder. These studies suggest similar relationships to those in schizophrenia where illness stage, and symptom severity, may be moderators of cortisol dysfunction.

One important factor that may contribute to elevated cortisol levels in psychotic disorders is psychotropic medication. A recent meta-analysis also shows that, while medicated schizophrenia patients had elevated cortisol compared to healthy controls, the cortisol levels of patients that were medication-free were greater still (Girshkin et al. 2014). This meta-analysis also found no increase in cortisol level in first-episode medication-naïve participants compared to healthy controls, consistent with reports of normal pituitary volume in first-episode participants before after antipsychotic treatment (Gruner et al. 2012). These findings in first-episode samples suggest that antipsychotic medication may have less impact on cortisol levels in the early stage of illness. Other studies have identified an increase in cortisol from baseline after a 12-day period of withdrawal from psychotropic treatment (Naber et al. 1985; Albus et al. 1985). Unlike antipsychotics, mood stabilisers, including lithium carbonate, have been shown to increase cortisol levels in a dose dependent manner (Platman and Fieve 1968; Bschor et al. 2011; Eroglu et al. 1979).

The manipulation of cortisol levels via the administration of glucocorticoid antagonists provides interesting results at odds with the growing list of similarities among schizophrenia and bipolar disorder: in a series of elegant investigations using mifepristone (a glucocorticoid antagonist), both schizophrenia and bipolar disorder individuals demonstrated a rise in cortisol and decrease in brain-derived neurotrophic factor (BDNF) in response to administration, compared to controls; however, it was only the schizophrenia participants whose changes in cortisol were associated with peripheral BDNF levels (Mackin et al. 2007). Another study by this group found that, following administration of mifepristone, the cortisol awakening response of bipolar disorder participants increased from baseline, and predicted improvement in spatial working memory over the course of treatment (Watson et al. 2012; Young et al. 2004). There was no similar effect of mifepristone on cognition in schizophrenia, despite the effects of mifepristone in increasing plasma levels of cortisol in schizophrenia (Gallagher et al. 2005). Together, these investigations implicate differential neurobiological mechanisms underlying the aberrant stress responses reflected in their abnormal heightened cortisol levels. Further work is clearly required to disentangle the similarities from differences with respect to abnormal HPA function in the psychotic and mood disorders.

The biological mechanisms of elevated cortisol levels in schizophrenia and bipolar disorder may be associated with the predisposition to psychosis, environmental effects or an interaction of the two (Wang et al. 2011; Aina 2013; Perroud et al. 2011). Predisposing genetic factors may include common variants on single nucleotide polymorphisms (SNPs) in genes associated with cortisol metabolism (SRD5A2) (Steen et al. 2010), the regulation of cortisol (glucocorticoid receptor, NR3C1) (Schatzberg et al. 2014), dopamine catabolism (catechol-o-methyltransfease COMT; dopamine D4 receptor gene (DRD4) (Jabbi et al. 2007), inhibitory neurotransmittors (GABA α6 receptor subunit gene; GABRA6) (Uhart et al. 2004) and stress-vulnerability (serotonin transporter-linked polymorphic region; 5-HTTLPR) (Miller et al. 2013). Similarly, environmental factors such as substance abuse (Lopez-Larson et al. 2011; Gavrieli et al. 2011), sleep deprivation (Spiegel et al. 1999), dietary changes (Cheng and Li 2012), lower socioeconomic status (Rudolph et al. 2014) and a lower level of education (Karlamangla et al. 2013) may contribute to the increased cortisol. While it remains unclear whether elevated cortisol levels are a risk factor for these disorders or a consequence of onset, recent studies suggest that it may be an interaction of the two (Wang et al. 2011; Aina 2013; Perroud et al. 2011).

Notably, the meta-analysis by Girshkin et al. (2014) revealed a positive relationship between duration of illness and cortisol levels in schizophrenia; that is, the magnitude of increase in cortisol level in established schizophrenia was greater than that for first-episode psychosis, relative to healthy controls. Increasing cortisol with illness chronicity may be accounted for by several factors such as an inability to habituate to stimuli that, therefore, are perceived as salient (potentially threatening), and therefore tax the HPA system (Kirschbaum et al. 1995; Braunstein-Bercovitz et al. 2001). This constant experience of events perceived to be salient may create lasting neural changes resulting hypersensitivity to external stimuli (J. Wang et al. 2005; Tognin et al. 2012; Zimmermann et al. 2011; McEwen 2000; Rao et al. 1989; Starkman et al. 1992; Lupien et al. 1998; Tessner et al. 2007; McGowan et al. 2009), which over time, may amass to a state of extreme sensitivity to both internal and externally driven stress (De Kloet et al. 1998; Holsboer 2000). The heightened levels of cortisol in bipolar disorder and schizophrenia not seen in earlier stages of illness may thus be due to a lack of accumulated stressful experiences, in combination with the effects of continued medication.

4 Inflammation and Stress

Stress-mediated immune activation is well known to affect HPA axis function, and can result in chronic inflammation capable of altering neural networks and brain morphology (van Winkel et al. 2008b; McEwen 2007). Notably, one of the most commonly implicated genetic markers of schizophrenia is in the Major Histocompatibility Complex (MHC) locus, which encodes more than 400 genes critical to immune functions (Corvin and Morris 2014); MHC-mediated immune molecules are highly expressed in neurons and regulate key aspects of brain development (McAllister 2014).

Inflammation has been linked to psychosis (Bergink et al. 2014), as well as depression, mania and cognitive impairment (Laan et al. 2009; Wadee et al. 2002; Larson and Dunn 2001). Interestingly, monocyte genomic profiling of mRNA has shown that inflammatory markers in bipolar disorder were overlapping with almost all those associated with schizophrenia (Drexhage et al. 2010). Positron Emission Tomography (PET) imaging studies in schizophrenia patients experiencing psychosis have also identified an increase of activated microglia in the hippocampus, suggesting focal neuroinflammation of this region of the brain (Doorduin et al. 2009).

The immunosuppressive and anti-inflammatory actions of antipsychotic drugs such as chlorpromazine suggest that inflammation has a role in altering central nervous system (CNS) function (Drzyzga et al. 2006). Similarly, antidepressants such as imipramine work by suppressing pro-inflammatory responses (cytokine production) and increasing BDNF (Sairanen et al. 2005; Kenis and Maes 2002). In healthy individuals pro-inflammatory cytokines modulate apoptotic and neurotrophic processes as well as prevent morphological brain changes (de Vries et al. 1997). However, evidence for immune dysregulation in schizophrenia and bipolar disorder suggests that inflammatory mechanisms shift from being neuroprotective to neurotoxic (Potvin et al. 2008; Kapczinski et al. 2011). As interactions between immune, gene and neural networks are complex, it remains unclear to what extent the pro-inflammatory response might contribute to the neuropathology of psychosis; however, a number of interesting models are emerging as new evidence shapes our understanding of these relationships (Girgis et al. 2014; Corvin and Morris 2014; Bergink et al. 2014).

4.1 Pro-Inflammatory Markers in Schizophrenia and Bipolar Disorder

Cytokines are pro-inflammatory markers which accompany immune responses to stress and have diverse roles in immunomodulation and cellular function (Kunze et al. 2013). Elevated levels of cytokines are evident in the peripheral blood of schizophrenia and bipolar disorder patients, relative to healthy controls (Kunze et al. 2013; Kim et al. 2007). Specifically, increased levels of serum interleukin (IL)-1, IL-2, IL6 and tumour necrosis factor-alpha (TNF-α) have been found in schizophrenia (Potvin et al. 2008; Theodoropoulou et al. 2001), and elevated plasma cytokine levels of IL-2, IL-4, IL-6, IL-8 and TNF-α are also evident in bipolar disorder (Brietzke et al. 2011; O'Brien et al. 2006).

Chronically elevated levels of cytokines can cause increased oxidative stress and alter neuronal function (Brietzke et al. 2011; Schafers and Sorkin 2008), and may also contribute to the grey matter loss seen in schizophrenia and bipolar disorder (Viviani et al. 2004). Pro-inflammatory cytokines can also stimulate excess secretion of corticotrophin-releasing hormone, which has an inverse relationship with the adaptive stress response (Sauvage and Steckler 2001). Thus, long-term increase in cytokines can further exacerbate the molecular changes associated with stress-induced HPA dysfunction.

In a balanced immune system, cytokines provide neurotrophic support to neurons and play a role in learning and memory via their effect on the hippocampus (Wilson et al. 2002). However, in bipolar disorder, cytokines have been shown to be positively correlated with symptom severity during active manic and depressive episodes (Kim et al. 2007; O'Brien et al. 2006) and with paranoid–hallucinatory symptoms in schizophrenia (Muller et al. 1999). As such, altered levels of cytokines implicate inflammatory processes in the pathophysiology of schizophrenia and bipolar disorder. The timing of inflammation (particularly whether before or after the onset of illness), and the precise effects on the brain are yet to be fully explicated. Perhaps not surprisingly, on the basis of the current evidence, preliminary studies of the use of common anti-inflammatory pharmacological agents to improve psychotic symptoms show promise as adjunct treatments for schizophrenia (Sommer et al. 2014).

5 Early Life Stress and Psychosis

Childhood trauma and other types of adversity are now well established as significant risk factors for the development of several mental disorders, including schizophrenia and bipolar disorder (Janssen et al. 2004; Kessler et al. 2010; Read and Bentall 2010; Cutajar et al. 2010; Etain et al. 2010; Hyun et al. 2000; Read et al. 2005; Matheson et al. 2013; Varese et al. 2012; Schafer and Fisher 2011). Childhood adversity refers to a number of experiences in the early stages of life including maltreatment (encompassing physical, sexual or emotional abuse and various forms of neglect), parental loss or divorce, parental substance abuse and poverty (Rosenberg et al. 2007). While sexual abuse has been reported as a significant risk factor for psychosis alone (Cutajar et al. 2010), a recent meta-analysis shows no evidence that any particular type of trauma is a stronger predictor of psychosis than the others (Varese et al. 2012). However, one recent study reports higher rates of emotional neglect in psychotic patients, with higher rates of physical abuse and neglect differentiating individuals with schizophrenia from those with bipolar disorder (Larsson et al. 2013).

A recent review of the neurobiological and clinical features of maltreatment across a variety of mental disorders concludes that sufficient evidence points towards the utility of examining subtypes of cross-disorder clinical cases who share a history of childhood trauma, as a phenotypic specialisation of environmental adversity, or *'ecophenotype'* (Teicher and Samson 2013). For example, neuroimaging studies of maltreated individuals within various clinical categories included in this review, report alterations in the size and integrity of the corpus callosum, hippocampus, cerebellum, and primary sensory cortices; as well as in sub-cortical region including the striatum/basal ganglia; and neocortical regions including the anterior cingulate cortex (ACC), the orbitofrontal cortex and the dorsolateral prefrontal cortex. More recently, Teicher et al. (2013) showed that maltreatment was associated with *decreased* connectivity among regions involved in emotion regulation and in theory of mind skills (left ACC, right medial frontal gyrus, right occipital pole and left temporal pole), and *enhanced* centrality among regions involved in emotion perception, self-referential thinking, self-awareness [right superior temporal gyrus (STG), right anterior insula] (Craig 2009). These regions are commonly implicated in the neuropathology of both schizophrenia and bipolar disorder, such that further examination of the effects of stress on these brain systems is warranted.

Only few studies have investigated brain abnormalities associated with childhood adversity in psychotic disorders. In individuals with first-episode of psychosis, exposure to childhood trauma has been associated with worse cognitive function and smaller amygdala (Aas et al. 2012), or decreased hippocampal volume (Hoy et al. 2012). Sexual abuse has been specifically associated with reduced (total) grey matter volume in psychotic patients compared to healthy participants (Sheffield et al. 2013), whereas, in line with Carrion et al. (2001), psychotic cases with a history of childhood trauma showed bilateral reduction of the PFC relative to cases without trauma history. Results of a functional neuroimaging investigation of the effects of childhood adversity on brain networks for working memory, in a combined sample of patients with schizophrenia and bipolar disorder, demonstrate failure to deactivate the posterior cingulate cortex (PCC) in patients with a history of severe childhood trauma; in contrast, psychotic patients without a history of childhood trauma show aberrant brain activation of the visuo-motor/attentional network (pre-post central gyrus, cuneus/visual areas) (Quidé et al. 2014). Interestingly, Quidé et al. also demonstrated the effects of childhood trauma on brain regions involved in salience and threat detection (amygdala, thalamus), as well as directed attention (amygdala, cuneus/lingual gyrus), and emotion regulation (amygdala, thalamus, superior temporal gyrus). Findings from neuroimaging studies thus converge with the implications of neurobiological investigations (reviewed above), in both suggesting that long-term changes in brain systems responsible for salience and threat detection may coincide with chronic HPA axis dysregulation. Determining the primary antecedent of these mechanisms requires further investigation of psychotic samples with due diligence in relation to the characterization of pre-term birth complications and early adverse life events.

6 Genetic Interactions with Early Traumatic Experiences

Several investigations have recently examined the effects of trauma in the context of common genetic variation, as reviewed by van Winkel et al. (2013). Study of the additive interaction of genes and trauma demonstrate worse cognitive functioning in first-episode patients with a history of physical childhood trauma (neglect or abuse) carrying the short-version of the serotonin transporter (5-HTTLPR) gene (Aas et al. 2012); in addition, four studies have shown increased symptoms in psychotic individuals homozygous for the Met allele of the *COMT* Val^{158}Met genotype in response to daily stress or in those with a history of childhood trauma (van Winkel et al. 2008a; Collip et al. 2011; Peerbooms et al. 2012; Green et al. 2014). These and other emerging studies of epigenetic processes highlight the likely interaction of genetic variations with traumatic experiences that may be also affected by epigenetic processes.

Epigenetic mechanisms refer to functionally relevant modifications to the genome that do not involve changes in nucleotide sequence; in contrast to examinations of stable DNA sequences, epigenetic processes are highly dynamic and can be modified by environmental factors (Weaver 2007). While the epigenetic mechanisms influencing the expression of genes in the human brain are yet to be fully understood, a series of elegant studies in rodents have shown methylation changes to sites of DNA in association with maternal rearing behaviours; these studies also demonstrated the ongoing genetic heritability, and potential reversibility, of epigenetically determined brain changes set in motion by early social experiences (Weaver et al. 2004, 2007).

A growing number of investigations of DNA methylation in humans have demonstrated epigenetic regulation of genes relevant to the function of the HPA axis, in the context of childhood trauma (Uher and Weaver 2014). For example, a recent study demonstrates increased risk of developing stress-related psychiatric disorders in association with allele-specific, childhood trauma-dependent DNA demethylation in the FK506 binding site protein 5 (FKBP5) gene, known to regulate glucocorticoid response functions (Klengel et al. 2013); in this study demethylation at this site was also associated with stress-dependent gene transcription alterations, leading to long-term dysregulation of the HPA system and a global effect on the function of immune cells and brain areas associated with stress regulation. Another earlier study showed aberrant methylation of a neuron-specific glucocorticoid receptor (NR3C1) promoter in the human brain, in association with childhood trauma (McGowan et al. 2009). Notably, increased methylation of this site has also been recently associated with the number and severity of childhood maltreatment (sexual, physical, emotional) in patients with bipolar disorder (Perroud et al. 2014). Evidence for aberrant DNA methylation patterns in schizophrenia patients have been reported for genes implicated in molecular genetic and neurobiological investigations of this disorder (Wockner et al. 2014). Interestingly, there has also been recent demonstration of common genomic sites of hypomethylation among twins with bipolar disorder and schizophrenia (Dempster et al. 2013); although this study did not examine associations with childhood trauma, their results imply common biological features of schizophrenia and bipolar disorder that likely result from common environmental effects on existing neurobiological vulnerabilities. These studies converge to suggest a crucial role of epigenetic mechanisms in modulating neurocognitive development that will be important for our understanding and treatment of psychotic disorders (Champagne et al. 2009).

7 Summary and Conclusions

Current neurodevelopmental models of psychosis implicate early dysfunction in biological systems regulating hypothalamic–pituitary–adrenal axis and immune function, with long-term effects on the development of the brain networks responsible for higher order cognitive processes and stress reactivity in later life. There is also increasing evidence of childhood trauma in psychosis, and its impact on the development of brain systems regulating stress, at the same time as new epigenetic methods facilitating the study of environmental effects on gene expression. The evidence is converging to suggest that exposure to stress at critical periods in life may be a critical factor in the development of the brain dysfunction that *represents* psychosis vulnerability, rather than stress merely precipitating the expression of an independent 'biological vulnerability' to psychosis.

Acknowledgments MJG was supported by an Australian Research Council Future Fellowship (FT0991511), the Netherlands Institute for Advanced Study in the Humanities and Social Sciences (NIAS), and the National Health and Medical Research Council (NHMRC, APP1061875; APP630471) during the preparation of this chapter.

References

Aas M, Navari S, Gibbs A, Mondelli V, Fisher HL, Morgan C et al (2012) Is there a link between childhood trauma, cognition, and amygdala and hippocampus volume in first-episode psychosis? Schizophr Res 137(1–3):73–79

Aina O (2013) Adrenal psychosis, A diagnostic challenge. *Endocrinol Metab Synd* 2:115 doi: 10.4172/2161-1017.10001

Albus M, Ackenheil M, Muller-Spahn F (1985) Do norepinephrine serum levels predict the effects of neuroleptic withdrawal on psychopathology? Pharmacopsychiatry 18(1):67–68

APA (2013). Diagnostic and statistical manual of mental disorders, 5th edn. American Psychiatric Association, Washington DC, p 991

Bebbington P, Wilkins S, Jones P, Foerster A, Murray R, Toone B et al (1993) Life events and psychosis. Initial results from the Camberwell collaborative psychosis study. Br J Psychiatry 162(1):72–79

Belanoff JK, Gross K, Yager A, Schatzberg AF (2001) Corticosteroids and cognition. J Psychiatr Res 35(3):127–145

Bergink V, Gibney SM, Drexhage HA (2014) Autoimmunity, inflammation, and psychosis: a search for peripheral markers. Biol Psychiatry 75(4):324–331

Berk M, Malhi GS, Cahill C, Carman AC, Hadzi-Pavlovic D, Hawkins MT et al (2007) The bipolar depression rating scale (BDRS): its development, validation and utility. Bipolar Disord 9(6):571–579

Bolanos SH, Khan DA, Hanczyc M, Bauer MS, Dhanani N, Brown ES (2004) Assessment of mood states in patients receiving long-term corticosteroid therapy and in controls with patient-rated and clinician-rated scales. Ann Allergy Asthma Immunol 92(5):500–505

Braehler C, Holowka D, Brunet A, Beaulieu S, Baptista T, Debruille JB et al (2005) Diurnal cortisol in schizophrenia patients with childhood trauma. Schizophr Res 79(2–3):353–354

Braunstein-Bercovitz H, Dimentman-Ashkenazi I, Lubow RE (2001) Stress affects the selection of relevant from irrelevant stimuli. Emotion 1(2):182–192

Brietzke E, Stabellini R, Grassis-Oliveira R, Lafer B (2011) Cytokines in bipolar disorder: recent findings, deleterious effects but promise for future therapeutics. CNS Spectr 16(7):157–168

Brown ES, Suppes T, Khan DA, Carmody TJ 3rd (2002) Mood changes during prednisone bursts in outpatients with asthma. J Clin Psychopharmacol 22(1):55–61

Bschor T, Ritter D, Winkelmann P, Erbe S, Uhr M, Ising M et al (2011) Lithium monotherapy increases ACTH and cortisol response in the DEX/CRH test in unipolar depressed subjects. A study with 30 treatment-naive patients. PLoS One 6(11):1–8

Canton G, Fraccon IG (1985) Life events and schizophrenia. A replication Acta Psychiatr Scand 71(3):211–216

Carrion VG, Weems CF, Eliez S, Patwardhan A, Brown W, Ray RD et al (2001) Attenuation of frontal asymmetry in pediatric posttraumatic stress disorder. Biol Psychiatry 50(12):943–951

Champagne DL, de Kloet ER, Joels M (2009) Fundamental aspects of the impact of glucocorticoids on the (immature) brain. Semin Fetal Neonatal Med 14(3):136–142

Cheng LC, Li LA (2012) Flavonoids exhibit diverse effects on CYP11B1 expression and cortisol synthesis. Toxicol Appl Pharmacol 258(3):343–350

Cohen AN, Hammen C, Henry RM, Daley SE (2004) Effects of stress and social support on recurrence in bipolar disorder. J Affect Disord 82(1):143–147

Collip D, van Winkel R, Peerbooms O, Lataster T, Thewissen V, Lardinois M et al (2011) COMT Val158Met-stress interaction in psychosis: role of background psychosis risk. CNS Neurosci Ther 17(6):612–619

Corvin A, Morris DW (2014) Genome-wide Association Studies: Findings at the major histocompatibility complex locus in psychosis. Biol Psychiatry 75(4):276–283

Craddock N, O'Donovan MC, Owen MJ (2005) The genetics of schizophrenia and bipolar disorder: dissecting psychosis. J Med Genet 42(3):193–204

Craig AD (2009) How do you feel–now? The anterior insula and human awareness. Nat Rev Neurosci 10(1):59–70

Cutajar MC, Mullen PE, Ogloff JR, Thomas SD, Wells DL, Spataro J (2010) Schizophrenia and other psychotic disorders in a cohort of sexually abused children. Arch Gen Psychiatry 67(11):1114–1119

De Kloet ER, Vreugdenhil E, Oitzl MS, Joels M (1998) Brain corticosteroid receptor balance in health and disease. Endocr Rev 19(3):269–301

de Vries HE, Kuiper J, de Boer AG, Van Berkel TJ, Breimer DD (1997) The blood-brain barrier in neuroinflammatory diseases. Pharmacol Rev 49(2):143–155

Demjaha A, MacCabe JH, Murray RM (2012) How genes and environmental factors determine the different neurodevelopmental trajectories of schizophrenia and bipolar disorder. Schizophr Bull 38(2):209–214

Dempster E, Viana J, Pidsley R, Mill J (2013) Epigenetic studies of schizophrenia: progress, predicaments, and promises for the future. Schizophr Bull 39(1):11–16

Deuschle M, Weber B, Colla M, Müller M, Kniest A, Heuser I (1998) Mineralocorticoid receptor also modulates basal activity of hypothalamus-pituitary-adrenocortical system in humans. Neuroendocrinology 68(5):355–360

Doom JR, Cicchetti D, Rogosch FA, Dackis MN (2013) Child maltreatment and gender interactions as predictors of differential neuroendocrine profiles. PNEC 38(8):1442–1454

Doorduin J, de Vries EF, Willemsen AT, de Groot JC, Dierckx RA, Klein HC (2009) Neuroinflammation in schizophrenia-related psychosis: a PET study. J Nucl Med 50(11): 1801–1807

Drexhage RC, van der Heul-Nieuwenhuijsen L, Padmos RC, van Beveren N, Cohen D, Versnel MA et al (2010) Inflammatory gene expression in monocytes of patients with schizophrenia: overlap and difference with bipolar disorder. A study in naturalistically treated patients. Int J Neuropsychopharmacol 13(10):1369–1381

Drzyzga L, Obuchowicz E, Marcinowska A, Herman ZS (2006) Cytokines in schizophrenia and the effects of antipsychotic drugs. Brain Behav Immun 20(6):532–545

Eichenbaum H, Otto T, Cohen NJ (1992) The hippocampus—what does it do? Behav Neural Biol 57(1):2–36

Eroglu L, Atamer-Simsek S, Yazici O, Keyer-Uysal M, Yuksel S (1979) A study of the relationship between serum lithium and plasma cortisol levels in manic depressive patients. Br J Clin Pharmacol 8(1):89–90

Etain B, Mathieu F, Henry C, Raust A, Roy I, Germain A et al (2010) Preferential association between childhood emotional abuse and bipolar disorder. J Trauma Stress 23(3):376–383

Gallagher P, Watson S, Smith MS, Ferrier IN, Young AH (2005) Effects of adjunctive mifepristone (RU-486) administration on neurocognitive function and symptoms in schizophrenia. Biol Psychiatry 57(2):155–161

Gallagher P, Watson S, Smith MS, Young AH, Ferrier IN (2007) Plasma cortisol-dehydroepiandrosterone (DHEA) ratios in schizophrenia and bipolar disorder. Schizophr Res 90(1–3):258–265

Garner B, Phassouliotis C, Phillips LJ, Markulev C, Butselaar F, Bendall S et al (2011) Cortisol and dehydroepiandrosterone-sulphate levels correlate with symptom severity in first-episode psychosis. J Psychiatr Res 45(2):249–255

Gavrieli A, Yannakoulia M, Fragopoulou E, Margaritopoulos D, Chamberland JP, Kaisari P et al (2011) Caffeinated coffee does not acutely affect energy intake, appetite, or inflammation but prevents serum cortisol concentrations from falling in healthy men. J nutr 141(4):703–707

Girgis RR, Kumar SS, Brown AS (2014) The cytokine model of schizophrenia: emerging therapeutic strategies. Biol Psychiatry 75(4):292–299

Girshkin L, Matheson SL, Shepherd AM, Laurens KR, Carr VJ, Schofield PR et al (2014) Morning cortisol level in schizophrenia and bipolar disorder: a meta-analysis (in preparation)

Green MJ, Chia TY, Cairns MJ, Wu J, Tooney PA, Scott RJ et al (2014) Catechol-O-methyltransferase (COMT) genotype moderates the effects of childhood trauma on cognition and symptoms in schizophrenia. J Psychiatr Res 49(1):43–50

Gruner P, Christian C, Robinson DG, Sevy S, Gunduz-Bruce H, Napolitano B et al (2012) Pituitary volume in first-episode schizophrenia. Psychiatry Res 203(1):100–102

Herman JP, Ostrander MM, Mueller NK, Figueiredo H (2005) Limbic system mechanisms of stress regulation: hypothalamo-pituitary-adrenocortical axis. Prog Neuropsychopharmacol Biol Psychiatry 29(8):1201–1213

Heuser I, Deuschle M, Weber A, Kniest A, Ziegler C, Weber B et al (2000) The role of mineralocorticoid receptors in the circadian activity of the human hypothalamus-pituitary-adrenal system: effect of age. Neurobiol Aging 21(4):585–589

Hill SK, Reilly JL, Keefe RS, Gold JM, Bishop JR, Gershon ES et al (2013) Neuropsychological impairments in schizophrenia and psychotic bipolar disorder: findings from the bipolar-schizophrenia network on intermediate phenotypes (B-SNIP) study. Am J Psychiatry 170(11):1275–1284

Holsboer F (2000) The corticosteroid receptor hypothesis of depression. Neuropsychopharmacology 23(5):477–501

Hoy K, Barrett S, Shannon C, Campbell C, Watson D, Rushe T et al (2012) Childhood trauma and hippocampal and amygdalar volumes in first-episode psychosis. Schizophr Bull 38(6):1162–1169

Hyun M, Friedman SD, Dunner DL (2000) Relationship of childhood physical and sexual abuse to adult bipolar disorder. Bipolar Disord 2(2):131–135

Jabbi M, Kema IP, Van Der Pompe G, Te Meerman GJ, Ormel J, Den Boer JA (2007) Catechol-o-methyltransferase polymorphism and susceptibility to major depressive disorder modulates psychological stress response. Psychiatr Genet 17(3):183–193

Jacobson L, Sapolsky R (1991) The role of the hippocampus in feedback regulation of the hypothalamic-pituitary-adrenocortical axis. Endocr Rev 12(2):118–134

Janssen I, Krabbendam L, Bak M, Hanssen M, Vollebergh W, de Graaf R et al (2004) Childhood abuse as a risk factor for psychotic experiences. Acta Psychiatr Scand 109(1):38–45

Joels M, Sarabdjitsingh RA, Karst H (2012) Unraveling the time domains of corticosteroid hormone influences on brain activity: rapid, slow, and chronic modes. Pharmacol Rev 64(4):901–938

Kapczinski F, Dal-Pizzol F, Teixeira AL, Magalhaes PV, Kauer-Sant'Anna M, Klamt F et al (2011) Peripheral biomarkers and illness activity in bipolar disorder. J Psychiatr Res 45(2):156–161

Karlamangla AS, Friedman EM, Seeman TE, Stawksi RS, Almeida DM (2013) Daytime trajectories of cortisol: demographic and socioeconomic differences–findings from the national study of daily experiences. PNEC 38(11):2585–2597

Kenis G, Maes M (2002) Effects of antidepressants on the production of cytokines. Int J Neuropsychopharmacol 5(4):401–412

Kessler RC, Amminger GP, Aguilar-Gaxiola S, Alonso J, Lee S, Ustun TB (2007) Age of onset of mental disorders: a review of recent literature. Curr Opin Psychiatry 20(4):359–364

Kessler RC, McLaughlin KA, Green JG, Gruber MJ, Sampson NA, Zaslavsky AM et al (2010) Childhood adversities and adult psychopathology in the WHO world mental health surveys. Br J Psychiatry 197(5):378–385

Kim YK, Jung HG, Myint AM, Kim H, Park SH (2007) Imbalance between pro-inflammatory and anti-inflammatory cytokines in bipolar disorder. J Affect Disord 104(1–3):91–95

Kirschbaum C, Prussner JC, Stone AA, Federenko I, Gaab J, Lintz D et al (1995) Persistent high cortisol responses to repeated psychological stress in a subpopulation of healthy men. Psychosom Med 57(5):468–474

Klaassens ER (2010) Bouncing back-trauma and the HPA-axis in healthy adults. Eur J Psychotraumatology 1. doi:10.3402/ejpt.v1i0.5844

Klengel T, Mehta D, Anacker C, Rex-Haffner M, Pruessner JC, Pariante CM et al (2013) Allele-specific FKBP5 DNA demethylation mediates gene-childhood trauma interactions. Nat Neurosci 16(1):33–41

Kunze U, Bohm G, Groman E (2013) Influenza vaccination in Austria from 1982 to 2011: a country resistant to influenza prevention and control. Vaccine 31(44):5099–5103

Laan W, Smeets H, de Wit NJ, Kahn RS, Grobbee DE, Burger H (2009) Glucocorticosteroids associated with a decreased risk of psychosis. J Clin Psychopharmacol 29(3):288–290

Larson SJ, Dunn AJ (2001) Behavioral effects of cytokines. Brain Behav Immun 15(4):371–387

Larsson S, Andreassen OA, Aas M, Rossberg JI, Mork E, Steen NE et al (2013) High prevalence of childhood trauma in patients with schizophrenia spectrum and affective disorder. Compr Psychiatry 54(2):123–127

Lichtenstein P, Yip BH, Bjork C, Pawitan Y, Cannon TD, Sullivan PF et al (2009) Common genetic determinants of schizophrenia and bipolar disorder in Swedish families: a population-based study. Lancet 373(9659):234–239

Liddle PF (1987) The symptoms of chronic schizophrenia. A re-examination of the positive-negative dichotomy. Br J Psychiatry 151(2):145–151

Lopez-Larson MP, Bogorodzki P, Rogowska J, McGlade E, King JB, Terry J et al (2011) Altered prefrontal and insular cortical thickness in adolescent marijuana users. Behav Brain Res 220(1):164–172

Lupien SJ, de Leon M, de Santi S, Convit A, Tarshish C, Nair NP et al (1998) Cortisol levels during human aging predict hippocampal atrophy and memory deficits. Nat Neurosci 1(1):69–73

Mackin P, Gallagher P, Watson S, Young AH, Ferrier IN (2007) Changes in brain-derived neurotrophic factor following treatment with mifepristone in bipolar disorder and schizophrenia. Aust N Z J Psychiatry 41(4):321–326

Manji HK, Quiroz JA, Payne JL, Singh J, Lopes BP, Viegas JS et al (2003) The underlying neurobiology of bipolar disorder. World Psychiatry 2(3):136–146

Mason O, Startup M, Halpin S, Schall U, Conrad A, Carr V (2004) Risk factors for transition to first episode psychosis among individuals with 'at-risk mental states'. Schizophr Res 71(2):227–237

Matheson SL, Shepherd AM, Pinchbeck RM, Laurens KR, Carr VJ (2013) Childhood adversity in schizophrenia: a systematic meta-analysis. Psychol Med 43(2):225–238

McAllister AK (2014) Major histocompatibility complex I in brain development and schizophrenia. Biol Psychiatry 75(4):262–268

McCormick CM, Lewis E, Somley B, Kahan TA (2007) Individual differences in cortisol levels and performance on a test of executive function in men and women. Physiol Behav 91(1):87–94

McEwen BS (2000) The neurobiology of stress: from serendipity to clinical relevance. Brain Res 886(1–2):172–189

McEwen BS (2007) Physiology and neurobiology of stress and adaptation: central role of the brain. Physiol Rev 87(3):873–904

McEwen BS, Sapolsky RM (1995) Stress and cognitive function. Curr Opin Neurobiol 5(2):205–216

McEwen BS, Seeman T (1999) Protective and damaging effects of mediators of stress. Elaborating and testing the concepts of allostasis and allostatic load. Ann NY Acad Sci 896:30–47

McGowan PO, Sasaki A, D'Alessio AC, Dymov S, Labonte B, Szyf M et al (2009) Epigenetic regulation of the glucocorticoid receptor in human brain associates with childhood abuse. Nat Neurosci 12(3):342–348

McGuffin P, Rijsdijk F, Andrew M, Sham P, Katz R, Cardno A (2003) The heritability of bipolar affective disorder and the genetic relationship to unipolar depression. Arch Gen Psychiatry 60(5):497–502

Miller R, Wankerl M, Stalder T, Kirschbaum C, Alexander N (2013) The serotonin transporter gene-linked polymorphic region (5-HTTLPR) and cortisol stress reactivity: a meta-analysis. Mol Psychiatry 18(9):1018–1024

Moskvina V, Craddock N, Holmans P, Nikolov I, Pahwa JS, Green E et al (2009) Gene-wide analyses of genome-wide association data sets: evidence for multiple common risk alleles for schizophrenia and bipolar disorder and for overlap in genetic risk. Mol Psychiatry 14(3):252–260

Muck-Seler D, Pivac N, Mustapic M, Crncevic Z, Jakovljevic M, Sagud M (2004) Platelet serotonin and plasma prolactin and cortisol in healthy, depressed and schizophrenic women. Psychiatry Res 127(3):217–226

Muller N, Riedel M, Hadjamu M, Schwarz MJ, Ackenheil M, Gruber R (1999) Increase in expression of adhesion molecule receptors on T helper cells during antipsychotic treatment and relationship to blood-brain barrier permeability in schizophrenia. Am J Psychiatry 156(4):634–636

Munck A, Guyre PM, Holbrook NJ (1984) Physiological functions of glucocorticoids in stress and their relation to pharmacological actions. Endocr Rev 5(1):25–44

Murray RM, Sham P, Van Os J, Zanelli J, Cannon M, McDonald C (2004) A developmental model for similarities and dissimilarities between schizophrenia and bipolar disorder. Schizophr Res 71(2–3):405–416

Naber D, Albus M, Burke H, Muller-Spahn F, Munch U, Reinertshofer T et al (1985) Neuroleptic withdrawal in chronic schizophrenia: CT and endocrine variables relating to psychopathology. Psychiatry Res 16(3):207–219

Naber D, Sand P, Heigl B (1996) Psychopathological and neuropsychological effects of 8-days' corticosteroid treatment. A prospective study. PNEC 21(1):25–31

O'Brien SM, Scully P, Scott LV, Dinan TG (2006) Cytokine profiles in bipolar affective disorder: focus on acutely ill patients. J Affect Disord 90(2–3):263–267

O'Donovan MC, Craddock N, Norton N, Williams H, Peirce T, Moskvina V et al (2008) Identification of loci associated with schizophrenia by genome-wide association and follow-up. Nat Genet 40(9):1053–1055

Peerbooms O, Rutten B, Collip D, Lardinois M, Lataster T, Thewissen V et al (2012) Evidence that interactive effects of COMT and MTHFR moderate psychotic response to environmental stress. Acta Psychiatr Scand 125(3):247–256

Perroud N, Dayer A, Piguet C, Nallet A, Favre S, Malafosse A et al (2014) Childhood maltreatment and methylation of the glucocorticoid receptor gene NR3C1 in bipolar disorder. Br J Psychiatry 204:30–35

Perroud N, Paoloni-Giacobino A, Prada P, Olie E, Salzmann A, Nicastro R et al (2011) Increased methylation of glucocorticoid receptor gene (NR3C1) in adults with a history of childhood maltreatment: a link with the severity and type of trauma. Transl Psychiatry 1(12):e59

Phillips LJ, McGorry PD, Garner B, Thompson KN, Pantelis C, Wood SJ et al (2006) Stress, the hippocampus and the hypothalamic-pituitary-adrenal axis: implications for the development of psychotic disorders. Aust N Z J Psychiatry 40(9):725–741

Pillai AG, de Jong D, Kanatsou S, Krugers H, Knapman A, Heinzmann JM et al (2012) Dendritic morphology of hippocampal and amygdalar neurons in adolescent mice is resilient to genetic differences in stress reactivity. PLoS One 7(6):e38971

Platman SR, Fieve RR (1968) Lithium carbonate and plasma cortisol response in the affective disorders. Arch Gen Psychiatry 18(5):591–594

Pope HG Jr, Lipinski JF Jr (1978) Diagnosis in schizophrenia and manic-depressive illness: a reassessment of the specificity of 'schizophrenic' symptoms in the light of current research. Arch Gen Psychiatry 35(7):811–828

Potvin S, Stip E, Sepehry AA, Gendron A, Bah R, Kouassi E (2008) Inflammatory cytokine alterations in schizophrenia: a systematic quantitative review. Biol Psychiatry 63(8):801–808

Power C, Li L, Hertzman C (2008) Cognitive development and cortisol patterns in mid-life: findings from a British birth cohort. PNEC 33(4):530–539

Quevedo K, Johnson A, Loman M, Lafavor T, Gunnar M (2012) The confluence of adverse early experience and puberty on the cortisol awakening response. Int J Behav Dev 36(1):19–28

Quidé Y, Shepherd AM, Gould IC, Rowland JE, O'Reilly N, Girshkin L et al (2014) Effects of childhood trauma on brain function during working memory in psychosis. (in preparation)

Rao VP, Krishnan KR, Goli V, Saunders WB, Ellinwood EH Jr, Blazer DG et al (1989) Neuroanatomical changes and hypothalamo-pituitary-adrenal axis abnormalities. Biol Psychiatry 26(7):729–732

Read J, Bentall R (2010) Schizophrenia and childhood adversity. Am J Psychiatry 167(6): 717–718

Read J, van Os J, Morrison AP, Ross CA (2005) Childhood trauma, psychosis and schizophrenia: a literature review with theoretical and clinical implications. Acta Psychiatr Scand 112(5):330–350

Rosenberg SD, Lu W, Mueser KT, Jankowski MK, Cournos F (2007) Correlates of adverse childhood events among adults with schizophrenia spectrum disorders. Psychiatr Serv 58(2):245–253

Rudolph KE, Gary SW, Stuart EA, Glass TA, Marques AH, Duncko R et al (2014) The association between cortisol and neighborhood disadvantage in a U.S. population-based sample of adolescents. Health Place 25:68–77

Sairanen M, Lucas G, Ernfors P, Castren M, Castren E (2005) Brain-derived neurotrophic factor and antidepressant drugs have different but coordinated effects on neuronal turnover, proliferation, and survival in the adult dentate gyrus. J Neurosci 25(5):1089–1094

Sapolsky RM, Krey LC, McEwen BS (1986) The neuroendocrinology of stress and aging: the glucocorticoid cascade hypothesis. Endocr Rev 7(3):284–301

Sauvage M, Steckler T (2001) Detection of corticotropin-releasing hormone receptor 1 immunoreactivity in cholinergic, dopaminergic and noradrenergic neurons of the murine basal forebrain and brainstem nuclei–potential implication for arousal and attention. Neuroscience 104(3):643–652

Schafer I, Fisher HL (2011) Childhood trauma and psychosis—what is the evidence? Dialogues Clin Neurosci 13(3):360–365

Schafers M, Sorkin L (2008) Effect of cytokines on neuronal excitability. Neurosci Lett 437(3):188–193

Schatzberg AF, Keller J, Tennakoon L, Lembke A, Williams G, Kraemer FB et al (2014) HPA axis genetic variation, cortisol and psychosis in major depression. Mol Psychiatry 19(2):220–227

Sheffield JM, Williams LE, Woodward ND, Heckers S (2013) Reduced gray matter volume in psychotic disorder patients with a history of childhood sexual abuse. Schizophr Res 143(1):185–191

Sommer IE, van Westrhenen R, Begemann MJH, de Witte LD, Leucht S, Kahn RS (2014) Efficacy of anti-inflammatory agents to improve symptoms in patients with schizophrenia: an update. Schizophr Bull 40(1):181–191

Spiegel K, Leproult R, Van Cauter E (1999) Impact of sleep debt on metabolic and endocrine function. Lancet 354(9188):1435–1439

Starkman MN, Gebarski SS, Berent S, Schteingart DE (1992) Hippocampal formation volume, memory dysfunction, and cortisol levels in patients with cushing's syndrome. Biol Psychiatry 32(9):756–765

Steen NE, Tesli M, Kahler AK, Methlie P, Hope S, Barrett EA et al (2010) SRD5A2 is associated with increased cortisol metabolism in schizophrenia spectrum disorders. Prog Neuropsychopharmacol Biol Psychiatry 34(8):1500–1506

Sullivan PF, Kendler KS, Neale MC (2003) Schizophrenia as a complex trait: evidence from a meta-analysis of twin studies. Arch Gen Psychiatry 60(12):1187–1192

Tamminga CA, Ivleva EI, Keshavan MS, Pearlson GD, Clementz BA, Witte B et al (2013) Clinical phenotypes of psychosis in the bipolar-schizophrenia network on intermediate phenotypes (B-SNIP). Am J Psychiatry 170(11):1263–1274

Teicher MH, Anderson CM, Ohashi K, Polcari A (2013) Childhood maltreatment: altered network centrality of cingulate, precuneus, temporal pole and insula. Biol Psychiatry. doi:10.1016/j.biopsych.2013.09.016

Teicher MH, Samson JA (2013) Childhood maltreatment and psychopathology: a case for ecophenotypic variants as clinically and neurobiologically distinct subtypes. Am J Psychiatry 170(10):1114–1133

Tessner KD, Walker EF, Dhruv SH, Hochman K, Hamann S (2007) The relation of cortisol levels with hippocampus volumes under baseline and challenge conditions. Brain Res 1179(1):70–78

Theodoropoulou S, Spanakos G, Baxevanis CN, Economou M, Gritzapis AD, Papamichail MP et al (2001) Cytokine serum levels, autologous mixed lymphocyte reaction and surface marker analysis in never medicated and chronically medicated schizophrenic patients. Schizophr Res 47(1):13–25

Tognin S, Rambaldelli G, Perlini C, Bellani M, Marinelli V, Zoccatelli G et al (2012) Enlarged hypothalamic volumes in schizophrenia. Psychiatry Res 204(2–3):75–81

Uhart M, McCaul ME, Oswald LM, Choi L, Wand GS (2004) GABRA6 gene polymorphism and an attenuated stress response. Mol Psychiatry 9(11):998–1006

Uher R, Weaver ICG (2014) Epigenetic traces of childhood maltreatment in peripheral blood: a new strategy to explore gene-environment interactions. Br J Psychiatry 204:3–5

van Winkel R, Henquet C, Rosa A, Papiol S, Fananas L, De Hert M et al (2008a) Evidence that the COMT(Val158Met) polymorphism moderates sensitivity to stress in psychosis: an experience-sampling study. Am J Med Genet B Neuropsychiatr Genet 147B(1):10–17

van Winkel R, Stefanis NC, Myin-Germeys I (2008b) Psychosocial stress and psychosis. A review of the neurobiological mechanisms and the evidence for gene-stress interaction. Schizophr Bull 34(6):1095–1105

van Winkel R, van Nierop M, Myin-Germeys I, van Os J (2013) Childhood trauma as a cause of psychosis: linking genes, psychology, and biology. Can J Psychiatry 58(1):44–51

Varese F, Smeets F, Drukker M, Lieverse R, Lataster T, Viechtbauer W et al (2012) Childhood adversities increase the risk of psychosis: a meta-analysis of patient-control, prospective- and cross-sectional cohort studies. Schizophr Bull 38(4):661–671

Viviani B, Bartesaghi S, Corsini E, Galli CL, Marinovich M (2004) Cytokines role in neurodegenerative events. Toxicol Lett 149(1–3):85–89

Wadee AA, Kuschke RH, Wood LA, Berk M, Ichim L, Maes M (2002) Serological observations in patients suffering from acute manic episodes. Hum Psychopharmacol 17(4):175–179

Walker E, Compton M, Shapiro D (2010) The role of glucocorticoids in the emergence of psychosis: potential genetic and epigenetic mechanisms. Schizophr Res 117(2):174

Wang J, Rao H, Wetmore GS, Furlan PM, Korczykowski M, Dinges DF et al (2005) Perfusion functional MRI reveals cerebral blood flow pattern under psychological stress. Proc Natl Acad Sci USA 102(49):17804–17809

Wang T, Chen M, Liu L, Cheng H, Yan YE, Feng YH et al (2011) Nicotine induced CpG methylation of Pax6 binding motif in StAR promoter reduces the gene expression and cortisol production. Toxicol Appl Pharmacol 257(3):328–337

Watson S, Gallagher P, Porter RJ, Smith MS, Herron LJ, Bulmer S et al (2012) A randomized trial to examine the effect of mifepristone on neuropsychological performance and mood in patients with bipolar depression. Biol Psychiatry 72(11):943–949

Weaver IC (2007) Epigenetic programming by maternal behavior and pharmacological intervention. Nature versus nurture: let's call the whole thing off. Epigenetics 2(1):22–28

Weaver IC, Cervoni N, Champagne FA, D'Alessio AC, Sharma S, Seckl JR et al (2004) Epigenetic programming by maternal behavior. Nat Neurosci 7(8):847–854

Weaver IC, D'Alessio AC, Brown SE, Hellstrom IC, Dymov S, Sharma S et al (2007) The transcription factor nerve growth factor-inducible protein a mediates epigenetic programming: altering epigenetic marks by immediate-early genes. J Neurosci 27(7):1756–1768

Wilson CJ, Finch CE, Cohen HJ (2002) Cytokines and cognition–the case for a head-to-toe inflammatory paradigm. J Am Geriatr Soc 50(12):2041–2056

Wingenfeld K, Wolf S, Krieg JC, Lautenbacher S (2011) Working memory performance and cognitive flexibility after dexamethasone or hydrocortisone administration in healthy volunteers. Psychopharmacology 217(3):323–329

Wockner LF, Noble EP, Lawford BR, Young RM, Morris CP, Whitehall VLJ et al (2014) Genome-wide DNA methylation analysis of human brain tissue from schizophrenia patients. Transl Psychiatry Psychiatry 4:e339

Young AH, Gallagher P, Watson S, Del-Estal D, Owen BM, Ferrier IN (2004) Improvements in neurocognitive function and mood following adjunctive treatment with mifepristone (RU-486) in bipolar disorder. Neuropsychopharmacology 29(8):1538–1545

Zimmermann P, Bruckl T, Nocon A, Pfister H, Binder EB, Uhr M et al (2011) Interaction of FKBP5 gene variants and adverse life events in predicting depression onset: results from a 10-year prospective community study. Am J Psychiatry 168(10):1107–1116

Stress, Substance Abuse, and Addiction

Tiffany M. Duffing, Stefanie G. Greiner, Charles W. Mathias and Donald M. Dougherty

Abstract Experiencing stressful life events is reciprocally associated with substance use and abuse. The nature of these relationships varies based on the age of stress exposure and stage of substance use involvement. This chapter reviews the developmental and biological processes involved in the relationship of stress exposure and substance use initiation, substance use maintenance and relapse, and response to substance abuse treatment. Special emphasis is given to describing the various stress-related mechanisms involved in substance use and abuse, highlighting the differences between each of these phases of drug use and drawing upon current research to make suggestions for treatments of substance use disorder (SUD) patients. Stress is inherent to the experience of life and, in many situations, unavoidable. Through ongoing research and treatment development, there is the potential to modify the relationship of stress with ongoing substance use and abuse.

Keywords Stress · Substance use · Substance abuse · Substance dependence

Contents

1 Stress, Substance Abuse, and Addiction	238
2 Substance Use Initiation	239
2.1 Early Developmental Stressors	240
2.2 Genes and Epigenetic Processes Involved in Substance Use Initiation	242

T. M. Duffing · S. G. Greiner
Fielding Graduate University, Santa Barbara, CA, USA

C. W. Mathias · D. M. Dougherty (✉)
Division of Neurobehavioral Research, NRLC MC 7793, Department of Psychiatry, The University of Texas Health Science Center, 7703 Floyd Curl Drive, San Antonio, TX 78229-3900, USA
e-mail: doughertyd@uthscsa.edu

2.3 Stress Response System and Substance Use Initiation 243
3 Substance Use Maintenance and Relapse 245
  3.1 Neurochemical and Neuroanatomical Mechanisms.................................. 245
  3.2 Early Life Stressors 246
  3.3 Stress Response System and Substance Use Maintenance and Relapse 247
  3.4 Reward Pathways and Substance Use Maintenance and Relapse 250
  3.5 Gender... 251
  3.6 Stress-Related Predictors of Relapse 252
4 Treatment... 253
  4.1 Predictors of Treatment Outcome... 254
  4.2 Treatment Modalities.. 255
  4.3 Co-occurring/Integrated Treatment 256
5 Conclusions.. 257
References.. 258

1 Stress, Substance Abuse, and Addiction

Phenotypic expression of genetic vulnerability to substance abuse disorders (SUD) is likely triggered by stressful events and environments. Stress during early development (including prenatal development) leads to epigenetic and anatomical changes in the brain that increase vulnerability to addiction and influence how children process subsequent stressful stimuli. These early developmental stressors have the potential to compound future stress exposure and are reported more often among drug using adolescents, compared to their nondrug using peers. Later, adolescent stress exposure is also independently linked to an increased likelihood of adolescent drug use as well as a younger age onset of substance use. Those who initiate substance use during adolescence further increase their risk for adult substance use disorder. The risk of substance use among adults is associated not only with current stress and substance use exposure, but is also affected by accumulated lifetime stress and prior substance use. Looking at the larger picture, stress at any point in the life span has the potential to independently or collectively modify various internal mechanisms such as the stress or reward systems, thereby increasing the risk of substance abuse. Continuous substance use leads to adaptations of neurochemistry and anatomy similar to the alternations created by stress. Each of these life span factors of stress and substance use plays a role in substance use initiation, addiction, and relapse (which then impacts treatment outcomes). This chapter will describe the various stress-related mechanisms involved in substance use and abuse, highlighting the differences between each of these phases of drug use and drawing upon current research to make suggestions for treatments of substance use disorder (SUD) patients.

Prior to the evaluation of the relationship between stress and substance abuse, there are certain elements and limitations of the laboratory model which require acknowledgement. Animal models of initiation, addiction, and relapse have allowed researchers to examine the physiological and behavioral aspects of stress and addiction in a controlled laboratory setting by utilizing strictly controlled experimental design. Stress in animal models is applied in species-specific ways by applying aversive stimuli such as footpad shocks or tail pulls (Schramm-Sapyta et al. 2009), or creating situations of isolation or social defeat (Neisewander et al. 2012). Drug craving and seeking is modeled with conditioned place preference methods, in which drugs are administered at one location in the cage (Schramm-Sapyta et al. 2009) and animals which explore the drug administration area after receiving stressors are considered to be seeking the drugs in response to stress. Researchers have also utilized animal models to learn about aversive effects of drugs and how those effects impact future use and abuse of substances (Schramm-Sapyta et al. 2009).

The absence of human social stressors is both a benefit to the methodology and a limitation to the applicability of such studies, yet reviewers believe that animal models of isolation or competition can still provide relevant information on the social influences on drug abuse (Neisewander et al. 2012). Most animal studies do not allow 24-h availability for self-administration and they limit the dosage that animal subjects can administer, which does not lend itself to generalization for human addiction models (Koob and Kreek 2007). However, the physiology of stress and addiction in humans does appear to be quite similar to stress and addiction in primates (Schwandt et al. 2007) and other mammals (Briand and Blendy 2010; Schramm-Sapyta et al. 2009). One of the unique strengths of some preclinical studies is the relatively short life cycle of rodents, which allows for more rapid tests of stress and substance use interaction than may be conducted in humans. This difference facilitates a more rapid, advanced understanding of the mechanisms involved in adolescent drug initiation, abuse, and addiction (Schramm-Sapyta et al. 2009).

Studying human stress and pathophysiology of addiction in the laboratory is also challenging for several reasons. Addicted individuals' mental state typically includes increased levels of anxiety and negative emotions (Sinha 2008a). It is often difficult to disentangle which stress symptoms are products of substance use, prior history of accumulated stress, or even effects of study participation. Stress symptoms are expressed during active use, in the early stages of drug abstinence, and during acute withdrawal, making it especially difficult to isolate stress effects from physiological changes accompanying addiction (Sinha 2008a). While clinical studies consistently identify significant variations in stress levels, response to stress stimuli, and the affects of substance abuse, exactly which individual factor or combination of factors proceeds or causes the other has not been consistently established.

2 Substance Use Initiation

This section will examine the role of stress in substance use initiation through a review of relevant and current preclinical and clinical studies. Animal subjects continue to effectively demonstrate the behaviors associated with stress and either

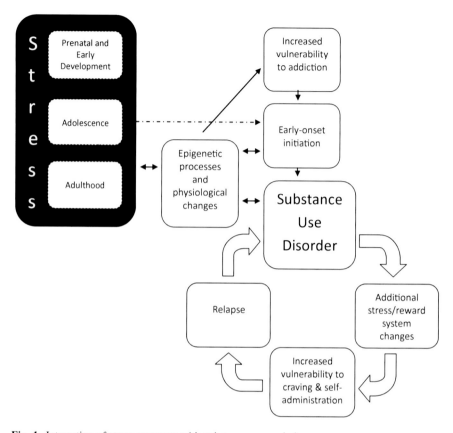

Fig. 1 Interaction of stress exposure with substance use and abuse

seeking out of rewards or avoiding aversive stimuli via substance use. Stress in early developmental stages is thought to result in genetic alterations and is shown to influence the age at which adolescents first engage insubstance use, which is then predictive of addiction in adulthood (Fig. 1). Finally, dysregulation of the hypothalamic–pituitary–adrenal (HPA) axis has been implicated in substance use initiation.

2.1 Early Developmental Stressors

Prenatal Stress. Evidence from preclinical and clinical research shows that early life stress predisposes individuals to be at higher risk of developing substance abuse and other psychological disorders later in life. Research with C57BL/6 J mice (Campbell et al. 2009) has shown that early environmental trauma results in heavy alcohol consumption. Adult mice that were subjected to prenatal stress

(PNS) consumed more alcohol during learning trials than mice that were not exposed to PNS. PNS mice were hyperresponsive to stress after birth, as well as had greater reactivity to novel stimuli and psychomotor stimulants (Campbell et al. 2009). This parallels research with children whose mothers experienced stressors during pregnancy. These children have an increased risk for neuropsychiatric disorders, including substance abuse disorders (Campbell et al. 2009).

Early Childhood Stress. Stress in early development can also increase drug self-administration (Becker et al. 2011). In ethanol self-administration experiments (Becker et al. 2011), mice exposed to early social isolation, between weaning and adulthood, had greater ethanol intake than group-housed mice. Social isolation during adulthood, however, did not increase ethanol intake compared to group-housed mice. In addition to isolation, chronic variable stress caused greater ethanol intake, but did not increase the ethanol intake any more in the mice that were isolated early in life.

Several studies (Anda et al. 2002; Verona and Sachs-Ericsson 2005) identify early life stress as a risk for adult substance dependence, independent of genetic factors. The cumulative number of stressors, timing, severity, and frequency of the early life stressors are all thought to contribute to increased risk for alcoholism and drug dependence (Enoch 2011). Although not every child that experiences early life stress develops a substance use disorder, the combination of genetic predisposition and early life stress can lead to early onset of substance use (Enoch 2011). The specific physiological mechanisms involved in this relationship are discussed later.

Adolescent Stress. Animal models show that adolescents can be more vulnerable to drug addiction than adults due to the reinforcing and aversive qualities of drugs that differ in adolescence compared to adult subjects (Schramm-Sapyta et al. 2009). In humans, early substance use often leads to either future adult substance use or a perpetuating cycle where stressful experiences, coupled with an inability to positively respond, lead to an increased risk of substance abuse (Enoch 2011). McCarty et al. 2012) report that youth experiencing stressful events by the sixth grade are at a significantly greater risk of initiating substances by the end of the eighth grade. A rare longitudinal study (Englund et al. 2008) illustrated the prospective risks of early life stressors by following low-income subjects from birth to 28 years. They concluded that adult alcohol abuse may be the result of a combination of stressful events in childhood and early adolescence that lead to externalizing behaviors among males, including binge drinking. For girls, interestingly, high achievers at age 12 had more likelihood of being heavy drinkers at age 23 compared to their non-high-achieving counterparts (Englund et al. 2008). Schmid et al. (2009) examined the potential role of the dopamine transporter gene (*DAT1*) in adolescent drug use and adult addiction. They found that early onset of smoking or drinking moderated the effect of *DAT1* on adult alcoholism; those with the *DAT1* VTNR 10r allele who initiated drinking and smoking before age 15 were most likely to be heavy drinkers as adults.

2.2 Genes and Epigenetic Processes Involved in Substance Use Initiation

In the classic vulnerability-stress model for psychological disorders and stress-related medical conditions, genetic predispositions were seen as vulnerabilities to substance abuse or other psychological disorders, and added environmental stressors were believed to interact with genetic predisposition to create influence. However, more recent research (Nestler 2011) has supported the notion that stress can actually alter genetic expressions and can affect neurodevelopment (Campbell et al. 2009) by altering brain anatomy and activity (Koob and Kreek 2007). For example, animals exposed to prenatal stress showed long-term changes in the dopamine and glutamate within the limbic system, leading researchers to suggest that this plays a role in the willingness to self-administer large doses of drugs (Campbell et al. 2009). Rhesus monkeys reared with three age-matched peers rather than a mother for the first 6 months of life experienced morphological brain changes (e.g., enlarged cerebellar vermis, dorsomedial prefrontal cortex, and dorsal anterior cingulate cortex) relative to mother-reared monkeys after (Spinelli et al. 2009). These stress-induced changes in brain structure may be utilize the high concentrations of glucocorticoid activity; previous research has reported that early life stress can lower the level of glucocoriticoid receptor nRNA expression and lowered glucocorticoid feedback sensitivity (Enoch 2011).

Severe childhood stressors and cumulative adversity through development initiate epigenetic adaptations via enzymes that can affect how relaxed or condensed a segment of chromatin is wrapped around a section of DNA (Nestler 2011). The environment can influence the "writers and erasers" (Nestler 2011, p. 78) of the epigenetic code, thereby causing alterations in stress and reward pathways that facilitate emotional distress susceptibility and influence the reinforcing elements of addictive substances (Sinha 2009). There is also an apparent similarity between exposure to stress and exposure to drugs in long-term changes in gene expression and activation of transcription factors for genes strongly associated with both drug addiction and stress (Briand and Blendy 2010). For example, acute stress exposure of various kinds (forced swim, footshock, restraint) leads to increases in cAMP response element-binding (CREB) protein (Briand and Blendy 2010). CREB protein regulates the transcription of target genes such as corticotropin-releasing factor (CRF), brain-derived neurotrophic factor (BDNF), and dynorphin, which are involved in stress and addiction. Chronic stress has been shown to inhibit CREB function (Briand and Blendy 2010); leading to decreases of CREB in the hippocampus, striatum, and frontal cortex, which likely indicates compensation to the chronic stress.

It is possible that a less active sympathetic nervous system may serve as a biomarker for future substance use disorders (Brenner and Beauchaine 2011). The altered mesolimbic dopamine function transforms the overall operation of the reward system and is thought to alter the way the body responds to various rewards, including those associated with substance use (Brenner and Beauchaine

2011). More specifically, reduced pre-ejection period reactivity to incentives is associated with increased risk of future alcohol use (Brenner and Beauchaine 2011). Theserecent developments inunderstanding the epigenetic impacts of stress support the view that stress is vulnerability in the development of SUDs, and suggest that stress may create biological vulnerabilities to SUDs.

2.3 Stress Response System and Substance Use Initiation

Stress, and the subsequent activation of the stress response system, is a known risk factor for drug addiction and relapse (Enoch 2011; McCarty et al. 2012; Sapolsky et al. 2000; Sinha 2001, 2008b, 2011; Sinha et al. 2005). It is apparent that each person who experiences stress does not engage in drug use nor develop a substance use disorder. However, recurrent or constant activation of the system leads to maladaptive changes which result in increased risk for physiological and psychological problems (Sapolsky et al. 2000). These increased vulnerabilities are then further compounded by other substance use risk factors. One primary area of research regarding the stress response system involves how stress and substance use, present and past, initiate and affect HPA axis activity.

Although the majority of the current literature appears to evaluate how HPA axis activity is involved in the relationship between stress and substance *abuse or addiction* (described in a subsequent section), there is a growing body of research evaluating the connection in regards to substance *initiation*. It has been established that one way in which the human body naturally responds to stress is by activating the HPA axis through the secretion of the corticotropin-releasing hormone (CRH) (Goeders 2003). This activation of this stress response system then modifies salivary cortisol secretion levels (Sapolsky et al. 2000) and is commonly utilized to measure stress response changes of the HPA axis (Fox et al. 2006; Junghanns et al. 2007). In addition to measuring variations in the HPA axis as a method for biological markers or predictors, knowledge of this information can also open doors to evaluating why there are different cortisol responses to different drugs and what other factors are involved in this biological adaptation. Aside from immediate changes in cortisol levels due to stress, there are also adaptations in basal HPA axis from stress exposure. For instance, preclinical studies have shown that the quality of maternal care, to include physical and emotional interaction, experienced by offspring during early development influences future HPA axis stress responsivity as well as increase drug self-administration (see review in Enoch 2011). This research indicates a cumulative response to stress where current stress not only has the potential to create adaptations, but prior and recurring stress impacts the way in which the body responds to and adapts to stress and therefore creates a vulnerability to maladaptive behaviors, including substance abuse.

Clinical studies have further substantiated the connection between stress, HPA axis activity, and substance use. The first step in demonstrating this relationship in humans is evident in the variations in HPA axis activity, as measured by cortisol levels, upon stress exposure. Although there is still not a consensus on whether these cortisol levels are diminished (Carpenter et al. 2009) or elevated (Tarullo and Gunnar 2006; Preussner et al. 2004) with stress exposure, it is evident that cortisol levels are significantly different from controls who have not experienced prior psychosocial stressors. For example, children who experience maltreatment, compared to those who have not, have elevated basal cortisol levels as children and then lower cortisol levels as adults (Tarullo and Gunnar 2006). In a more recent longitudinal study, nonstress morning cortisol levels were measured from childhood to adulthood in individuals who had experienced confirmed familial sexual abuse (Trickett et al. 2010). This study also demonstrated that individuals who had experienced childhood stress had attenuated cortisol levels during adolescents and lower cortisol levels during adulthood, compared to controls. Since cortisol levels are associated with many health concerns and dysfunction in other areas (Sapolsky et al. 2000), these stress-induced cortisol level adaptations are important in conceptualizing the stress/substance abuse process as a continuing cycle. While these studies specifically evaluated the putative connection between childhood stress and disrupted basal cortisol levels, they also further verified the fact that stress, particularly early childhood stress, not only affects HPA axis activity at the time of stress exposure, but potentially creates aversive adaptations which could lead to vulnerabilities in adulthood. Furthermore, since HPA axis activity and stress response have been shown to vary as a function of family history of alcoholism (Dai et al. 2007); it is possible that these HPA axis changes also impact future offspring.

In addition to the HPA axis related vulnerabilities for substance use initiation, it is suspected that HPA axis functioning may serve as a predictor for age of onset of drug use (Evans et al. 2011). In a study evaluating over 2,000 adolescents, researchers discovered that cortisol levels at the onset of and during social stress exposure significantly explained the variance in age of onset of alcohol use (Evans et al. 2011). This expanded previous research which demonstrated that early onset cannabis users had significantly lower morning cortisol levels, compared to non-users (Huizink et al. 2006). Additionally, the differences in morning cortisol levels may substantiate a widely known hypothesis which suggests drug using individuals are seeking to restore "normal" biological levels through the use of substances (discussed in Maintenance and Relapse).

More research in this area is necessary to confirm the association between HPA axis dysfunction and substance abuse, and further explore the mechanisms involved in this dynamic interplay of the stress response system and prolonged substance use. While this research has certainly facilitated the identification of personalized responses to drugs and stress experiences, the individualized way in which mechanisms interact and affect each other has inhibited consistent research replication. More specifically, it is difficult to demonstrate clear relationships between one factor and the next when the very nature of individualistic and drug-specific responses is a seemingly endless list of potential interactive relationships. What is more clear is that the next step in furthering our understanding of stress and substance abuse may lie in our ability to pinpoint exactly which factors, environmental and biological, moderate stress system adaptations, and how this knowledge can be implemented into treatment to produce more positive outcomes (discussed further in Treatment). Employing programs which incorporate known predictors, as well as potential factors that interfere with or reverse maladaptive changes in the stress response system, may be the key to effective substance abuse prevention programs.

3 Substance Use Maintenance and Relapse

In this section, we will address the neurochemical and neuroanatomical changes that accompany the transition from substance abuse to addiction and relapse and how stress affects this process. We also return to the theme of early life stressors, mentioning research that links early life stress to drug addiction and relapse. The research examining addiction and early life stress lends support to the theory that early life stress alters important neuroanatomy and neurochemistry that may change the rewarding and aversive effects of drugs, thereby altering the individual's vulnerability to addiction. We also discuss how adaptations of the stress response system specifically affect, and are impacted by, the cycle of addiction andrelapse.

3.1 Neurochemical and Neuroanatomical Mechanisms

In the transition from occasional use or limited access to compulsive use of drugs, animal models show a dysregulation of brain reward pathways (Koob and Kreek 2007). Despite the common finding that aversive events trigger dopamine release, it also appears that acute stress or aversive stimuli can also inhibit dopamine release (Ungless et al. 2010). When the aversive stimulus such as foot shock is removed, the dopamine neurons fire, which supports the concept that the offset of an aversive event can then release dopamine (Ungless et al. 2010). Glucocorticoids also sensitize the reward pathways as a result of HPA axis activation from both stress and drug use (Uhart and Wand 2009).

Large increases in adrenocorticotropic hormone (ACTH) and corticosterone during cocaine administration in rat models have been reported. Additionally, tobacco smokers experiencing stress also show elevated ACTH and HPA axis reactivity in addition to tobacco craving (McKee et al. 2011). Researchers have also shown that stress reduces the ability of smokers to resist smoking. Acute withdrawal states reactivate the HPA axis (Koob and Kreek 2007). Escalation in drug intake produces activation of the CRF in the extended amygdala (outside the hypothalamus). This is supported by functional magnetic resonance imaging (fMRI) studies that show stress and drug exposure both activate the mesolimbic and mesocortical dopamine projection areas (Briand and Blendy 2010). Chronic stress affects the dorsolateral striatum-dependent habit system, which thereby may accelerate the transition from initially goal-directed, involuntary drug use to more dependent and compulsive drug-seeking and taking behaviors (Schwabe et al. 2011). Schwabe et al. (2011) argue that stress or stress hormones may promote or induce the transition from voluntary drug use to dependence or abuse. Rats pretreated with corticosterone seek cocaine, and cocaine is also observed to increase levels of ACTH and corticosterone (Goeders 1997). The role of adrenocorticosteroids in cocaine reinforcement is further supported in findings which showed that adrenalectomized rats did not self-administer any cocaine (Uhart and Wand 2009), yet this effect could be reversed by administering corticosterone.

Uhart and Wand (2009) discuss stress and vulnerability to drug addiction in terms of allostatic load. When an individual lives with chronic stress, allostasis, or "the process of maintaining apparent reward function stability through changes in reward and stress system neurocircuitry that are maladaptive" (Uhart and Wand 2009, p. 44), becomes more difficult and the allostatic load increases. An overcapacity allostatic load results in overexposure to glucocorticoids, other stress peptides, and proinflammatory cytokines (Uhart and Wand 2009).

Stress also causes a downregulation of neuropeptides designed to assist in stress management. For example, neuropeptide Y (NPY) is known to have anxiolytic properties, which regulate the stress response (Witt et al. 2011; Xu et al. 2012). NPY plasma levels are predicted by genetic variations, and patients with stress disorders have suppressed plasma NPY levels (Xu et al. 2012). In a study examining stress, NPY and risk for substance abuse relapse (which is related to managing stressful stimuli), Xu and colleagues (2012) characterized abstinent substance-dependent (SD) patients and healthy controls on NPY diplotypes (HH: high expression; HLLL: intermediate to low expression). All subjects were exposed to stress and subsequently presented with alcohol/drug cues and neutral relaxing cues via individualized guided imagery. Results from this 90-day prospective study showed that HH subjects and SD subjects showed lower stress-induced NPY. Additionally, lower NPY levels predicted higher number of days and greater quantities of drug use posttreatment (Xu et al. 2012).

In summary, stress contributes to the transition from substance abuse to addiction by altering neurochemical pathways that are critical in producing and regulating the stress response. It is possible that an interaction between the downregulation of neuropeptides such as NPY, increased production of corticosteroids, and dysregulation of the dopaminergic pathways contribute to the transition from impulse-control issues of substance abuse to the compulsive drug-seeking behaviors seen in addiction.

3.2 Early Life Stressors

Animal models have shown that early life stressors alter behavioral responses to psychomotor stimulants, which is likely due to changes in the dopamine and glutamate systems within the limbic structures (Campbell et al. 2009). Prenatal stress is believed to decrease sensitivity to alcohol, which is correlated with alcoholism vulnerability (Campbell et al. 2009). Maternal deprivation (MD) in early childhood is believed to alter the reward system significantly enough that one 24-h episode of MD is shown to produce physiological changes similar to depression and anxiety, increasing the time it takes for adolescent mice to meet acquisition criteria for cocaine self-administration (Martini and Valverde 2012).

Early stress has been shown to influence addiction and relapse as well as initiation of substance use in humans. For example, family history predicted relapse in a study examining four types of psychological disorders and prognoses (Milne et al. 2009). The authors suggested that knowledge of family drug use could be informative when determining prognosis for the addicted individual. Having a drug-addicted parent predicts poor parenting due to increased chance of drug abuse by the adult child of the addicted parent (Locke and Newcomb 2004). However, in other research, the childhood stress of abuse predicted drug use and other externalizing behaviors above and beyond parental history of substance abuse (Verona and Sachs-Ericsson 2005). In a retrospective study, Anda et al. (2002) found that individuals who reported parental alcohol abuse were more likely to acknowledge problems with alcohol abuse themselves. In addition, the potential to have nine other risk factors (examples include mentally ill, suicidal, or criminal behavior in parents) was higher in the respondents who reported parental alcohol abuse during their childhood (Anda et al. 2002). Further, early stress influences brain development. Individuals who experience childhood maltreatment and subsequent alcohol abuse beginning in adolescence have been shown to have lower hippocampal volume, which is associated with longer duration of drinking (Enoch 2011).

3.3 Stress Response System and Substance Use Maintenance and Relapse

Stress is related to fewer percent days of abstinence, lower rates of complete abstinence, a significantly shorter time to relapse, and more drinks consumed upon relapse (Breese et al. 2005; Higley et al. 2011; McKee et al. 2011). It has also been reported as the primary reason for relapse among previously abstinent substance users (Sinha 2008a) and is also indicated in provoking subjective drug cravings (Sinha 2011). While overcoming stress can build upon an individual's perceived and actual accomplishments and overall resilience, chronic or uncontrollable stress may lead to maladaptive behaviors as well as potentially instigate adaptations of behavioral, cognitive, and physiological systems (Sinha 2008a; Uhart and Wand 2009). More specifically, these changes affect glucocorticoid gene expression, serotonin function, mesolimbic dopamine transmission, and other stress-related systems which contribute to the differences in an individual's stress response process (Sinha 2009). As previously mentioned, there is also evidence to suggest that specific genes involved in the stress response system may contribute to a genetic predisposition for substance dependence (Clarke et al. 2012). These biological adaptations are hypothesized to be a product of the interaction of continued stress, individualgenetic susceptibility, history andseverity of drug dependence, baseline stress levels, the type ofstressor experienced, the duration for which the stress is experienced, and potentially additional undiscovered elements (Sinha 2008a, b, 2011). It is this perpetuating system of adverse adaptations which is thought to disrupt the HPA axis and emotion regulatory mechanism, inhibit the regulation and normalizing of the stress response system, and thereby prevent the homeostasis necessary for healthy living. Stress has the potential to affect future coping, current and future biological responses, and ultimately affect drug maintenance and relapse. It is this interactive, cyclic relationship which makes understanding all of the involved predictors and mechanisms of stress and substance use especially difficult to distinguish.

Although it has become quite obvious that stress is involved in the transition between various phases of drug addiction, whether stress actually causes continuous drug use is not as clear. Preclinical studies have demonstrated that various forms of stress are associated with increased drug self-administration. For instance, Higley et al. 1991) found that rhesus monkeys exposed to peer-rearing consumed significantly more alcohol than their mother-reared peers. Social isolation of the mother-reared monkeys during adulthood was later associated with more alcohol intake. Further, Kosten and Kehoe 2010) reported that neonatal isolation was significantly associated with continued self-administration of drugs.

Clinical studies have corroborated, albeit inconsistently, the relationship between stress and increased or continued substance use. While Pratt and Davidson (2009) did not find an association between stress and induced or greater drinking in alcohol-dependent individuals, Clarke et al. (2012) found a significant association between levels of alcohol consumption and activation of the stress response system upon acute stress exposure. Furthermore, Thomas et al. (2011) found that alcohol-dependent individuals exposed to a psychosocial stressor were twice as likely to drink all of the alcohol available when compared to nonstressed controls. This study furthered prior research which demonstrated that alcoholics, but not social drinkers, consumed greater amounts of alcohol after engaging in laboratory social stress exposure scenarios (Miller et al. 1974). Research examining possible neurochemical and/or functional differences in the brains of alcohol-dependent and social drinkers, while engaged in stressful situations, would help further our understanding of the progression from initiation to maintenance stages. For example, it would be helpful to evaluate potential biological differences between social drinkers and alcohol-dependent drinkers while also examining variations that correspond with the environment in which drinking typically occurs (i.e., social situations versus alone). Another area of potential stress system research could investigate whether social drinkers' NPY levels are higher and whether their glucocorticoid levels are more stable than alcohol-dependent subjects.

A great deal of research has examined the interactive relationship between the HPA axis and substance abuse and relapse. Pre-clinical studies have shown that dysregulation of the HPA axis, as a result of stress, have a specific influence on drug self-administration (Goeders 2003). The measurement of corticosterone, a substance typically released as a final step in the HPA axis activation, is one method which has been utilized in preclinical studies to demonstrate the impact of the HPA axis and stress on substance use. For instance, when corticosterone was injected, rats became more sensitive to cocaine doses (Goeders 2003). When another study attempted to regulate the HPA axis with a corticosterone synthesis inhibitor, drug acquisition decreased (Campbell and Carroll 2001). Further, rats completely cease self-administration of cocaine following bilateral adrenalectomy (Goeders and Guerin 1996). These preclinical studies illustrate one potential mechanism of stress and substance use and highlight the critical impact a dysregulated HPA axis has on drug use behaviors.

In addition to the HPA axis being evaluated as a predictor of substance use, researchers have also been investigating the bidirectional effects of different drugs on HPA axis regulation. For instance, exposure to stress and the use of various stimulants has been associated with decreased (Gerra et al. 2003), increased (Sinha et al. 2000; King et al. 2010), and no change in (Harris et al. 2005) salivary cortisol levels as a function of the type of stimulant ingested. More recent research has replicated this dysfunctional HPA axis activity in relation to alcohol, cocaine, and nicotine relapse outcomes (see review in Sinha 2011). This growing body of research regarding HPA activity and changes helps substantiate the individual nature in which people will respond to stress as well as the heterogeneity of substances abused. Additional research has even indicated that relapse risk is not necessarily exclusively associated to provoked (i.e., stressed) HPA axis activity; rather, cortisolresponse to corticotropin (also known as adrenal sensitivity) at morning resting levels is also significantly associated with relapse risk (Sinha et al. 2011). This suggests that the HPA axis dysfunction is occurring across situations rather than solely disrupted during stress exposure. The variable reactions of the HPA axis suggests that exposure to stress interacts with individual differences in biological response and it is likely that these responses are dependent on numerous individualized environmental and biological factors. Furthermore, since HPA axis activity during a resting state is also predictive of relapse, it lends additional support to the idea that there is more at play than current stress exposure.

Taken together, research shows that not only does the function of the HPA axis affect the ability to cope with stress and substance use, but substance use also independently affects the function of the HPA axis. In other words, research corroborates the notion that stress can increase risk for substance use just as substance use increases risk for stress. Unfortunately, all of the mechanisms involved in the relationship between HPA axis activity, stress, and substance use are not clear. However, there are two primary hypotheses which attempt to explain the link between HPA axis dysfunction and continued substance use. First is the widely recognized self-medication hypothesis which posits that substance abusers are actually attempting to reduce or eliminate negative symptoms such as anxiety or depression or integrate some perceived control into their life (Goeders 2003; Khantzian 1997). The second hypothesis of drug addiction proposes that an individual continues to seek out substances in an attempt to produce an internal state of arousal which would then counter the established HPA axis hypoactivity (Goeders 2003; Majewska 2002).

3.4 Reward Pathways and Substance Use Maintenance and Relapse

Although it is consistently recognized that stress increases dopamine release, the various effects that stress has on different dopamine neurons and at different phases is less known (Ungless et al. 2010). However, it appears from several studies that the mesolimbic dopamine system becomes downregulated between acquisition and addiction, and that stress-induced changes in glucocorticoid activity may be responsible for this downregulation (Uhart and Wand 2008). Rodent models show that movement from acquisition to maintenance in cocaine use results in brain reward dysfunction, evidenced by elevated reward thresholds and decreased dopamine signaling (Koob and Kreek 2007). To complicate matters further, the downregulation of the mesolimbic reward system is often accompanied by negative affect induced by the increased allostatic load (Uhart and Wand 2008).

Specific areas of the limbic system are involved in stress and reward processes. The central amygdala projects to the bed nucleus of the stria terminalis (BNST) which is involved in drug reward modulation by stress (Briand and Blendy 2010). The BNST projects to the ventral tegmental area (VTA), thereby activating the mesolimbic and mesocortical dopamine pathways (Briand and Blendy 2010). Several compoundsare known to enactchanges in these areas when stress and drug areinduced (Briand and Blendy 2010). The cAMP response element-binding protein (CREB) is involved in stress response, drug exposure, and reinstatement of drug reward (Kreibich et al. 2009) as shown in forced swim studies. This form of stress increases CREB in the nucleus accumbens, amygdala, dentate gyrus, and the neocortex (Briand and Blendy 2010). Abusive drug use also increases CREB in reward pathway areas (Briand and Blendy 2010). Brain-derived neurotrophic factor (BDNF) is identified as a potential downstream target of CREB and exposure to stress alters the BDNF mRNA and protein levels in the brain (Briand and Blendy 2010), either increasing or decreasing depending upon the type of stress and the brain region. Decreases in BDNF appear to increase anxiety behaviors and lead to increased vulnerability to stressors (Advani et al. 2009). As mentioned previously, the neuropeptide CRF is involved in mediating autonomic, neuroendocrine, and behavioral responses to stress. Increased stress leads to increased CRF, and drugs of abuse cause changes in CRF activity and how CRF mediates drug seeking and reward (Briand and Blendy 2010).

3.5 Gender

Research evaluating the impact of gender on stress-related aspects of substance use is not as robust as the more commonly recognized substance use gender differences. More specifically, it is clear that there are gender sensitive differences involved in the age of onset, motivations for drug use, progression to addiction, and substance abuse treatment outcomes (Brady and Randall 1999; Chen and Jacobson 2012; Randall et al. 1999). It is also thought that gender-specific hormones are at least partially responsible for these differences (Fox et al. 2013; Wetherington 2007) and that sex chromosomes may also have a strong influence (Becker et al. 2007). So while it has been established that there are diverse, gender-specific responses to substance use, it also is plausible that these responses interact with the gender effects in stress responsivity. For instance, Kosten et al. (2005) have conducted numerous animal studies which have demonstrated that early life stress, specifically neonatal isolation, shapes neurodevelopment and future stress response. In this case, gender was significantly associated with differences in stress sensitivity and the development of context-induced and unconditioned fear. Female rats presented with enhanced responses and higher foot shock sensitivity, context-induced fear, as well as a greater overall unconditioned fear. Neonatal isolation is not only thought to modify gender- specific aversive learning but is also hypothesized to affect neuronal reorganization in a gender-sensitive manner (Kosten et al. 2005).

Clinical studies have further supported the impact of gender on stress and substance use by investigating gender differences among stress-exposed drug-using populations. It has been demonstrated that cocaine-dependent women exposed to stress manipulations reported higher stress, nervousness, and pain ratings. When compared to men, women also endured the physical stress activity for less time than men (Back et al. 2005). Further, differences in the reinforcing effects of stress, neural activations, and reactions to drug and stress cues have been associated with gender (Potenza et al. 2012). In another study, women, but not men, who experienced childhood abuse, demonstrated an increased risk for drug relapse (Hyman et al. 2008). These gender sensitive variations indicate greater subjective stress reactivity for women and suggest that stress exposure may be perceived and experienced differently among drug-dependent men and women.

Gender also moderates drug craving and stress arousal in guided imagery studies and clinical settings. In a study where both men and women received progesterone, only women reported decreased negative emotion with increased relaxed mood (Fox et al. 2013). However in the same study, cue-induced cravings, cortisol responses, and inhibitory control did not show significant gender sensitivity (Fox et al. 2013). So, while it is clear that some stress and substance use mechanisms may operate in a gender sensitive manner, research still lacks a comprehensive explanation for how these differences will vary. There is also a lack of clear answers to the gender-specific stress mechanisms involved in initiating drug use and drug dependence. Based on the present research, and gaps in what we still do not know, it is critical to acknowledge that gender does not serve as a universal predictive factor for SUD or relapse. More research is necessary in order to broaden our understanding of these deviations as well as facilitate the creation and implementation of appropriate gender- specific clinical treatments.

3.6 Stress-Related Predictors of Relapse

In addition to the putative mechanisms involved in substance abuse relapse, there are several specific, stress-related factors which have been identified as predictors of, or at least strongly associated with, relapse. While some associations such as psychopathology (Dodge et al. 2005; Greenfield et al. 1998), experiencing stress or cues (Breese et al. 2005; Coffey et al. 2010; McKee et al. 2011; Sinha et al. 2000, 2011), cravings (Epstein et al. 2010; Higley et al. 2011; Rohsenow and Monti 1999; Sinha et al. 2011), and gender (Hyman et al. 2008; Potenza et al. 2012) have been consistently supported by both clinical and preclinical research (Koob 2008; Liu and Weiss 2002), there are also additional factors associated with relapse that are not as widely known or consistently researched. For instance, an upregulation of the HPA axis (at baseline, resting, and with cue exposure) of individuals abstaining from cocaine, alcohol, opiates, and nicotine has been shown to be associated with relapse (Sinha 2011). Blunted stress- and cue-induced cortisol levels among alcoholics and nicotine abstainers are also predictive of relapse (Adinoff et al. 2005; Al'absi et al. 2005; Breese et al. 2005; Junghanns et al. 2003; Sinha et al. 2001, 2011). These studies combined suggest that screening for HPA axis reactivity upon treatment admission may be helpful in identifying those at higher risk for relapse. Additionally, a higher morning serum BDNF level has also shown to be predictive of cocaine relapse (D'Sa et al. 2011). Preclinical studies have also evaluated the involvement of BDNF and found an association to drug reinstatement (Schoenbaum et al. 2007). There has even been an association established through neuroimaging studies exploring gray matter and neurological activity. These studies have shown that lower BDNF volumes in multiple areas of the brain, to include the amygdala, are significantly associated with relapse, in some cases predicting time to relapse (Wrase et al. 2008; Sinha 2011). Increased activity in certain areas of the cingulate cortex has also been associated with relapse risk (Li and Sinha 2008). While it is not practical to suggest the integration of neuroimaging, or even HPA axis screening, intotreatment admission in order to determine relapse risk, these studies do help highlight neural andbiological adaptations that occur in relation to stress and prolonged substance use. As Sinha (2011) pointed out, a continuation and furthering of these types of studies may be the key to establishing a relapse risk biological profile capable of being utilized in clinical settings.

Despite the current research which establishes relationships between stress, substance abuse, and additional predictors, the mechanisms that underlie these relationships are still relatively vague and inconclusive. Furthermore, the literature has yet to adequately establish practical recommendations for how to incorporate these known relationships into substance abuse treatment and relapse prevention interventions. One caveat to this disparity is the relatively recent move to effectively treat patients with co-occurring mental health and substance abuse disorders. The following section will discuss research and the current state of knowledge regarding stress-related risks, predictors, and diagnosis associated with substance abuse treatment.

4 Treatment

Since stress alters brain chemistry and function during addiction and relapse, stress reduction techniques may be critical in substance abuse treatment: by reversing the effects of stress and thereby raising the likelihood of long-term abstinence. For example, preclinical research utilizes environmental enrichment (Solinas et al. 2010) to show that certain environmental conditions, such as larger housing and the inclusion of toys or other stimulating objects in the cages can have beneficial effects (Rawas et al. 2009). Environmental enrichment is shown to function in opposition to stress, and can bring about long-term changes that prevent drug addiction by changing the structure and function of key areas involved in stress and stress regulation. For example, mice raised in enriched environments have been shown to be less sensitive to the reinforcing aspects of cocaine (Solinas et al. 2009) and heroin (Rawas et al. 2009) than mice raised in standard environments. Solinas et al. (2010) proposed that environmental enrichment can also play a role in transforming the brain back to preaddicted states of hippocampal fitness and to repair disrupted learning and memory functions that can then decrease the likelihood of relapse. It may be hypothesized, for example, that environmental enrichment may help to raise levels of NPY in the amygdala and related cortical areas, which would then assist the addicted individual in coping with stress by minimizing the stress response in the brain and lowering craving. One limitation in this line of research is that studies seem to focus on the effects of environmental enrichment during early developmental periods on later potential for addiction. For instance, environmental enrichment can diminish cocaine self-administration among animals exposed to cocaine (Solinas et al. 2008). While early positive environments and experiences can serve as buffers for adolescents and adults who abuse drugs (Solinas et al. 2009), enriching the environment of a drug-addicted individual could be important in avoiding relapse (Solinas et al. 2008). Therefore, this type of research may expand our understanding of the value of enriched environments as well as increase the efficacy of treatment and relapse prevention among humans.

Clinical studies robustly demonstrate that stress negatively impacts substance abuse treatment outcomes (Higley et al. 2011; Rooke et al. 2011; Sinha 2001, 2011; Sinha et al. 2011; Sinha and Jastreboff 2013; Tate et al. 2006, 2008). Emotional and physiological stressors, usually produced as a result of psychosocial experiences or hardships, are generally the most recognized forms of stress. They are typically associated as interfering with an abusers ability to obtain and sustain drug abstinence (Fox et al. 2013b; Sinha 2008a, b). In addition to these types of stressors, research also recognizes the impact of environmental stimuli or cues, historically associated with the drug of choice, as also having the potential to create an internal state of stress (Fox et al. 2013b; Goeders 2003). Given the potential for stress to arise from drug or environment cues as well as psychosocial influences and impact relapse (Higley et al. 2011; McKee et al. 2011), it seems obvious that successful treatment modalities and interventions would need to not only address the actual substance use and anticipated cycle, but also incorporate services which help moderate the other factors creating stress. For instance, if the stress is a product of additional psychopathology or other psychosocial influences, successful treatment must address the dual diagnosis and/or other psychosocial needs in order to advance both objectives—increasing remission success and decreasing additional symptomology. Overall, studies with outcome measures for both stress (regardless of how this is defined) and substance use are relatively scarce and those which do demonstrate at least moderate promise lack consistent replication. However, there is a growing body of research evaluating the impact of and appropriate treatment for co-occurring disorders—an indirect representation of the relationship between stress and substance abuse. Nevertheless, it is generally known that stress has negative implications for relapse. How these specific risks can be mitigated or incorporated into treatment program development is strikingly indistinguishable.

4.1 Predictors of Treatment Outcome

There are numerous stress-related elements which have been significantly associated with substance abuse treatment outcomes. The most obvious elements to consider include those discussed in the relapse predictors section (i.e., chronic stress exposure and psychopathology). However, research has also specifically evaluated factors associated with successful treatment outcomes. While some of these factors cannot be controlled, knowledge of the predictors can help clinicians properly screen substance abuse patients and help in the development of individualized treatment programs which incorporate research-supported stress management interventions. For instance, Higley et al. (2011) demonstrated that the level of stress-induced cravings recorded during treatment were associated with time to relapse, increased use during relapse, and less time of overall abstinence. The increased cravings levels of those with poor outcomes suggest that some individuals may be more prone to enhanced physiological responses and maintain a unique biological vulnerability to relapse. This association indicates that treatment programs which distinguish the clients with the highest craving responses may be more informed and equipped to incorporate high-risk interventions or relapse prevention measures. Looking at this relationship in another way and acknowledging that stress promotes drug cravings (Breese et al. 2005; Sinha et al. 2000), prioritizing stress management skills during treatment could also reduce experienced cravings and therefore may moderate relapse rates, although more research is needed to investigate these potential outcomes.

Another possible biological marker to treatment outcomes is the reactivity of the HPA axis. Higher salivary cortisol levels have been predictive of residential treatment program dropout (Daughters et al. 2009). While attempting to manage cortisol levels, by decreasing perceived stress and therefore stress responsivity, is one potential treatment objective, measuring cortisol levels as part of a screening process could also be useful in determining the dropout risk. In addition to increased craving response and cortisol levels, avoidant coping style, a lack of overall coping, self-efficacy, less education, and more frequent exposure to other substance users have also been indicated in unsuccessful treatment outcomes (Demirbas et al. 2012; Laurent et al. 1997; Rooke et al. 2011). Other studies suggest that self-reported social exclusion, personal control, and social regulation all affect a person's perceived level of stress, which in return further degrades that person's physical and mental health and potentially increases the risk for relapse (Cole et al. 2011). Treatments that target this perceived stress through the teaching of coping skills and psychoeducation, regarding the impact of powerlessness, may be more equipped to mitigate the detrimental cycle of stress and relapse. Taking these studies together, it seems that treatment programs which have knowledge of and incorporate established treatment outcome vulnerabilities may be more equipped to develop relevant treatment programs to facilitate successful treatment completion.

4.2 Treatment Modalities

Cognitive-Behavioral Therapy (CBT) and CBT-oriented interventions have become the standard for treating SUDs (see reviews in Dutra et al. 2008; Hayes et al. 2011; McHugh et al. 2010). A meta-analysis of 34 randomized controlled trials on the effectiveness of CBT for drug abuse and dependence found an overall moderate effect size ($d = 0.45$) with CBT and a contingency management intervention combination showing the highest effect sizes ($d = 1.02$) (Dutra et al. 2008; McHugh et al. 2010). While each intervention does not work equally well across diagnostic combinations, CBT has shown a relatively strong ability to reduce overall stress and substance use (Brewer et al. 2009; Drake et al. 2004; Sannibale et al. 2013). Although there is not much discussion or research on the *direct* treatment of stress, the apparent consensus is that substance abuse treatment cannot produce long-term positive outcomes without addressing compounding stress-related factors such as psychopathology and ongoing psychosocial needs (e.g., homelessness or unhealthy interpersonal relationships). The teaching and implementation of CBT interventions such as Relapse Prevention (RP), Motivational Interviewing (MI), case management, and skills training of coping, communication, and mindfulness are just a few examples of the potential to concurrently address stress and substance use (Dutra et al. 2008; McHugh et al. 2010). With the use of these interventions, stress may be reduced indirectly as a mechanism of (a) reduced cue reactivity, cravings and substance use, as well as (b) increased distress tolerance or emotion regulation through appropriate skill implementation. Furthermore, the implementation of learned skills such as healthy coping have the additional potential to directly reduce stress in that skills can help moderate stress as well as facilitate the avoidance of situations most likely to lead to stress. Research is clear that CBT and CBT-oriented interventions are effective at treating SUD; whether they are treating stress directly or indirectly is less obvious.

4.3 Co-occurring/Integrated Treatment

Co-occurring treatment, which concurrently addresses both substance use disorders and mental health disorders, is becoming the gold standard for treating dual diagnosis. While initial studies were inconsistent on the benefit of combining mental health and substance abuse treatment (see review in Drake et al. 1998) and some researchers still advise caution (Conrod and Stewart 2005; Torchalla et al. 2012), the majority of the research solidifies the idea that providing comprehensive, concurrent, and multidisciplinary services to those struggling with dual diagnosis is the most effective treatment option available to date (Cleminshaw et al. 2005; Drake et al. 1998, 2008; Durell et al. 1993). The efficacy of co-occurring treatment has become particularly salient with co-occurring posttraumatic stress disorder and substance dependence (McGovern et al. 2011; Mills et al. 2013). Since the mainstream acknowledgement of co-occurring treatment, researchers have begun to distinguish comprehensive integrated treatment (IT) as the ideal form of co-occurring treatment (Drake et al. 1998, 2004). Comprehensive IT typically refers to a coordinated, concurrent effort of mental health and substance abuse providers to provide motivational interventions, assertive outreach, intensive case management, individual counseling, and family interventions for ideally 18 months or longer (Drake et al. 1998). It is thought that it is this all-inclusive, long-term approach which facilitates treatment engagement, social support, and a true dual-diagnosis focus which not only leads to a reduction in substance use and mental health symptomology, but also extends substance use remission (Drake et al. 1998). Another potential strength of co-occurring treatment is the apparent compatibility to interweave diagnosis-specific CBT interventions into both the substance abuse and mental health treatment objectives.

It is clear that a perfect treatment of stress and substance abuse has not been established. While CBT, CBT-oriented interventions, and comprehensive integrated treatment certainly show promise and are moving in a more positive direction, there is still more to be discovered. For instance, Hein et al. (2012) highlighted the importance of client-modulated treatment dosage in ensuring that treatment services are given a true opportunity to produce desired outcomes. It is also clear that screening for treatment risk and incorporating relevant preventative measures can help reduce treatment dropout risk (Choi et al. 2013). This may be the most important component to facilitating successful treatment outcomes, since those which drop out of treatment do not receive all of the elements thought to facilitate symptom reduction and drug use remission. Even among methods which have shown initial positive indications, consistent replication is lacking. Research investigating the relationship and involved mechanisms of stress and substance abuse have certainly helped progress the overall understanding of these two interactive factors; the critical next step is determining how to incorporate this information into practical, evidenced-based treatment programs.

5 Conclusions

Chronic and acute stress exposure leads to maladaptive epigenetic and physiological changes that each has the potential, individually and cooperatively, to increase the vulnerability for substance use initiation, addiction, and relapse. While chronic stress has the potential to collectively increase substance abuse risks, acute stress exposure during each phase of life can also independently increase these vulnerabilities. Current research suggests that adaptations to numerous stress-related neurological mechanisms, such as the stress response and reward pathway systems, play an integral role in the transition from occasional to compulsive drug behaviors. These changes then produce unique responses, which then create further individualized risks for future stress and substance abuse. While there is ample evidence regarding a relationship between stress and substance abuse, it is not clear exactly what scenarios or combinations of stress and other factors (to include gender) will produce precisely which outcomes. The causal relationships between stress or substance use have not been established and it is possible that the direction of these effects may be reciprocal. It is likely that there are numerous other factors and systems involved in the stress by substance use/abuse interaction that are yet to be identified or adequately substantiated. The mechanisms that underlie these known vulnerabilities are even less clear. Despite the lack of understanding surrounding how all of these systems interact and affect each other, co-occurring and integrated treatments have shown initial success in acknowledging and incorporating some of these known vulnerabilities with specific CBT interventions. Even so, there are still vast risks factors associated with relapse and treatment outcomes which have not been consistently integrated into empirically based treatments.

The association between stress and substance abuse and addiction has been the focus of a great amount of research in the last two decades; however, there is still a need for further research in order to expand our overall knowledge and develop relevant clinical applications. There are several themes described in thischapter that could benefit from future research. First,there is great potential to expand our understanding of the identified life spanvariations in the relationship between stress and substance abuse through prospective studies that could look at the interaction of stress and substance abuse over time and across developmental transitions (i.e., early development, adolescence, adulthood). An examination of the impact of stress and substance abuse in specific life stages could help clarify the divergent mechanisms involved during each phase of life. These future studies could also help clarify the impact of genetics and the difference between stress exposure during one developmental phase and the compounding effect of early development stress with adolescent or adult stress. Another area that warrants greater attention is how to successfully integrate known risk factors for relapse and treatment outcome into practical, empirically based treatment programs. While knowledge of these vulnerabilities is certainly insightful in our broad understanding of the relationship between stress and substance abuse, the lack of practical and research-based treatment applications has prevented some of the risk factors from being clinically exploited. Finally, although there is a growing body of research demonstrating the success of comprehensive integrated treatment, clinical treatment-centered research has not adequately incorporated the factors shown in preclinical studies to enhance treatment outcomes. For example, one potential direction for incorporating enriched environments into research is to integrate measures which assess for housing satisfaction, community or family involvement, and regular participation in stimulating activities. An awareness of how an enriched environment impacts treatment outcomes could alter the way we envision and develop future residential and community treatment programs. Finally, stress is inherent to life and, in many situations, unavoidable. Through ongoing research and treatment development, there is the potential to modify the relationship of stress with ongoing substance use and abuse.

Acknowledgments Research reported in this chapter was supported by the National Institute on Drug Abuse (R01-DA026868), National Institute on Alcohol Abuse and Alcoholism (R01-AA014988), and the National Institute of Mental Health (R01-MH077684; R01-MH081181) of the National Institutes of Health. The content is solely the responsibility of the authors and does not necessarily represent the official views of the National Institutes of Health. Dr. Dougherty also acknowledges the support of the William & Marguerite Wurzbach Distinguished Professorship, and the University of Texas Star Program.

References

Adinoff B, Junghanns K, Kiefer F et al (2005) Suppression of the HPA axis stress-response: implications for relapse. Alcohol Clin Exp Res 29:1351–1355

Advani T, Koek W, Hensler JG (2009) Gender differences in the enhanced vulnerability of BDNF ± mice to mild stress. Int J Neuropsychopharmacol 12:583–588

Al'Absi M, Hatsukami DK, Davis G (2005) Attenuated adrenocorticotropic responses to psychological stress are associated with early smoking relapse. Psychopharmacology 181:107–117

Anda RF, Whitfield CL, Felitti VJ et al (2002) Adverse childhood experiences, alcoholic parents, and later risk of alcoholism and depression. Psychiatric Services 53(8):1001–1009

Back SE, Brady KT, Jackson JL et al (2005) Gender differences and stress reactivity among cocaine dependent individuals. Psychopharmacology 180:169–176

Becker HC, Lopez MF, Doremus-Fitzwater TL (2011) Effects of stress on alcohol drinking: a review of animal studies. Psychopharmacology (Berlin) 218(1):121–155. doi:10.1007/s00213-011-2443-9

Becker JB, Monteggia LM, Perro-Sinal TS et al (2007) Stress and disease: is being female a predisposing factor? J Neurosci 27(44):11851–11855

Brady KT, Randall CL (1999) Gender differences in substance use disorders. Psychiatr Clin North Am 22:241–252

Breese GR, Chu K, Dayas CV et al (2005) Stress enhancement of craving during sobriety: a risk for relapse. Alcohol Clin Exp Res 29:185–195

Brenner SL, Beauchaine TP (2011) Pre-ejection period reactivity and psychiatric comorbidity prospectively predict substance use initiation among middle-schoolers: a pilot study. Psychophysiology 48(11):1587–1595. doi:10.1111/j.469.8986.2011.0123.x

Brewer JA, Sinha R, Chen JA et al (2009) Mindfulness training and stress reactivity in substance abuse: results from a randomized, controlled stage 1 pilot study. Subst Abuse 30(4):306–317

Briand LA, Blendy JA (2010) Molecular and genetic substrates linking stress and addiction. Brain Res 219. doi:10.1016/j.brainres.2009.11.002

Campbell UC, Carroll ME (2001) Effects of ketoconazole on the acquisition of intravenous cocaine self-administration under different feeding conditions in rats. Psychopharmacology 154:311–318

Campbell UC, Szumlinski KK, Kippin TE (2009) Contribution of early environmental stress to alcoholism vulnerability. Alcohol 43(7):547–554. doi:10.1016/j.alcohol.2009.09.029

Carpenter LL, Tyrka AR, Ross NS et al (2009) Effect of childhood emotional abuse and age on cortisol responsivity in adulthood. Biol Psychiatry 66(1):69–75

Chen P, Jacobson KC (2012) Developmental trajectories of substance use from early adolescence to young adulthood: gender and racial/ethnic differences. J Adolescent Health 50(2):154–163

Choi S, Adam SM, MacMaster SA et al (2013) Predictors of residential treatment retention among individuals with co-occurring substance abuse and mental health disorders. J Psychoactive Drugs 45(2):122–131

Clarke T, Dempster E, Docherty SJ et al (2012) Multiple polymorphisms in genes of the adrenergic stress system confer vulnerability to alcohol abuse. Addict Biol 17(1):202–208. doi:10.1111/j.1369-1600.2010.00263.x

Cleminshaw HK, Shepler R, Newman I (2005) The integrated co-occurring treatment (ICT) model. J Dual Diagnosis 1(3):85–94

Coffey SF, Schumacher JA, Baillie LE et al (2010) Craving and physiological reactivity to trauma and alcohol cues in posttraumatic stress disorder and alcohol dependence. Exp Clin Psychopharm 18(4):340–349

Cole J, Logan TK, Walker R (2011) Social exclusion, personal control, self-regulation, and stress among substance abuse treatment clients. Drug Alcohol Depend 113:13–20

Conrod PJ, Steward SH (2005) A critical look at dual-focused cognitive-behavioral treatments for comorbid substance use and psychiatricdisorders: strengths, limitations, and future directions. J Cog Psych: Intern Q 19(3):261–284

D'Sa C, Fox HC, Hong AK et al (2011) Increased serum brain-derived neurotrophic factor (BDNF) is predictive of cocaine relapse outcomes: a prospective study. Biol Psychiatry 70(8):706–711

Dai X, Thavundayil J, Santella S et al (2007) Response of the HPA-axis to alcohol in stress as a function of alcohol dependence and family history of alcoholism. Psychoneuroendocrinology 32(3):293–305

Daughters SB, Richards JM, Gorka SM et al (2009) HPA axis response to psychological stress and treatment retention in residential substance abuse treatment: a prospective study. Drug Alcohol Depen 105:202–208

Demirbas H, Ilhan IO, Dogan YB (2012) Ways of problem solving as predictors of relapse in alcohol dependent male inpatients. Addict Behav 37(1):131–134

Dodge R, Sindelar J, Sinha R (2005) The role of depression symptoms in predicting drug abstinence in outpatient substance abuse treatment. J Subst Abuse Treat 28:189–196

Drake RE, Mercer-McFadden C, Meuser KT et al (1998) Review of integrated mental health substance abuse treatment for patients with dual disorders. Schizophr Bull 24(4):589–608

Drake RE, Mueser KT, Brunette MF et al (2004) A review of treatments for people with severe mental illnesses and co-occurring substance use disorders. Psychiatr Rehabil J 27(4):360–375

Drake RE, O'Neal EL, Wallach MA (2008) A systematic review of psychosocial research on psychosocial interventions for people with co-occurring severe mental and substance use disorders. J Subst Abuse Treat 34:123–138

Durell J, Lechtenberg B, Corse S et al (1993) Intensive care management of persons with chronic mental illness who abuse substances. Hosp Community Psych 44:415–416

Dutra L, Stathopoulou G, Basden S et al (2008) A meta-analytic review of psychosocial interventions for substance use disorders. Am J Psych 165(2):179–187

Englund MM, Egeland B, Oliva EM et al (2008) Childhood and adolescent predictors of heavy drinking and alcohol use disorders in early adulthood: a longitudinal developmental analysis. Addiction 103(S1):23–35

Enoch MA (2011) The role of early life stress as a predictor for alcohol and drug dependence. Psychopharmacology 214:17–31

Epstein DH, Marrone GF, Heishman SJ (2010) Tobacco, cocaine, and heroin: raving and use during daily life. Addict Behav 35:318–324

Evans BR, Greaves-Lord K, Eusar AS et al (2011) The relation between hypothalamic-pituitary-adrenal (HPA) axis activity and age of onset of alcohol use. Addiction 107:312–322. doi:10.1111/.1360-0443.2011.03568.x

Fox HC, Sofuoglu M, Morgan PT et al (2013a) The effects of exogenous progesterone on drug craving and stress arousal in cocaine dependence: impact of gender and cue type. Psychoneuroendocrinology (electronic print first). doi:10.1016/j/{sumeiem/2-12.12.022

Fox HC, Tuit KL, Sinha R (2013b) Stress system changes associated with marijuana dependence may increase craving for alcohol and cocaine. Hum Psychopharmacol 28:40–53. doi:10.1002/hup.2280

Fox HC, Wilker EH, Dreek MJ et al (2006) Reliability of salivary cortisol assessments and cocaine dependent individuals. J Psychopharmacol 20(5):650–655

Gerra G, Bassignana S, Zaimovic A et al (2003) Hypothalamic-pituitary-adrenal axis responses to stress in subjects with 3, 4-methylenediosy-methamphetamine ('ecstasy') use history: correlation with dopamine receptor sensitivity. Psych Res 120(2):115–124

Goeders NE (2003) The impact of stress on addiction. Eur Neuropsychopharmacol 12:435–441. doi:10.1016/j.euroneuro.2003.08.004

Goeders NE (1997) A neuroendocrine role in cocaine reinforcement. Psychoneuroendocrinology 22(4):237–259

Goeders NE, Guerin GF (1996) Effects of surgical and pharmacological adrenalectomy on the initiation and maintenance of intravenous cocaine self-administration in rats. Brain Res 722:145–152

Greenfield SF, Weiss RD, Muenz LR et al (1998) The effect of depression on return to drinking: a prospective study. Arch Gen Psych 55:259–265

Harris DS, Reus VI, Wolkowitz OM (2005) Repeated psychological stress testing in stimulant-dependent patients. Prog Neuropsychopharmacol Biol Psychiatry 29(5):669–677

Hayes SC, Villatte M, Levin M et al (2011) Open, aware, and active: contextual approaches as an emerging trend in the behavioral and cognitive therapies. Annu Rev Clin Psychol 7:141–168

Hein DA, Morgan-Lopez AA, Campbell AN et al (2012) Attendance and substance use outcomes for the seeking safety program: sometimes less is more. J Consult Clin Psychol 80(1):29–42

Higley AE, Crane NA, Spadoni AD et al (2011) Craving in response to stress induction in a human laboratory paradigm predicts treatment outcome in alcohol-dependent individuals. Psychopharmacology (Berlin, Germany) 218(1):121. doi:10.1007/s00213-011-2355-80

Higley JD, Hasert MF, Suomi SJ et al (1991) Nonhuman primate model of alcohol abuse: effects of early experience, personality, and stress on alcohol consumption. Proc Natl Acad Sci USA 88(16):7261–7265

Huizink AC, Ferdinand RF, Ormel J et al (2006) Hypothalamic-pituitary-adrenal axis activity and early onset of cannabis use. Addiction 101:1581–1588

Hyman SM, Paliwal P, Chaplin TM et al (2008) Severity of childhood trauma is predictive of cocaine relapse outcomes in women but not men. Drug Alcohol Depend 92:208–216

Junghanns KK, Backhous J, Tietz U (2003) Impaired serum cortisol stress response is a predictor of early relapse. Alcohol 38:189–193

Junghanns KK, Horbach R, Ehrenthal D et al (2007) Cortisol awakening response in abstinent alcohol-dependent patients as a marker of HPA-axis dysfunction. Psychoneuroendocrinology 32(8):1133–1137

Khantzian EJ (1997) The self-medication hypothesis of substance use disorders: a reconsideration and recent applications. Harv Rev Psychiatry 4:231–244

King G, Alicata D, Cloak C et al (2010) Psychiatric symptoms and HPA axis function in adolescent methamphetamine users. J Neuroimmune Pharmacol 5(4):582–591

Koob GF (2008) A role for brain stress systems in addiction. Neuron 59:11–34

Koob GF, Kreek MJ (2007) Stress, dysregulation of drug reward pathways, and the transition to drug dependence. Am J Psychiatry 164(8):1149–1159

Kosten TA, Kehoe P (2010) Immediate and enduring effects of neonatal isolation on maternal behavior in rats. Int J Dev Neurosci 28(1):53–61

Kosten TA, Miserendino MJ, Bombace JC et al (2005) Sex-selective effects of neonatal isolation on fear conditioning and foot shock sensitivity. Behav Brain Res 2(28):235–244

Kreibich AS, Briand L, Cleck JN et al (2009) Stress-induced potentiation of cocaine reward: a role for CRF_{R1} and CREB. Neuropsychopharmacology 34:2609–2617

Laurent L, Catanzaro SJ, Callan MK (1997) Stress, alcohol-related expectancies and coping preferences: a replication with adolescents of the Cooper et al. (1992) model. J Stud Alcohol 58:644–651

Li CS, Sinha R (2008) Inhibitory control and emotional stress regulation: neuroimaging evidence for frontal-limbic dysfunction in psychostimulant addiction. Neurosci Biobehav Rev 32(3):581–597

Liu X, Weiss F (2002) Additive effect of stress and drug cues on reinstatement of ethanol seeking: exacerbation by history or dependence and role of concurrent activation of corticotropin-releasing factor and opioid mechanisms. J Neurosci 22:7856–7861

Locke T, Newcomb M (2004) Child maltreatment, parent alcohol- and drug-related problems, polydrug problems, and parenting practices: a test of gender differences and four theoretical perspectives. J Fam Psychol 18(1):120–134

Majewska MD (2002) HPA axis and stimulant dependence: an enigmatic relationship. Psychoneuroendocrinology 27(1–2):5–12

Martini M, Valverde O (2012) A single episode of maternal deprivation impairs the motivation for cocaine in adolescent mice. Psychopharmacology 219:149–158

Mccarty CA, Rhew IC, Murowchick E et al (2012) Emotional health predictors of substance use initiation during middle school. Psychol Addict Behav 26(2):351–357. doi:10.1037/a0025630

McGovern MP, Lambert-Harris C, Alterman AI et al (2011) A randomized controlled trial comparing integrated cognitive behavioral therapy versus individual addiction counseling for co-occurring substance use and posttraumatic stress disorders. J Dual Diagn 7(4):207–227. doi:10.1080/15504263.2011.620425

McHugh RK, Hearon BA, Otto MW (2010) Cognitive-behavioral therapy for substance use disorders. Psychiatr Clin North Am 33(3):511–525

McKee S, Sinha R, Weinberger AH et al (2011) Stress decreases the ability to resist smoking and potentiates smoking intensity and reward. J of Psychopharmacol 25(4):490–502. doi:10.1177/0269881110376694

Miller PM, Hersen M, Eisler RM et al (1974) Effects of social stress on operant drinking of alcoholics and social drinkers. Behav Res Ther 12(2):67–72

Mills KL, Teesson M, Back SE et al (2013) Integrated exposure-based therapy for co-occurring posttraumatic stress disorder and substance dependence. JAMA 308(7):690–699

Milne BJ, Caspi A, Harrington H et al (2009) Predictive value of family history on severity of illness: the case for depression, anxiety, alcohol dependence, and drug dependence. Arch Gen Psychiatry 66(7):738–747. doi:10.1001/archgenpsychiatry.2009.55

Neisewander JL, Peartree NA, Pentkowski NS (2012) Emotional valence and context of social influences on drug abuse-related behavior in animal models of social stress and prosocial interaction. Psychopharmacology 224:33–56

Nestler EJ (2011) Hidden switches in the mind. Sci Am 305(6):76–83

Potenza MN, Hong KA, Lacadie CM et al (2012) Neural correlates of stress-induced and cue-induced drug craving: influences of sex and cocaine dependence. Am J Psychiatry 169(4):406–414

Pratt WM, Davidson D (2009) Role of the HPA axis and the A118G polymorphism of the mu-opioid receptor in stress-induced drinking behavior. Alcohol 44:358–365

Preussner JC, Champagne F, Meaney MJ et al (2004) Dopamine release in response to a psychological stress in humans and its relationship to early life maternal care: a positron emission tomography study using [^{11}C]Raclopride. J Neurosci 24(11):2825–2831

Randall CL, Robers JS, Del Boca FK et al (1999) Telescoping of landmark events associated with drinking: a gender comparison. J Stud Alcohol 60:252–260

Rawas RE, Thiriet N, Lardeux V et al (2009) Environmental enrichment decreases the rewarding but not the activating effects of heroin. Psychopharmacology 203:561–570

Rohsenow DJ, Monti PM (1999) Does urge to drink predict relapse after treatment? Alcohol Res Health 23:225–232

Rooke SE, Norberg MM, Copeland J (2011) Successful and unsuccessful cannabis quitters: comparing group characteristics and quitting strategies. Subst Abuse Treat Prev Policy 6(30):1–9

Sannibale C, Teesson M, Creamer M et al (2013) Randomized controlled trial of cognitive behavior therapy for comorbid post-traumatic stress and alcohol use disorders. Addiction 108:1397–1410

Sapolsky RM, Romero LM, Munch AU (2000) How do glucocorticoids influence stress responses? Integrating permissive, suppressive, stimulatory, and preparation actions. Endocr Rev 21:55–89

Schmid B, Blomeye D, Becker K et al (2009) The interaction between the dopamine transporter gene and age at onset in relation to tobacco and alcohol use among 19-year-olds. Addict Bio 14:489–499. doi:10.1111/j.1369-1700.2009.00171.x

Schoenbaum G, Stalnaker TA, Shaham Y (2007) A role for BDNF in cocaine reward and relapse. Nat Neurosci 10(8):935–936

Schramm-Sapyta NL, Walker DQ, Caster JM et al (2009) Are adolescents more vulnerable to drug addiction than adults?Evidence from animal models. Psychopharmacology 207:1–21. doi:10.1007/s00213-009-1585-5

Schwabe L, Dickenson A, Wolf OT (2011) Stress, habits, and drug addiction: a psychoneuroendocrinological perspective. Exp Clin Psychopharmacol 19(1):53–63. doi:10.1037/a0022212

Schwandt ML, Barr CS, Suomi SJ et al (2007) Age-dependent variation in behavior following acute ethanol administration in male and female adolescent rhesus macaques (*Macaca mulatta*). Alcohol Clin Exp Res 31(2):228–237

Sinha R (2001) How does stress increase risk of drug abuse and relapse? Psychopharmacology 158:343–359

Sinha R (2008a) Modeling stress and drug craving in the laboratory: implications for addiction treatment development (review). Addict Biol 14:84–98. doi:10.1111/j.1369-1600.2008.00134.x

Sinha R (2008b) Chronic stress, drug use, and vulnerability to addiction. Ann NY Acad Sci 1141:105–130. doi:10.1196/annals.1441.030

Sinha R (2009) Stress and addiction: a dynamic interplay of genes, environment and drug intake. Biol Psychiatry 66(2):100–101. doi:10.1016/j.bbiopsych.2009.05.003

Sinha R (2011) New findings on biological factors predicting addiction relapse vulnerability. Curr Psychiatry Rep 13(5):398–405. doi:10.1007/s.11920-011-0224-0

Sinha R, Fox H, Hong KA et al (2001) Effects on adrenal sensitivity, stress- and cue-induced craving, and anxiety on subsequent alcohol relapse and treatment outcomes. Arch Gen Psychiatry 68(9):942–952

Sinha R, Fuse T, Aubin LR et al (2000) Psychological stress, drug-related cues and cocaine cravings. Psychopharmacology 152(2):140–148

Sinha R, Jastreboff AM (2013) Stress as a common risk factor for obesity and addiction. Biol Psychiatry 73(9):827–835. doi:10.1016/j.biopsych.2013.01.032

Sinha R, Lacadie C, Skudlarski P et al (2005) Neural activity associated with stress-induced cocaine craving: a functional magnetic imaging study. Psychopharmacology 183(2):171–180

Sinha R, Shaham Y, Heilig M (2011) Translational and reverse translational research on the role of stress in drug craving and relapse. Psychopharmacology 218:69–82. doi:10.1007/s00213-011-2263-z

Solinas M, Chauvet C, Thiriet N et al (2008) Reversal of cocaine addiction byenvironmental enrichment. Proc Natl Acad Sci USA 105(44):17145–17150

Solinas M, Thiriet N, Chauvet C et al (2010) Prevention and treatment of drug addiction by environmental enrichment. Prog Neurobiol 92(4):572–592. doi:10.1016/j.pneurobio.2010.08.002

Solinas M, Thiriet N, Rawas RE et al (2009) Environmental enrichment during early stages of life reduces the behaviors, neurochemical, and molecular effects of cocaine. Neuropsychopharmacology 34:1102–1111

Spinelli S, Chefer S, Suomi SJ et al (2009) Early life stress induces long-term morphologic changes in primate brain. Arch Gen Psychiatry 66:658–665. doi:10.1001/archgenpsychiatry.2009.52

Tarullo AR, Gunnar MR (2006) Child maltreatment and the developing HPA axis. Horm Behav 50:632–639

Tate SR, Brown SA, Glasner SV et al (2006) Chronic life stress, acute stress events, and substance availability in relapse. Addict Res Theory 14(3):303–322

Tate SR, Wu J, McQuaid JR et al (2008) Comorbidity of substance dependence and depression: role of life stress and self-efficacy in sustaining abstinence. Psychol Addict Behav 22(1):47–57

Thomas SE, Bacon AK, Randall PK et al (2011) An acute psychosocial stressor increases drinking in non-treatment-seeking alcoholics. Psychopharmacology 218(1):19–28

Torchalla I, Nosen L, Rostam H et al (2012) Integrated treatment programs for individuals with concurrent substance sue disorders and trauma experiences: a systematic review and meta-analysis. J Subst Abuse Treat 42:65–77

Trickett PK, Noll JG, Susman EJ et al (2010) Attenuation of cortisol across development for victims of sexual abuse. Dev Psychopathol 22:165–175

Uhart M, Wand GS (2009) Stress, alcohol and drug interaction: an update of human research. Addict Biol 14(1):43–64. doi:10.1111/j.1369-1600.2008.00131.x

Ungless MA, Argilli E, Bonci A (2010) Effects of stress and aversion on dopamine neurons: implications for addiction. Neurosci Biobehav Rev 35:151–156. doi:10.1016/jneubiorev.2010.04.006

Verona E, Sachs-Ericsson N (2005) The intergenerational transmission of externalizing behaviors in adult participants: the mediating role of childhood abuse. J Consult Clin Psychol 73:1135–1145

Wetherington CL (2007) Sex-gender differences in drug abuse: a shift in the burden of proof? Exp Clin Psychopharmacol 15(5):411–417

Witt SH, Buchmann AF, Blomeyer D et al (2011) An interaction between a neuropeptide Y gene polymorphism and early adversity modulates endocrine stress responses. Psychoneuroendocrinology 36(7):1010–1020

Wrase J, Makris N, Braus DG et al (2008) Amygdala volume associated with alcohol abuse relapse and craving. Am J Psychiatry 165(9):1179–1184

Xu K, Hong A, Zhou Z et al (2012) Genetic modulation of plasma NPY stress response is suppressed in substance abuse: association with clinical outcomes. Psychoneuroendocrinology 37:554–564

Role of Stress, Depression, and Aging in Cognitive Decline and Alzheimer's Disease

Mak Adam Daulatzai

Abstract Late-onset Alzheimer's disease (AD) is a chronic neurodegenerative disorder and the most common cause of progressive cognitive dysfunction and dementia. Despite considerable progress in elucidating the molecular pathology of this disease, we are not yet close to unraveling its etiopathogenesis. A battery of neurotoxic modifiers may underpin neurocognitive pathology via deleterious heterogeneous pathologic impact in brain regions, including the hippocampus. Three important neurotoxic factors being addressed here include aging, stress, and depression. Unraveling "upstream pathologies" due to these disparate neurotoxic entities, vis-à-vis cognitive impairment involving hippocampal dysfunction, is of paramount importance. Persistent systemic inflammation triggers and sustains neuroinflammation. The latter targets several brain regions including the hippocampus causing upregulation of amyloid beta and neurofibrillary tangles, synaptic and neuronal degeneration, gray matter volume atrophy, and progressive cognitive decline. However, what is the fundamental source of this peripheral inflammation in aging, stress, and depression? This chapter highlights and delineates the inflammatory involvement—i.e., from its inception from gut to systemic inflammation to neuroinflammation. It highlights an upregulated cascade in which gut-microbiota-related dysbiosis generates lipopolysaccharides (LPS), which enhances inflammation and gut's leakiness, and through a Web of interactions, it induces stress and depression. This may increase neuronal dysfunction and apoptosis, promote learning and memory impairment, and enhance vulnerability to cognitive decline.

Keywords Aging · Stress · Depression · Gut dysbiosis · Neuroinflammation · Cognitive decline · Alzheimer's disease

M.A. Daulatzai (✉)
Sleep Disorders Group, EEE Department, Melbourne School of Engineering,
The University of Melbourne, Building 193, 3rd Floor, Room no. 3/344, Parkville,
VIC 3010, Australia
e-mail: makd@unimelb.edu.au

Contents

1 Introduction	266
2 Neurotoxic Modifiers: Aging, Stress, and Depression	267
2.1 Aging: The Old Bag of Tricks	267
2.2 Stress	269
2.3 Stress-Related Depression	272
3 Discussion	275
3.1 Aging and Neurodegeneration	275
3.2 Depression: Prodromal to AD	276
3.3 Gut Dysbiosis, Depression, and Cognitive Decline	277
4 Conclusion	280
References	281

1 Introduction

The number of individuals 65 year and older is projected to exceed 71.5 million in the year 2030 (Ferri et al. 2005). This would lead to a dramatic increase in the number of individuals suffering from aging-related diseases including Alzheimer's disease (AD). AD is a chronic neurodegenerative disorder marked by a progressive loss of memory and cognitive function. About 6 % of the elderly will develop AD, while about 94 % of aging seniors will have to cope and live with some memory and cognitive dysfunction. The dementia cases will nearly double every 20 years, costing about 65.7 million in 2030 and 115.4 million in 2050 (Ferri et al. 2005). It is essential, therefore, that we understand the etiopathogenesis mechanism(s) underlying aging-related decline in memory/cognition. This would aid in developing appropriate therapeutic strategies to retard and ameliorate cognitive impairment.

Mental health disorders, including depression, are on an increase globally. Lifestyle and environmental changes may be responsible in driving the increased prevalence (WHO 2009). Numerous factors are implicated to underpin the epidemic increase in mental health disorders (Reynolds et al. 1999; Wium-Andersen et al. 2012). These may include, but are not limited to, socioeconomic changes, alterations in dietary habits, chronic alcohol consumption in excess, sedentary lifestyle, paucity of adequate sunlight exposure, decrease in social support, and loneliness (Iannotti and Wang 2013; Nousen et al. 2013; Peterson et al. 2013). These factors may be responsible for upregulating stress and the development of depression.

The gastrointestinal (GI) commensal microbiota and the human host have an intimate, bidirectional interaction that is symbiotic in health. Such mutually beneficial interactions between the gut microbiota and the host influence GI physiology and systemic immunity, defense against pathogens, intestinal sensorimotor function, GI secretion, mucosal barrier function, detoxification of xenobiotics, energy harvest, general anabolism, and behavior. Thus, the GI microbiota influences the host's homeostasis through modulation of gene expression and immunological, physiological, and psychological functions. Conversely, the host modulates the composition and activity of its gut microbiome. In an array of physiological perturbations of the host, there is an alteration in the quality and quantity of the gut microbiota, generally referred to as dysbiosis. The dysbiotic gut microbiome thus has the propensity to trigger and exacerbate a number of disease states.

In recent years, we have become aware of several heterogeneous conditions, as consequences of intestinal dysbiotic microbiota; some of these include Guillain-Barré disease, seronegative spondyloarthropathies, non-celiac gluten sensitivity, irritable bowel syndrome, inflammatory diseases of gut, diabetes mellitus (DM2), and obesity. This pathogenic interconnection has led to substantial research on gut and its resident microbiota (Amin et al. 2007, 2008, 2009; Cani et al. 2012; Everard and Cani 2013; Forsythe and Kunze 2013; Vindigni et al. 2013). Further, the neuropsychological consequences owing to alterations in gut microbiota have been amply emphasized. Indeed, accumulating evidence supports that the gut microbiota impacts brain chemistry, activates neural pathways, and consequently has an influence on behavior/mental disorders (Konturek et al. 2011; Dinan and Cryan 2013; Foster and McVey Neufeld 2013; Park et al. 2013).

This chapter aims to delineate the upstream neurotoxic risk factors, interrelate these to aging, stress, and depression, and comment as to how these may underpin synaptic neuronal injury and dysfunction, leading to neurocognitive impairment. A range of neurotoxic insults impacts different brain regions including the hippocampus (Rossler et al. 2002; West et al. 1994, 2000). There is a vast literature on various aspects of the hippocampal pathology in a host of conditions, including aging, stress, depression, and AD. Selected facets of hippocampal pathophysiology, caused by selected neurotoxic factors, will be highlighted.

2 Neurotoxic Modifiers: Aging, Stress, and Depression

2.1 Aging: The Old Bag of Tricks

Age-related changes occur in several body systems and their contributions to various diseases are well documented. Several studies have explored the neurotoxic effect of age in inducing the neuropathological and clinical manifestation of cognitive dysfunction in the elderly (Hof 1997; Sarkar and Fisher 2006; Small 2001). In particular, inflammation and oxidative damage are characteristic features of neuropathology in age-associated diseases (Burke and Barnes 2006; Wilson et al. 2006; Daulatzai 2010a, 2012a, d, 2013a, b). With increasing age, the hippocampus undergoes early changes and long-term potentiation (LTP) function declines (Abe et al. 1999; Jacobson et al. 2013). Various studies have highlighted structural and functional alterations in the hippocampal network in aging. These alterations have been correlated with memory dysfunction (Craik and Simon 1980), including spatial memory and navigation (Newman and Kaszniak 2000), contextual source memory (Henkel et al. 1998), and recollection (Jennings and Jacoby 1997; Robitsek et al. 2008). Recent electrophysiological data have shed light on some of the possible neural mechanisms responsible for memory decline in the aged rodent hippocampus (Wilson et al. 2006). The hippocampus undergoes structural, inflammatory, atrophic, and electrophysiological changes during aging (Cerbai et al. 2012; Liu et al. 2012). The number of CA1 neurons decreases, which also shows a high level of cellular loss in aged rats (Cerbai et al. 2012). This would necessarily lead to a functional imbalance in the dentate gyrus (DG) and CA1–CA3 of the hippocampus (Wilson et al. 2006).

Functional and structural neuroimaging evidence has repeatedly demonstrated that the hippocampal volume, hippocampal activation, and neurocognitive performance are correlated; indeed, they may predict cognitive performance in old age and disease states (Jack et al. 2000a, b; Mormino et al. 2009; Rabinovici and Jagust 2009). Such structure–function correlative studies in the brain were mostly cross-sectional comparisons. However, a recent interesting study on hippocampus evaluated intra-individual fMRI signal and change in memory performance over two decades. The results found a positive relationship between activation change in the hippocampus and change in memory performance, thus validating the correlation between reduced hippocampal activation and declining memory performance (Persson et al. 2012).

Ultrahigh-resolution microstructural diffusion tensor imaging (msDTI) work showed changes in diffusion properties within hippocampal subfield gray matter in aging (Yassa et al. 2011). A correlation has been emphasized between the DG and CA3 functional rigidity and the same regions' fractional anisotropy (i.e., directional diffusion). Directional diffusion in gray matter is taken as an index of dendritic integrity (Jespersen et al. 2007; Yassa et al. 2010). Therefore, structural dendritic changes in the DG and CA3 regions are selectively vulnerable to the aging process and may therefore promote cognitive impairments (Yassa et al. 2011). Worse performance in total recall in elderly without dementia was strongly associated with a decrease in hippocampal volume (Reitz et al. 2009). This was shown to be due to a functional imbalance in the hippocampal DG and CA3 network.

Subjects without hypertension, diabetes mellitus, or hyperlipidemia underwent T2-weighted MRI brain screening. The prevalence of hippocampal atrophy and white matter lesions increased significantly with age, as did cerebral microbleeds (Chowdhury et al. 2011). Cerebral amyloid angiopathy was considered to underlie age-related brain changes (Chowdhury et al. 2011). Several studies have confirmed that the hippocampus shrinks with age (West 1993; Allen et al. 2005; Greenberg et al. 2008; Jernigan et al. 2001; Mu et al. 1999; Raz et al. 2004; Scahill et al. 2003; Schuff et al. 1999; Walhovd et al. 2005, 2011). Further, evaluations of the hippocampus in humans have shown that the DG is indeed quite vulnerable to aging (Mueller et al. 2008; Varela-Nallar et al. 2010). Similarly, non-human primates (Gazzaley et al. 1996; Small et al. 2002, 2004) and rodents (Small et al. 2004; Moreno et al. 2007) have also shown the DG vulnerability to aging (Chawla and Barnes 2007; Varela-Nallar et al. 2010). A significant relationship exists between the development of AD and premorbid hippocampal volume. Compared with 46 %

of those whose hippocampal volumes were in the lowest range, only 15 % of subjects with normal hippocampal volume developed AD (Jack et al. 1999). Individuals with mild or moderate AD show significantly greater gray matter loss (mean global rates) in the medial temporal lobe (Fox et al. 2001). MRI studies have provided evidence that a decrease in the volume of the hippocampus is linked to memory decline which may range from encoding new memories to retrieval of preexisting memories (Jespersen et al. 2007; Kramer et al. 2007; Reitz et al. 2009; Yassa et al. 2011).

Recent studies in aged rodents provide evidence for molecular and synaptic changes in the entorhinal cortex (ERC), where the perforant path (PP) commences (Stranahan et al. 2010) and in the hippocampus (Geinisman et al. 1992). One of the primary targets of PP input is the DG. Indeed, there is substantial documentation that DG also shows vulnerability to the aging process (West 1993; Gazzaley et al. 1996; Small et al. 2002; Moreno et al. 2007; Penner et al. 2010). Electrophysiologically recorded presynaptic potential at the PP-DG synapse (Barnes et al. 2000; Dieguez and Barea-Rodriguez 2004), as well as excitatory postsynaptic potentials (EPSPs) in the DG, shows reductions in aged rats (Barnes 1979; Barnes and McNaughton 1980; Barnes et al. 2000). This clearly suggests that physiological changes in the PP, DG, and CA3 may cause a reduction in dendritic integrity (Jespersen et al. 2007; Yassa et al. 2010), thus contributing to the hippocampal functional deficits. The PP degradation and a loss of functional connectivity between the ERC and DG may promote the age-related memory decline (Yassa et al. 2011). Finally, age-related alterations in both presynaptic and postsynaptic potentiation mechanisms in the aged rats reflect substantial changes in neuronal signal processing capabilities and local circuit function in the hippocampus; these may be pivotal in exacerbating poor spatial memory acquisition, retention, and eventual global cognitive dysfunction (Deupree et al. 1993).

2.2 Stress

Stress affects all humans. Selye (1936) defines stress as an acute threat to the homeostasis. It may be physical—therefore real or perceived—therefore psychological. However, these are triggered by events within or without of organism. Importantly, stress may evoke adaptive responses in order to maintain the internal milieu and thus ensure homeostasis and survival (Alfonso et al. 2006; Arnsten 2009; Heine et al. 2005; Kasselman et al. 2007).

Psychological stress is known to play an important role in functional GI disorders (FGID), e.g., irritable bowel syndrome (IBS) and non-celiac gluten sensitivity, by precipitating exacerbation of symptoms. Stress affects visceral sensory function in humans. Studies in experimental animals suggest that stress-induced visceral hypersensitivity is centrally mediated by endogenous corticotropin-releasing factor (CRF) and involvement of structures of the emotion pathway, e.g., the amygdale (Mönnikes et al. 2001). CRF-signaling pathways contribute in the endocrine, behavioral, and visceral responses to stress. Chronic psychological stress, however, results in reduced host defense, initiates intestinal inflammation, and alterations in gut physiology (Caso et al. 2008; Konturek et al. 2011). Further, endogenous CRF release, low-grade inflammation, ultrastructural epithelial abnormalities, and altered microbiome–host interactions causing dysbiosis and greater microbial translocation may play a significant role in mediating gut epithelial abnormalities as well as small and large intestine dysfunctions (Gareau et al. 2008; Zhang et al. 2009a, b; Maes 2008; Maes et al. 2011).

Acute or chronic stress can have deleterious effects on the brain and cognition in humans and animals (Guenzel et al. 2013; Hinwood et al. 2012; Holmes and Wellman, 2009; Margarinos et al.1996; Mika et al., 2012). Chronic stress is a risk factor for the development of cognitive dysfunction in animals and humans (Vander Weele et al. 2013; Yun et al. 2010; Bondi et al. 2008). Stress is quite common in AD patients, and the impact of stress on hippocampus-dependent declarative memory processes is well characterized (de Quervain et al. 2003; Grigoryan et al. 2013). In a 12-year follow-up, 38 % elderly developed mild cognitive impairment (MCI), with the risk of MCI increased by about 2 % for each one unit increase on the stress scale. Higher level of chronic psychological distress has been shown to increase the incidence of MCI (Wilson et al. 2007).

2.2.1 Stress, Glucocorticoid, and Hippocampus

Stress causes the release of glucocorticoid hormones, which are implicated in hippocampal neurogenesis. Studies have confirmed the effect of glucocorticoids on retrieval of hippocampus-dependent spatial memory (Roozendaal et al. 2003; Anacker et al. 2013). Glucocorticoids (GCs) in stress could enhance neuronal injury and promote learning and memory impairments. Indeed, high concentrations of cortisol (100 µM) decreased proliferation and neuronal differentiation into MAP2-positive neurons (Anacker and Pariante et al. 2012). There is a compelling evidence that neurons extending from the CA1 region of the hippocampus and from the subiculum project to the prefrontal cortex (PFC); this is referred to as the hippocampal–prefrontal (H-PFC) pathway. This pathway is critically involved in aspects of cognition related to executive function and to emotional regulation. Stress disorder displays structural and functional coupling anomalies within the H-PFC circuit. Considering that such a disorder involves varying degrees of cognitive impairment and emotional dysregulation, dysfunctional H-PFC pathway might play an important role in pathophysiology of cognition (Godsil et al. 2013). The GCs response to stressful stimuli is regulated by the hypothalamic–pituitary–adrenal (HPA) axis, which triggers the adrenal cortex to release GCs (cortisol in primates and corticosterone in mice and rats). Epidemiological evidence further supports a role for stress as a risk factor for AD because elderly individuals prone to psychological distress are more likely to develop the disorder than age-matched, nonstressed individuals (Wilson et al. 2005). Rodent and primates studies suggest that chronic exposure to elevated GCs has neurotoxic effects and lowers the threshold for hippocampal neuronal degeneration and loss (Uno et al. 1989; Sapolsky et al. 1990). There is ample evidence implicating HPA axis dysfunction in AD, reflected by markedly elevated basal level of circulating cortisol (Swanwick et al. 1998) and a failure of cortisol suppression after dexamethasone challenge (Nasman et al. 1995). Also, plasma levels of cortisol—the stress hormone—are correlated with the rate of dementia progression in patients with AD (Csernansky et al. 2006).

An altered response of the HPA system occurs in patients with AD, and these alterations may increase GCs levels (Hatzinger et al. 1995). At present, the mechanism by which increased HPA axis activity could accelerate the AD process is unknown (Jeong et al. 2006); however, impairment in declarative memory in stress may be related to a disturbance of medial temporal lobe function—possibly due to reduced blood flow (de Quervain et al. 2003). Another possible contributing factor to the development of cognitive deficits in stress is the impact of stress on synaptic plasticity. Interestingly, the molecular correlate of stress in the transgenic mice (3 × Tg-AD mice) was found to be an altered ratio of Aβ42/40 in both cortex and the hippocampus (Grigoryan et al. 2013). During chronic psychosocial stress (6-week stress period), these mice displayed increased levels of Aβ oligomers, intraneuronal Aβ, and decreased brain-derived neurotrophic factor (BDNF) levels, relative to controls (Rothman et al. 2012). Chronic stress also increases the expression of β1-integrin (CD29), a protein implicated in microglial ramification. This involvement may represent an important pathobiological mechanism through which microglia mediate the behavioral effects of chronic psychological stress (Hinwood et al. 2013).

Glucocorticoids affect the activity of transactive response DNA-binding protein-43 (TDP-43). TDP-43 is an RNA and DNA-binding protein involved in transcriptional repression, RNA splicing, and RNA metabolism during the stress response (Wilson et al. 2011). The accumulation of TDP-43 and its 25 kDa C-terminal fragment (TDP-25) is a hallmark of several neurodegenerative disorders. In a recent investigation, transgenic mice were utilized that overexpress TDP-25 (Caccamo et al. 2013). GCs increased the levels of soluble TDP-25 and exacerbated cognitive deficits. The neurotoxic action of TDP-25 potentiated by GCs increased the neurotoxic pathology via oxidative damage. Further, altering the brain's redox state, i.e., restoring ratio of reduced to oxidized glutathione, blocked the glucocorticoid-TDP-25-related deleterious effects (Caccamo et al. 2013).

TDP-43 has been linked to the pathogenesis of frontotemporal lobar degeneration (FTLD), amyotrophic lateral sclerosis (ALS), and AD (Kadocura et al. 2009; Caccamo et al. 2013). TDP-43 pathology also occurs in aging (Wilson et al. 2013), as well as in AD patients with hippocampal sclerosis (Wilson et al. 2011). Hippocampal sclerosis is characterized by selective neuronal loss in CA1 region of the hippocampus. TDP-43-immunoreactive neuronal inclusions have been documented in 20–30 % of AD brains (Kadocura et al. 2009). In contrast, TDP-43 pathology is infrequent (3 % or less) in neurologically normal elderly.

These findings suggest that glucocorticoid elevations in response to chronic stress may mediate hippocampal and neocortical damage, which may play an important role in AD pathogenesis (Gould et al. 1992; Mayer et al. 2006; David et al. 2009; Anacker et al. 2011a, b, 2013).

2.3 Stress-Related Depression

The interactions between chronic stress and the molecular, cellular, and behavioral alterations may promote the development of depression (Kendler et al. 1999; Caspi et al. 2003). Indeed, stress can lead to hippocampal atrophy similar to that noted in depression (Sapolsky 2000). Further, chronic stress recapitulates many behavioral characteristics of depression that respond to antidepressants (Willner 2005). Hence, the pathophysiology of depression is intimately intertwined with stress (Zhang et al. 2012), while the latter also has an effect on the mechanisms of neuroplasticity (Shors et al. 1989; McEwen 1999, 2002).

Stress affects neuronal signaling involving several heterogeneous pathways that impact synaptic plasticity (Belliveau et al. 1990). Chronic stress has been shown to increase the MAPK phosphorylation (Pardon et al. 2005; Lee et al. 2006). Acute stress leads to the phosphorylation of both MAPK and CaMKII (Ahmed et al. 2006) as does the acute glucocorticoid treatment (Revest et al. 2005). These processes may have a critical impact on long-term potentiation (LTP) of the hippocampus (Yang et al. 2004). Several acute and chronic stressor paradigms have shown an increase in phosphorylation of CREB in the hippocampus (Pardon et al. 2005; Ahmed et al. 2006; Nair et al. 2007), consistent with an alteration in signaling pathways and synaptic activity. Both acute stress and chronic stress influence neuroplasticity and lead to reductions in hippocampal BDNF mRNA levels (Nibuya et al. 1995, 1999; Smith et al. 1995; Russo-Neustadt et al. 2001; Rasmusson et al. 2002; Franklin and Perrot-Sinal 2006), as does glucocorticoid (Smith et al. 1995; Schaaf et al. 2000). The expression in the cell adhesion molecule (CAM) NCAM is critical for LTP (Muller et al. 1996). Further, neuronal activity regulates the expression of NCAM at the synapse and this expression is required for the induction of synaptic plasticity (Kim et al. 2006, 2007).

Chronic stress associated with glucocorticoid increase can lead to neuronal atrophy, notably affecting dendrites in both the medial prefrontal cortex (mPFC) and the hippocampus. Further, stress also enhances extracellular glutamate in various brain regions including the PFC (Bagley and Moghaddam 1997). Of note is the correlation between an increase in glucocorticoid and an increase in glutamate. For example, glucocorticoid excess increases glutamate release in the hippocampal CA1 region (Venero and Borrell 1999) and in the CA3 following chronic stress (Lowy et al. 1993). Several studies show that excess glutamatergic synapse may contribute to cell damage and cell death (Sapolsky 2000, 2003). Interestingly, glutamate antagonists attenuate and block the deleterious effects of glucocorticoid increase on dendrites in the hippocampus (Magariños and McEwen 1995). An increase in expression of the glial glutamate transporter (GLT-1) occurs due to excess glucocorticoid exposure (Zschocke et al. 2005; Autry et al. 2006).

Consequently, GLT-1 may promote increased reuptake of excess extracellular glutamate in persistent stress; this may result in variable pathology including atrophy of the apical dendrites in hippocampal CA3 pyramidal neurons (Magariños and McEwen 1995). Chronic stress-induced excess in extrasynaptic glutamate may exert an imbalance between extrasynaptic and synaptic NMDA receptors and perturb the mechanisms of synaptic plasticity, neuronal homeostasis, and viability (Guzowski et al. 1999; Guzowski et al. 2000; Guzowski et al. 2001; Hardingham et al. 2002; Hardingham and Bading 2003; Pittenger et al. 2007).

Neuronal cell death is the final pathological consequence of many CNS diseases, including AD. Hence, neuronal death is an important characteristic pathology of AD. Apoptosis is a variety of cell death that is involved in diverse physiological and pathological processes, including AD (Yang et al. 2008). Therefore, the effects of stress-level glucocorticoids were investigated on neuronal apoptosis in the hippocampus and neocortex. Histological examination showed that dexamethasone treatment induced degeneration of neurons in the hippocampus (CA1, CA3) and cortex (Haynes et al. 2001; Yang et al. 2008). The neuronal cell body became short and deeply stained with dye. Nuclear staining with Hoechst 33,258 showed nuclear condensation and fragmentation in dead cells.

Significant evidence indicates that mitochondria mediate oxidative stress. Oxidative stress has profound effect on cellular viability and damage in a range of different pathologies (Wang et al. 2013a, b; Rodríguez-Martínez et al. 2013). Mitochondria have been found to be essential in controlling at least certain apoptosis pathways (Green and Reed 1998). The mechanisms by which they exert this function include release of caspase(s) activators as cytochrome c and apoptosis-inducing factor (Liu et al. 1996), and disruption of electron transport and oxidative phosphorylation (Adachi et al. 1997; Garcia-Ruiz et al. 1997). Cytochrome c has been reported to be released from mitochondria into the cytosol of many cell types undergoing apoptosis (Kluck et al. 1997; Kong et al. 2013). Once in the cytosol, cytochrome c presumably binds to Apaf-1 and procaspase-9 and forms a functional apoptosome. Various stimuli that induce apoptosis lead to the release of cytochrome c from mitochondria, which then play a key role in a common pathway of activation of caspases (Mancini et al. 1998; Mulugeta et al. 2007). Indeed, cytosolic cytochrome c can bind Apaf-1 and subsequently trigger the sequential activation of caspase-9 and caspase-3 (Mancini et al. 1998). Activation of caspase-3 has been shown to be a key step in the execution process of apoptosis, and its inhibition can block apoptotic cell death. Activated caspases cleave a variety of target proteins, thereby disabling important cellular processes via breaking down structural components and eventually causing cell death (Thornberry et al. 1997). Numerous studies have also documented the activation of caspases in the AD brain as well as the cleavage of critical cellular proteins (Rohn et al. 2001; Su et al. 2002). These studies suggest that it is the caspase-mediated cleavage of important cellular proteins, per se, that may be important for driving the apoptotic pathology in AD.

2.3.1 Late-Life Depression

Depression is a highly prevalent mental health problem in the elderly, and it is associated with cognitive deficits (Smits et al. 2012). Depression represents an important psychiatric comorbidity in AD. It has a profound negative impact on memory and cognition in seniors, and AD patients (Baba 2010; Baba et al. 2012; Castrén 2013; Nihonmatsu-Kikuchi et al. 2013). For example, elevated memory complaints correlated significantly with depression (Fischer et al. 2010). In 2,160 community-dwelling Medicare recipients, (aged 65 years or older), an association existed between depression and mild cognitive impairment (MCI), suggesting that depression accompanies cognitive impairment (Richard et al. 2013).

The pathogenesis of depression is not fully known; however, studies suggest a hippocampal involvement. The hippocampus is a highly stress-sensitive brain region and has been implicated in the pathogenesis of major depressive disorder (MDD) (Kessler 1997; Thomas et al. 2007; McKinnon et al. 2009). In the hippocampus, diverse inhibitory circuits differentially control physiologically relevant network activities (Mendez et al. 2012). Patients with MDD were most impaired on measures of the hippocampus-dependent memory (Zakzanis et al. 1998). Compared with healthy controls, MDD patients show smaller volume of the hippocampus (Kempton et al. 2011; Lee et al. 2011; MacQueen and Frodl 2011). Meta-analyses of magnetic resonance imaging studies concluded that the hippocampus is smaller bilaterally in people with MDD (compared with age- and sex-matched controls) (Campbell et al. 2004; Videbech and Ravnkilde 2004). An aggregate meta-analysis has confirmed that the MDD patients indeed possess smaller left and right hippocampus volumes than controls (McKinnon et al. 2009). Interestingly, a correlation has been reported between the total number of depressive episodes and decreased hippocampus volume (Videbech and Ravnkilde 2004). Additionally, smaller hippocampus volumes are linked to a number of stigmata including the severity of depression (Vakili et al. 2000; Saylam et al. 2006), age at the onset of depression (Hickie et al. 2005; Taylor et al. 2005; Janssen et al. 2007), non-responsiveness to treatment (Vakili et al. 2000; Hsieh et al. 2002; Frodl et al. 2004), illness burden (Sheline et al. 1996, 1999; MacQueen et al. 2003; MacMaster and Kusumakar 2004), level of anxiety (Rusch et al. 2001), and polymorphisms in the serotonin transporter gene 5-HTTLPR (Taylor et al. 2005) as well as the BDNF gene (at position 66—Val66Met) (Frodl et al. 2007). However, duration of illness (>2 years) or more than 1 episode of illness may also impact the hippocampus volume adversely (Sheline et al. 1999; McKinnon et al. 2009). Not unexpectedly, hippocampal volume is significantly smaller in MDD patients during depressive episodes than during remission (Kempton et al. 2011).

Increasing severity of depression enhances the severity of psychopathological and neurological impairments, and even mild levels of depression can produce significant functional impairment in AD. Epidemiological studies suggest that depression may increase the risk of AD. In this regard, studies have found higher serum Aβ40/Aβ42 ratio in depression, compared with the controls (Baba 2010; Baba et al. 2012). Importantly, the serum Aβ40/Aβ42 ratio was negatively correlated with the age of onset of MDD (Namekawa et al. 2013). There is evidence that neuroplasticity is impaired in depression (Batsikadze et al. 2013; Player et al. 2013). Thus, it is not surprising that depressive symptoms have been associated with an increased risk of dementia (Dal Forno et al. 2005; Hermida et al. 2012; Fuhrer et al. 2003) and possible reductions in regional cerebral glucose metabolism (rCMRglu) in the hippocampus (Kennedy et al. 2001; Videbech et al. 2002; Davies et al. 2003). An eight-year follow-up showed that higher depressive symptoms were correlated with longitudinal cerebral blood flow (rCBF) decreases in temporal regions (Dotson et al. 2009). This may conceivably impact the hippocampal volume loss.

3 Discussion

3.1 Aging and Neurodegeneration

Age is the greatest risk factor for AD, and the most common cause of dementia in the elderly is AD. An estimated 13 and 45 % of Americans over age 65 and 85, respectively, suffer from AD (Alzheimer's Association 2012). A plethora of factors may lead to AD including aging, stress, and depression among others (Daulatzai 2013a, b; Jack et al. 2002).

An evaluation of cognitive dysfunction in neurodegenerative disorders must commence with the anatomical–physiological correlates and cellular and molecular mechanisms underlying normal cognition (Fjell et al. 2013; Giannakopoulos et al. 1996; Gonzales et al. 1995). The structure implicated in learning and memory is the hippocampus. This neocortical structure represents one of the most extensively studied region of the brain Price et al. 2001. This section focuses on the dysfunction and pathophysiological alterations of the hippocampus due to neurotoxic insults, including aging.

The hippocampal formation, a structure crucial for learning and memory, is particularly vulnerable to the aging process in several species (Small et al. 2004). Transient hypoperfusion targets the CA1 subregion and causes hippocampal-dependent memory deficits. Owing to inter-connectivity between the hippocampal subregions, dysfunction in one subregion affects physiologic functions in other hippocampal subregions (Barnes 1994), thus affecting the entire circuit. Sporadic AD is an age-related disorder; aging affects hippocampal performance by impairing normal neuronal physiology via synaptic dysfunction. Accumulating evidence has provided support to the feed-forward model that encompasses pernicious association between stress, aging, diabetes, insulin resistance, dysglycemia, sleep apnea, obesity, hypertension, and inflammation—all promoting hippocampal dysfunction. These have been implicated in AD pathogenesis (Peila et al. 2002; Row 2007; Yaffe et al. 2011; Umegaki 2012; Daulatzai 2010a, 2012a, d, 2013a, b).

3.2 Depression: Prodromal to AD

Epidemiological studies point out that depression increases the risk of AD. Since depressive symptoms in old age are found to be associated with AD risk, the question is whether depression is an independent risk factor for AD or an early clinical sign of AD pathogenesis. To answer this, older Catholic nuns, priests, and brothers were clinically evaluated annually, and their brain autopsied at death. Many members of the clergy are known to indulge in chronic alcohol intake (Vander Velt and McAllister 1962; Anderson et al. 2004). On the basis of linear regression, depressive symptoms were related to cognitive dysfunction proximate to death (Wilson et al. 2003). Importantly, the association of depressive symptoms with clinical AD and cognitive impairment appeared to be independent of cortical amyloid plaques and neurofibrillary tangles (Wilson et al. 2003). However, there is documented association of upregulated generation of amyloid beta (Aβ) with depression also (see below).

Patients with depression are usually characterized clinically by some cognitive impairments (Hollon et al. 2005; Jarrett et al. 2013). The well-documented mechanism that might juxtapose depression with AD is the amyloid beta (Aβ) pathophysiology, among others (Tran et al. 2011; Nihonmatsu-Kikuchi et al. 2013). Several studies have documented upregulated Aβ metabolism in depression. Higher plasma Aβ40/42 ratio in late-onset depression suggests this to be prodromal manifestation of AD (Baba 2010; Baba et al. 2012; Namekawa et al. 2013). Even younger subjects with depression show pathological plasma Aβ40/42 ratio (Baba 2010; Baba et al. 2012; Namekawa et al. 2013). It needs to be emphasized that quantitated CSF levels of these measures were similar in patients with major depression or AD (Hock et al. 1998).

3.2.1 Depression and Inflammation

Inflammation is a physiological process to overcome harmful stimuli and repairs tissues. However, when chronic, inflammation can impart deleterious consequences in terms of morbidity and mortality. Age-related changes in the immune system reflect immunosenescence.

Aging is a pro-inflammatory state. Accompanied with increased secretion of pro-inflammatory cytokines, older age represents a state of subacute chronic inflammation. Older subjects generally possess elevated levels of tumor necrosis factor (TNF-α), interleukin 1 (IL-1β), IL-6, and C-reactive protein (CRP) (Pedersen et al. 2000; Schram et al. 2007; Mooijaart et al. 2013; Michaud et al. 2013). There seems to be a relationship between inflammation and oxidative stress pathways and depressive disorder (Nunes et al. 2013). This potential pathophysiological mechanism may involve increased levels of pro-inflammatory cytokines, increased acute-phase proteins, increased oxidative stress, and decreased levels of antioxidants, thus reflecting a possible mechanistic framework underpinning depressive disorder (Nunes et al. 2013).

3.3 Gut Dysbiosis, Depression, and Cognitive Decline

The human gut contains ~1,000 different bacterial species with 99 % belonging to about 40 species (Neish 2009). The bacterial density increases progressively along the small bowel with a predominance of gram-negative aerobes and some obligate anaerobes (Zoetendal et al. 2006). GI enteric microbiome plays cardinal roles in physiological, nutritional, and immunological status of the host. Indeed, the composition and function of the gut microbiota has an impact on obesity and promote systemic inflammation and metabolic endotoxemia (Kopelman 2000; Clarke et al. 2012; Shen et al. 2013). Age affects the gut microbiota with a decrease in beneficial organisms such as anaerobes and bifidobacteria and an increase in enterobacteria (Woodmansey 2007; Guigoz et al. 2008; Hildebrandt et al. 2009; Jumpertz et al. 2011). Further, microbiota and microbiota-induced barrier dysfunction are modulated by various factors including age (Tran and Greenwood-Van Meerveld 2013), alcohol (Mutlu et al. 2009, 2012), fat consumption (Kopelman 2000; Jumpertz et al. 2011; de Wit et al. 2012), non-steroidal anti-inflammatories (NSAIDs) (Aabakken and Osnes 1989; Aabakken 1999), macronutrients such as protein (Tiihonen et al. 2010; Björklund et al. 2012), and indeed proinflammatory cytokines IFN-γ, IL-1β, and IL-6 (Al-Sadi et al. 2009; Daulatzai 2014a, b).

Endogenous LPS is a component of gram-negative bacterial cell walls and is continuously produced by the death of intestinal gram-negative bacteria; this significant proinflammatory effector then migrates into intestinal capillaries (Camilleri et al. 2012; Neal et al. 2006). It is the active component of this endotoxin that binds to LPS-binding protein (LBP), CD14, TLR4, and lymphocyte antigen 96, among other receptors. LPS infusion in mice resulted in increased fasting levels of glucose and insulin, as well as weight gain; the effects of this treatment on total body fat, steatosis, and adipose tissue were similar to those induced by a high-fat diet. The gut microbiota contributes to body fat deposition in mice (Bäckhed et al. 2004) since germ-free animals have a lower body fat content than do conventionally microbiome colonized animals (Bäckhed et al. 2004). Importantly, a 57 % increase in total body fat results, following the inoculation of germ-free mice with microbiota obtained from conventionally colonized adult animals (Bäckhed et al. 2004; Cani et al. 2007). Concomitant with these changes, macrophage numbers in the adipose tissue and levels of inflammatory markers increase both systemically and in the brain (Fig. 1). Hence, gut dysbiosis may be extremely important in upregulating stress metabolic dyshomeostasis, depression, and cognitive decline. Chronic dysbiosis in conjunction with inflamed gut may upregulate LPS permeability, increase in proinflammatory cytokines, systemic inflammation, and neuroinflammation.

Inflammation plays a pivotal role in its early pathology of AD (Engelhart et al. 2004; Griffin 2006; Churchill et al. 2006; Perry et al. 2007; Zhang et al. 2009a, b; Eikelenboom et al. 2010; 2012; Parachikova et al. 2007; Agostinho et al. 2010; Hoozemans et al. 2011). Diverse modulatory mechanisms (Daulatzai 2010a, b, 2011, 2012a, b, c, d, 2013a, b), including systemic and neuroinflammation, underpin cognitive decline and the development of AD. Emphasis has been placed on the matrix of

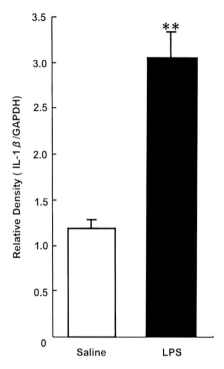

Fig. 1 Effect of lipopolysaccharide (LPS; 300 μg/kg ip) on IL-1β mRNA level in the hypothalamus 2 h after the injection. The amounts of IL-1β mRNA are expressed as ratios of densitometric measurements of the samples to the corresponding GAPDH as an internal standard. Values are means ±SE (n = 3 rats/group). (*Double asterick*) $P < 0.01$ (with permission, Hosoi et al. 2000). Similarly, systemic intraperitoneal injection of IL-β also promotes IL-β mRNA in the brain; however, subdiaphragmatic vagotomy blocked the IL-1β-induced increase of IL-1β mRNA in the brain stem and hippocampus (see Hansen et al. 1998)

Fig. 2 Plasma levels of LPS in healthy controls (21 ± 6 pg/m, n = 18), AD (61 ± 42 pg/ml, n = 18), and sALS (43 ± 18 pg/ml, n = 23) (with permission, Zhang et al. 2009a, b)

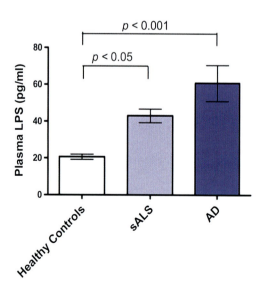

upstream interaction; these disparate key mechanisms/factors may have the "synergistic-additive impact" and promote an array of pathophysiological effects including memory dysfunction in AD (Bozzao et al. 2001; Daulatzai 2012c, d, 2013a, b).

Neuropathological investigations have revealed a variety of inflammation-related proteins including LPS, complement factors, acute-phase proteins, and pro-inflammatory cytokines in AD brains (Zhang et al. 2009a, b) (Fig. 2). These components of innate immunity promote crucial pathogenic cascade, involving systemic (Krstic et al. 2012) and neuroinflammation (Churchill et al. 2006; Perry et al. 2007; Eikelenboom et al. 2010), and are implicated in the etiopathogenesis of AD (Engelhart et al. 2004; Griffin 2006; Parachikovaa et al. 2007; Agostinho et al. 2010; Hoozemans et al. 2011; Eikelenboom et al. 2010, 2011, 2012).

Amyloid deposition is implicated in AD pathogenesis (Guo et al. 2002; Lee et al. 2008). Several pro-inflammatory cytokines including TNF-α, IL-1β, IL-6, or TGF-β can enhance amyloid precursor protein (APP) expression (Hirose et al. 1994; Buxbaum et al. 1992), upregulate β-secretase mRNA, protein, and enzymatic activity (Sastre et al. 2003), and thus increase Aβ formation (Blasko et al. 1999; Lee et al. 2008). A number of studies have confirmed that the systemic inflammation generated by LPS induces memory impairment (Shaw et al. 2001; Sparkman et al. 2005; Lee et al. 2008). The basis of this memory decline is the relationship between LPS-induced accumulation of Aβ and neuronal cell death; substantial increase of apoptotic cells was revealed in the hippocampus of LPS-treated mice (36.2 ± 3.6 %) relative to the controls (2.1 ± 0.8 %) (Lee et al. 2008). Chronic inflammation is linked to the onset and progression of AD-related pathologies in the brain, viz. deposition of Aβ plaques and neurofibrillary tangles (NFT). Peripheral inflammation was induced by using the bacterial endotoxin LPS in C57BL/6 J mice; hippocampus from LPS-treated mice contained significantly higher levels of Aβ1-42 (compared with saline controls) (Gasparini et al. 2004; Kahn et al. 2012), as was the cerebral cortex (Lee et al. 2008). Interestingly, even a single injection of LPS enhanced levels of both central and peripheral pro-inflammatory cytokines (Bossù et al. 2012; Kahn et al. 2012). The Morris water maze and contextual fear conditioning tests revealed cognitive deficits in LPS-treated mice (Kahn et al. 2012). Similarly, spatial memory was impaired in the mouse following sustained expression of IL-1β in the hippocampus (Moore et al. 2009). There are copious data showing that the pro-inflammatory mediators present in AD induce neuropathological cascade related to increases in Aβ generation (Eikelenboom et al. 1994; Turrin et al. 2001; Lee et al. 2010; Agostinho et al. 2010; Daulatzai 2010a, b, 2011, 2012a, b, c, d, 2013a, b).

Other than AD, substantial studies have also underscored an association between clinical depression and altered immune function. An increased translocation of gut's gram-negative bacterial LPS may be causally related to depression (Maes 2008). Indeed, depression is associated with inflammatory reaction indicated by increased production of pro-inflammatory cytokines, viz. IL-1β, IL-6, TNF-alpha, and interferon (IFN)-gamma (Maes et al. 2008; Nunes et al. 2013). Hippocampal neurogenesis, i.e., proliferation and differentiation of multipotent neural precursor cells, is an important source of neurons in adult brain. IL-1β impairs hippocampal neurogenesis. The hippocampal precursor cells showed a decrease in serotonergic neuronal differentiation in the presence of IL-1β; this effect was both dose dependent and could be blocked by an IL-1 receptor antagonist (IL-1ra) (Zhang et al. 2013). Further, following IL-1β treatment, lysate from the cultures of differentiated hippocampal neurons showed low levels of serotonin, Bcl-2, and phosphorylated extracellular-regulated kinase (pERK) (Zhang et al. 2013). These interactions may be one of the pathways contributing to the development of depression.

Clinical and experimental evidence document that increased peripheral cytokine levels and inflammation are associated with depression-like symptoms and neuropsychological disturbances in humans (Grigoleit et al. 2011). After endotoxin administration, the subjects showed a transient significant increase in the levels of anxiety and depression (Reichenberg et al. 2001). When LPS was given systemically to mice, it enhanced sickness response, depression-like behaviors, and expression of the relevant genes (Lawson et al. 2013). Current evidences support that inflammation and oxidative and nitrogen stress are intertwined, and their signaling mechanisms are important in upregulating depression pathophysiology (Biesmans et al. 2013; Ferreira Mello et al. 2013; Lawson et al. 2013; van Heesch et al. 2013). Finally, this insight suggests new therapeutic approaches for preventing/ameliorating depression caused by LPS and proinflammatory cytokine-induced mechanisms. Recently, vagus nerve stimulation (VNS) in lung disease has been shown to effectively attenuate the levels of proinflammatory cytokines including TNF-α, IL-1β, and IL-6 in bronchoalveolar lavage fluid (Chen et al. 2013). VNS also improves cardiac autonomic control and attenuates canine heart failure (Zhang et al. 2009a, b). VNS may also ameliorate LPS-induced inflammatory cascade via gut-brain axis stimulation. Indeed, such VNS effectiveness was experimentally tested in LPS-challenged (intraperitoneal injection) mice. The endotoxin induced intestinal tight junction injury with increased intestinal permeability and leakiness, as expected (Zhou et al. 2013). However, VNS of right cervical vagus nerve ameliorated the tight junction damage, decreased intestinal permeability, and reversed the decreased expression of tight junction proteins occludin and zonula occludens 1 (Zhou et al. 2013). These provide a strong case for VNS application in stress, depression, MCI, and AD (Christmas et al. 2013; Marras et al. 2013; Zhou et al. 2013). It should be noted that VNS is approved by the United States Food and Drug Administration as an adjunctive therapy for treatment-resistant depression (Groves and Brown 2005).

4 Conclusion

This chapter addressed the pathophysiological–neuromodulatory mechanism related to the gut dysbiosis which may upregulate gut–brain axis dysfunction—thus triggering depression and cognitive decline and promoting the development of AD.

Hence, future research must leverage manipulation of the gut microbiome for therapeutic purposes. Indeed, the treatment modality needs to exploit the known cascade of dysbiosis, gut leakiness, increased formation of LPS and proinflammatory cytokines, and inflammation. Stimulation of gut–brain axis may be an important conjoint strategy to overcome gut dysbiosis and gut-induced systemic inflammation, and attenuate neuroinflammation, depression, and cognitive dysfunction.

Acknowledgments I express my sincere gratitude to the following colleagues for their generosity in permitting me to use figures from their papers: Dr. Hiroshi Imaoka for Figure 1 and Dr. Mike McGrath for Fig. 2. Neela Khan of the Swinburne University, Victoria, Australia gave excellent help with EXCEL.

References

Aabakken L, Osnes M (1989) Non-steroidal anti-inflammatory drug-induced disease in the distal ileum and large bowel. Scand J Gastroenterol 163:48–55

Aabakken L (1999) Small-bowel side-effects of non-steroidal anti-inflammatory drugs. Eur J Gastroenterol Hepatol 11:383–388

Abe Y, Toyosawa K (1999) Age-related changes in rat hippocampal theta rhythms: a difference between type 1 and type 2 theta. J Vet Med Sci 61:543–548

Adachi S, Cross AR, Babior BM, Gottlieb RA (1997) Bcl-2 and the outer mitochondrial membrane in the inactivation of cytochrome c during Fas-mediated apoptosis. J Biol Chem 272:21878–21882

Agostinho P, Cunha RA, Oliveira C (2010) Neuroinflammation, oxidative stress and the pathogenesis of Alzheimer's disease. Curr Pharm Des 16:2766–2778

Ahmed T, Frey JU, Korz V (2006) Long-term effects of brief acute stress on cellular signaling and hippocampal LTP. J Neurosci 26:3951–3958

Alfonso J, Frick LR, Silberman DM, Palumbo ML, Genaro AM, Frasch AC (2006) Regulation of hippocampal gene expression is conserved in two species subjected to different stressors and antidepressant treatments. Biol Psych 59:244–251

Allen JS, Bruss J, Brown CK, Damasio H (2005) Normal neuroanatomical variation due to age: the major lobes and a parcellation of the temporal region. Neurobiol Aging 26:1245–1260

Al-Sadi R, Boivin M, Ma T (2009) Mechanism of cytokine modulation of epithelial tight junction barrier. Front Biosci 14:2765–2778

Alzheimer's Association (2012) Alzheimer's disease facts and figures. Alzheimers Dement 8:131–168

Amin PB, Diebel LN, Liberati DM (2007) Ethanol effects proinflammatory state of neutrophils in shock. J Surg Res 142:250–255

Amin PB, Diebel LN, Liberati DM (2008) The synergistic effect of ethanol and shock insults on Caco-2 cytokine production and apoptosis. Shock 29:631–635

Amin PB, Diebel LN, Liberati DM (2009) Dose-dependent effects of ethanol and E. coli on gut permeability and cytokine production. J Surg Res 157:187–192

Anacker C, Pariante CM (2012) Can adult neurogenesis buffer stress responses and depressive behaviour. Mol Psychiatry 17:9–10

Anacker C, Zunszain PA, Carvalho LA, Pariante CM (2011a) The glucocorticoid receptor: pivot of depression and of antidepressant treatment? Psychoneuroendocrinology 36:415–425

Anacker C, Zunszain PA, Cattaneo A, Carvalho LA, Garabedian MJ, Thuret S (2011b) Antidepressants increase human hippocampal neurogenesis by activating the glucocorticoid receptor. Mol Psych 16:738–750

Anacker C, Cattaneo A, Luoni A, Musaelyan K, Zunszain PA, Milanesi E (2013) Glucocorticoid-related molecular signaling pathways regulating hippocampal neurogenesis. Neuropsychopharmacology 38:872–883

Anderson PA, Grey SF, Nichols C, Parran TV Jr, Graham AV (2004) Is screening and brief advice for problem drinkers by clergy feasible? A survey of clergy. J Drug Educ 34:33–40

Arnsten AF (2009) Stress signalling pathways that impair prefrontal cortex structure and function. Nat Rev Neurosci 10:410–422

Autry AE, Grillo CA, Piroli GG, Rothstein JD, McEwen BS, Reagan LP (2006) Glucocorticoid regulation of GLT-1 glutamate transporter isoform expression in the rat hippocampus. Neuroendocrinology 83:371–379

Baba H (2010) A consideration for the mechanism of the transition from depression to dementia. Seishin Shinkeigaku Zasshi 112:1003–1008

Baba H, Nakano Y, Maeshima H, Satomura E, Kita Y, Suzuki T et al (2012) Metabolism of amyloid-β protein may be affected in depression. J Clin Psych 73:115–120

Bäckhed F, Ding H, Wang T, Hooper LV, Koh GY, Nagy A et al (2004) The gut microbiota as an environmental factor that regulates fat storage. Proc Natl Acad Sci USA 101:15718–15723

Bagley J, Moghaddam B (1997) Temporal dynamics of glutamate efflux in the prefrontal cortex and in the hippocampus following repeated stress: effects of pretreatment with saline or diazepam. Neuroscience 77:65–73

Barnes CA (1979) Memory deficits associated with senescence: A neurophysiological and behavioral study in the rat. J Comp Physiol Psychol 93:74–104

Barnes CA, McNaughton BL (1980) Physiological compensation for loss of afferent synapses in rat hippocampal granule cells during senescence. J Physiol 309:473–485

Barnes CA (1994) Normal aging: regionally specific changes in hippocampal synaptic transmission. Trends Neurosci 17:13–18

Barnes CA, Rao G, Houston FP (2000) LTP induction threshold change in old rats at the perforant path—granule cell synapse. Neurobiol Aging 21:613–620

Batsikadze G, Paulus W, Kuo MF, Nitsche MA (2013) Effect of serotonin on paired associative stimulation-induced plasticity in the human motor cortex. Neuropsychopharmacology 38:2260–2267

Belliveau JW, Rosen BR, Kantor HL, Rzedzian RR, Kennedy DN, McKinstry RC et al (1990) Functional cerebral imaging by susceptibility-contrast NMR. Magn Reson Med 14:538–546

Biesmans S, Meert TF, Bouwknecht JA, Acton PD, Davoodi N, De Haes P et al (2013) Systemic immune activation leads to neuroinflammation and sickness behavior in mice. Mediat Inflamm 2013:271359

Blasko I, Marx F, Steiner E, Hartmann T, Grubeck-Loebenstein B (1999) TNFalpha plus IFNgamma induce the production of Alzheimer beta-amyloid peptides and decrease the secretion of APPs. FASEB J 13:63–68

Bondi CO, Rodriguez G, Gould GG, Frazer A, Morilak DA (2008) Chronic unpredictable stress induces a cognitive deficit and anxiety-like behavior in rats that is prevented by chronic antidepressant drug treatment. Neuropsychopharmacology 33:320–331

Bossù P, Cutuli D, Palladino I, Caporali P, Angelucci F, Laricchiuta D et al (2012) A single intraperitoneal injection of endotoxin in rats induces long-lasting modifications in behavior and brain protein levels of TNF-α and IL-18. J Neuroinflammation 9:101

Bozzao A, Floris R, Baviera ME, Apruzzese A, Simonetti G (2001) Diffusion and perfusion MR imaging in cases of Alzheimer's disease: correlations with cortical atrophy and lesion load. AJNR Am J Neuroradiol 22:1030–1036

Burke SN, Barnes CA (2006) Neural plasticity in the ageing brain. Nat Rev Neurosci 7:30–40

Buxbaum JD, Oishi M, Chen HI, Pinkas-Kramarski R, Jaffe EA, Gandy SE et al (1992) Cholinergic agonists and interleukin 1 regulate processing and secretion of the Alzheimer beta/A4 amyloid protein precursor. Proc Natl Acad Sci USA 89:10075–10078

Caccamo A, Medina DX, Oddo S (2013) Glucocorticoids exacerbate cognitive deficits in TDP-25 transgenic mice via a glutathione-mediated mechanism: implications for aging, stress and TDP-43 proteinopathies. J Neurosci 33:906–913

Camilleri M, Madsen K, Spiller R (2012) Greenwood-Van Meerveld B, Verne GN. Intestinal barrier function in health and gastrointestinal disease. Neurogastroenterol Motil 24:503–512

Campbell J, Ciesielski CJ, Hunt AE, Horwood NJ, Beech JT, Hayes LA et al (2004) A novel mechanism for TNF-alpha regulation by p38 MAPK: involvement of NF-kappa B with implications for therapy in rheumatoid arthritis. J Immunol 173:6928–6937

Cani PD, Amar J, Iglesias MA, Poggi M, Knauf C, Bastelica D et al (2007) Metabolic endotoxemia initiates obesity and insulin resistance. Diabetes 56:1761–1772

Cani PD, Osto M, Geurts L, Everard A (2012) Involvement of gut microbiota in the development of low-grade inflammation and type 2 diabetes associated with obesity. Gut Microbes 3:279–288

Caso JR, Leza JC, Menchén L (2008) The effects of physical and psychological stress on the gastro-intestinal tract: lessons from animal models. Curr Mol Med 8:299–312

Caspi A, Sugden K, Moffitt TE, Taylor A, Craig IW, Harrington H et al (2003) Influence of life stress on depression: moderation by a polymorphism in the 5-HTT gene. Science 301:386–389

Castrén E (2013) Neuronal network plasticity and recovery from depression. JAMA Psych 70:983–989

Cerbai F, Lana D, Nosi D, Petkova-Kirova P, Zecchi S, Brothers HM et al (2012) The neuron-astrocyte-microglia triad in normal brain ageing and in a model of neuroinflammation in the rat hippocampus. PLoS One 7:e45250

Chawla MK, Barnes CA (2007) Hippocampal granule cells in normal aging: insights from electrophysiological and functional imaging experiments. Prog Brain Res 163:661–678

Chen C, Zhang Y, Du Z, Zhang M, Niu L, Wang Y, et al. (2013) Vagal efferent fiber stimulation ameliorates pulmonary microvascular endothelial cell injury by downregulating inflammatory responses. Inflammation 2013 Aug 4. (Epub ahead of print)

Chowdhury MH, Nagai A, Bokura H, Nakamura E, Kobayashi S, Yamaguchi S (2011) Age-related changes in white matter lesions, hippocampal atrophy, and cerebral microbleeds in healthy subjects without major cerebrovascular risk factors. J Stroke Cerebrovasc Dis 20:302–309

Christmas D, Steele JD, Tolomeo S, Eljamel MS, Matthews K (2013) Vagus nerve stimulation for chronic major depressive disorder: 12-month outcomes in highly treatment-refractory patients. J Affect Disord 2013 Jun 28. pii: S0165-0327(13)00459-X. doi: 10.1016/j.jad.2013.05.080. (Epub ahead of print)

Churchill L, Taishi P, Wang M, Brandt J, Cearley C, Rehman A et al (2006) Brain distribution of cytokine mRNA induced by systemic administration of interleukin-1beta or tumor necrosis factor alpha. Brain Res 1120:64–73

Clarke SF, Murphy EF, Nilaweera K, Ross PR, Shanahan F, O'Toole PW et al (2012) The gut microbiota and its relationship to diet and obesity: new insights. Gut Microbes 3:186–202

Craik FIM, Simon E (1980) New directions in memory and aging: proceedings of the George A Talland memorial conference, age differences in memory: the roles of attention and depth of processing. Lawrence Erlbaum Associates, Hillsdale, NJ. pp 95–112

Csernansky JG, Dong HX, Fagan AM, Wang L, Xiong CJ, Holtzman DM et al (2006) Plasma cortisol and progression of dementia in subjects with Alzheimer-type dementia. Am J Psych 163:2164–2169

Dal Forno G, Palermo MT, Donohue JE, Karagiozis H, Zonderman AB, Kawas CH (2005) Depressive symptoms, sex, and risk for Alzheimer's disease. Ann Neurol 57:381–387

Daulatzai MA (2010a) Early stages of pathogenesis in memory impairment during normal senescence and Alzheimer's disease. J Alzheimers Dis 20:355–367

Daulatzai MA (2010b) Conversion of Elderly to Alzheimer's Dementia: role of confluence of hypothermia and senescent stigmata—the plausible pathway. J Alzheimers Dis 21:1039–1063

Daulatzai MA (2011) Role of sensory stimulation in the amelioration of obstructive sleep apnea. Sleep Disord. Article ID 596879, doi:10.1155/2011/596879

Daulatzai MA (2012a) Memory and cognitive dysfunctions in Alzheimer's disease are Inextricably intertwined with neuroinflammation due to aging, obesity, obstructive sleep apnea, and other upstream risk factors. In: Costa E, Villalba E (eds) Horizons in neuroscience research. Nova Science Publishers Inc., NY, pp 69–106

Daulatzai MA (2012b) Pathogenesis of cognitive dysfunction in patients with obstructive sleep apnea: a hypothesis with emphasis on the nucleus tractus solitarius. Sleep Disord 2012:251096

Daulatzai MA (2012c) Neuroinflammation and dysfunctional nucleus tractus solitarius: their role in neuropathogenesis of Alzheimer's Dementia. Neurochem Res 37:846–868

Daulatzai MA (2012d) Quintessential risk factors: their role in promoting cognitive dysfunction and Alzheimer's disease. Neurochem Res 37:2627–2658

Daulatzai MA (2013a) Death by a thousand cuts in Alzheimer's disease: hypoxia—The prodrome. Neurotox Res 24:216–243

Daulatzai MA (2013b) Neurotoxic saboteurs: straws that break the Hippo's (Hippocampus) back drive cognitive impairment and Alzheimer's disease. Neurotox Res 24:407–459

Daulatzai MA (2014a) Chronic functional bowel syndrome enhances gut-brain axis dysfunction, neuroinflammation, cognitive impairment, and vulnerability to dementia. Neurochem Res 39:624–644

Daulatzai MA (2014b) Non-Celiac Gluten Sensitivity Triggers Gut Dysbiosis, Neuroinflammation, Gut-Brain Axis Dysfunction, and Vulnerability for Dementia. CNS Neurol Disord Drug Targets

David DJ, Samuels BA, Rainer Q, Wang JW, Marsteller D, Mendez I et al (2009) Neurogenesis-dependent and -independent effects of fluoxetine in an animal model of anxiety/depression. Neuron 62:479–493

Davies J, Lloyd KR, Jones IK, Barnes A, Pilowsky LS (2003) Changes in regional cerebral blood flow with venlafaxine in the treatment of major depression. Am J Psych 160:374–376

de Quervain DJ, Henke K, Aerni A, Treyer V, McGaugh JL, Berthold T et al (2003) Glucocorticoid-induced impairment of declarative memory retrieval is associated with reduced blood flow in the medial temporal lobe. Eur J Neurosci 17:1296–1302

de Wit N, Derrien M, Bosch-Vermeulen H, Oosterink E, Keshtkar S, Duval C et al (2012) Saturated fat stimulates obesity and hepatic steatosis and affects gut microbiota composition by an enhanced overflow of dietary fat to the distal intestine. Am J Physiol Gastrointest Liver Physiol 303:G589–G599

Deupree DL, Bradley J, Turner DA (1993) Age-related alterations in potentiation in the CA1 region in F344 rats. Neurobiol Aging 14:249–258

Dieguez DJ Jr, Barea-Rodriguez EJ (2004) Aging impairs the late phase of long-term potentiation at the medial perforant path-CA3 synapse in awake rats. Synapse 52:53–61

Dinan TG, Cryan JF (2013) Melancholic microbes: a link between gut microbiota and depression? Neurogastroenterol Motil 25:713–719

Dotson VM, Beason-Held L, Kraut MA, Resnick SM (2009) Longitudinal study of chronic depressive symptoms and regional cerebral blood flow in older men and women. Int J Geriatr Psych 24:809–819

Eikelenboom P, Zhan SS, Van Gool WA, Allsop D (1994) Inflammatory mechanisms in Alzheimer's disease. Trends Pharmacol Sci 15:447–450

Eikelenboom P, van Exel E, Hoozemans JJ, Veerhuis R, Rozemuller AJ, van Gool WA (2010) Neuroinflammation—an early event in both the history and pathogenesis of Alzheimer's disease. Neurodegener Dis. 7:38–41

Eikelenboom P, Veerhuis R, van Exel E, Hoozemans JJ, Rozemuller AJ, van Gool WA (2011) The early involvement of the innate immunity in the pathogenesis of late-onset Alzheimer'sdisease: neuropathological, epidemiological and genetic evidence. Curr Alzheimer Res 8:142–150

Eikelenboom P, Hoozemans JJ, Veerhuis R, van Exel E, Rozemuller AJ, van Gool WA (2012) Whether, when and how chronic inflammation increases the risk of developing late-onset Alzheimer's disease. Alzheimers Res Ther 4:15

Engelhart MJ, Geerlings MI, Meijer J, Kiliaan A, Ruitenberg A, van Swieten JC et al (2004) Inflammatory proteins in plasma and the risk of dementia: the Rotterdam study. Arch Neurol 61:668–672

Everard A, Cani PD (2013) Diabetes, obesity and gut microbiota. Best Pract Res Clin Gastroenterol 27:73–83

Ferreira Mello BS, Monte AS, McIntyre RS, Soczynska JK, Custódio CS, Cordeiro RC, et al. (2013) Effects of doxycycline on depressive-like behavior in mice after lipopolysaccharide (LPS) administration. J Psych Res 47:1521–1529

Ferri CP, Prince M, Brayne C, Brodaty H, Fratiglioni L, Ganguli M et al (2005) Global prevalence of dementia: a Delphi consensus study. Lancet 366:2112–2117

Fischer CE, Jiang D, Schweizer TA (2010) Determining the association of medical co-morbidity with subjective and objective cognitive performance in an inner city memory disorders clinic: a retrospective chart review. BMC Geriatr 10:89

Fjell AM, McEvoy L, Holland D, Dale AM, Walhovd KB (2013) Brain changes in older adults at very low risk for Alzheimer's disease. J Neurosci 33:8237–8242

Forsythe P, Kunze WA (2013) Voices from within: gut microbes and the CNS. Cell Mol Life Sci 70:55–69

Foster JA, McVey Neufeld KA (2013) Gut-brain axis: how the microbiome influences anxiety and depression. Trends Neurosci 36:305–312

Fox NC, Crum WR, Scahill RI, Stevens JM, Janssen JC, Rossor MN (2001) Imaging of onset and progression of Alzheimer's disease with voxel-compression mapping of serial magnetic resonance images. Lancet 358:201–205

Franklin TB, Perrot Sinal TS (2006) Sex and ovarian steroids modulate brain derived neurotrophic factor (BDNF) protein levels in rat hippocampus under stressful and non-stressful conditions. Psychoneuroendocrinology 31:38–48

Frodl T, Meisenzahl EM, Zetzsche T, Höhne T, Banac S, Schorr C et al (2004) Hippocampal and amygdala changes in patients with major depressive disorder and healthy controls during a 1-year follow-up. J Clin Psych 65:492–499

Frodl T, Schule C, Schmitt G, Born C, Baghai T, Zill P et al (2007) Association of the brain-derived neurotrophic factor Val66Met polymorphism with reduced hippocampal volumes in major depression. Arch Gen Psych 64:410–416

Fuhrer R, Dufouil C, Dartigues JF (2003) Exploring sex differences in the relationship between depressive symptoms and dementia incidence: prospective results from the PAQUID Study. J Am Geriatr Soc 51:1055–1063

Gareau MG, Silva MA, Perdue MH (2008) Pathophysiological mechanisms of stress-induced intestinal damage. Curr Mol Med 8:274–281

García-Ruiz C, Colell A, Marí M, Morales A, Fernández-Checa JC (1997) Direct effect of ceramide on the mitochondrial electron transport chain leads to generation of reactive oxygen species. Role of mitochondrial glutathione. J Biol Chem 272:11369–11377

Gasparini L, Rusconi L, Xu H, del Soldato P, Ongini E (2004) Modulation of beta-amyloid metabolism by non-steroidal anti-inflammatory drugs in neuronal cell cultures. J Neurochem 88:337–348

Gazzaley AH, Siegel SJ, Kordower JH, Mufson EJ, Morrison JH (1996) Circuit-specific alterations of N-methyl-D-aspartate receptor subunit 1 in the dentate gyrus of aged monkeys. Proc Natl Acad Sci USA 93:3121–3125

Geinisman Y, deToledo-Morrell L, Morrell F, Persina IS, Rossi M (1992) Age-related loss of axospinous synapses formed by two afferent systems in the rat dentate gyrus as revealed by the unbiased stereological dissector technique. Hippocampus 2:437–444

Giannakopoulos P, Hof PR, Kövari E, Vallet PG, Herrmann FR, Bouras C (1996) Distinct patterns of neuronal loss and Alzheimer's disease lesion distribution in elderly individuals older than 90 years. J Neuropathol Exp Neurol 55:1210–1220

Godsil BP, Kiss JP, Spedding M, Jay TM (2013) The hippocampal-prefrontal pathway: the weak link in psychiatric disorders? Eur Neuropsychopharmacol 23:1165–1181

González RG, Fischman AJ, Guimaraes AR, Carr CA, Stern CE, Halpern EF et al (1995) Functional MR in the evaluation of dementia: correlation of abnormal dynamic cerebral blood volume measurements with changes in cerebral metabolism on positron emission tomography with fludeoxyglucose F 18. Am J Neuroradiol 16:1763–1770

Gould E, Cameron HA, Daniels DC, Woolley CS, McEwen BS (1992) Adrenal hormones suppress cell division in the adult rat dentate gyrus. J Neurosci 12:3642–3650

Green DR, Reed JC (1998) Mitochondria and apoptosis. Science 281:1309–1312
Greenberg DL, Messer DF, Payne ME, Macfall JR, Provenzale JM, Steffens DC et al (2008) Aging, gender, and the elderly adult brain: an examination of analytical strategies. Neurobiol Aging 29:290–302
Griffin WS (2006) Inflammation and neurodegenerative diseases. Am J Clin Nutr 83:470S–474S
Grigoleit JS, Kullmann JS, Wolf OT, Hammes F, Wegner A, Jablonowski S et al (2011) Dose-dependent effects of endotoxin on neurobehavioral functions in humans. PLoS One 6:e28330
Grigoryan G, Biella G, Albani D, Forloni G, Segal M (2013) Stress impairs synaptic plasticity in triple-transgenic Alzheimer's disease mice: rescue by ryanodine. Neurodegener Dis 2013 Sep 4 (Epub ahead of print)
Groves Duncan A, Brown Verity J (2005) Vagal nerve stimulation: a review of its applications and potential mechanisms that mediate its clinical effects. Neurosci Biobehav Rev 29(3):493
Guenzel FM, Wolf OT, Schwabe L (2013) Stress disrupts response memory retrieval. Psychoneuroendocrinology 38:1460–1465
Guigoz Y, Doré J, Schiffrin EJ (2008) The inflammatory status of old age can be nurtured from the intestinal environment. Curr Opin Clin Nutr Metab Care 11:13–20
Guo JT, Yu J, Grass D, de Beer FC, Kindy MS (2002) Inflammation-dependent cerebral deposition of serum amyloid a protein in a mouse model of amyloidosis. J Neurosci 22:5900–5909
Guzowski JF, McNaughton BL, Barnes CA, Worley PF (1999) Environment-specific expression of the immediate-early gene Arc in hippocampal neuronal ensembles. Nat Neurosci 2:1120–1124
Guzowski JF, Lyford GL, Stevenson GD, Houston FP, McGaugh JL, Worley PF et al (2000) Inhibition of activity-dependent arc protein expression in the rat hippocampus impairs the maintenance of long-term potentiation and the consolidation of long-term memory. J Neurosci 20:3993–4001
Guzowski JF, Setlow B, Wagner EK, McGaugh JL (2001) Experience-dependent gene expression in the rat hippocampus after spatial learning: a comparison of the immediate-early genes Arc, c-fos, and zif268. J Neurosci 21:5089–5098
Hansen MK, Taishi P, Chen Z, Krueger JM (1998) Vagotomy blocks the induction of interleukin-1beta (IL-1beta) mRNA in the brain of rats in response to systemic IL-1beta. J Neurosci 18:2247–2253
Hardingham GE, Fukunaga Y, Bading H (2002) Extrasynaptic NMDARs oppose synaptic NMDRs by triggering CREB shut-off and cell death pathways. Nat Neurosci 5:405–414
Hardingham GE, Bading H (2003) The yin and yang of NMDA receptor signaling. Trends Neurosci 26:81–89
Hatzinger M, Z'Brun A, Hemmeter U, Seifritz E, Baumann F, Holsboer-Trachsler E et al (1995) Hypothalamic-pituitary-adrenal system function in patients with Alzheimer's disease. Neurobiol Aging 16:205–209
Haynes LE, Griffiths MR, Hyde RE, Barber DJ, Mitchell IJ (2001) Dexamethasone induces limited apoptosis and extensive sublethal damage to specific subregions of the striatum and hippocampus: implications for mood disorders. Neuroscience 104:57–59
Heine VM, Zareno J, Maslam S, Joels M, Lucassen PJ (2005) Chronic stress in the adult dentate gyrus reduces cell proliferation near the vasculature and VEGF and Flk-1 protein expression. Eur J Neurosci 21:1304–1314
Henkel LA, Johnson MK, De Leonardis DM (1998) Aging and source monitoring: cognitive processes and neuropsychological correlates. J Exp Psychol Gen 127:251–268
Hermida AP, McDonald WM, Steenland K, Levey A (2012) The association between late-life depression, mild cognitive impairment and dementia: is inflammation the missing link? Expert Rev Neurother 12:1339–1350
Hickie I, Naismith S, Ward PB, Turner K, Scott E, Mitchell P et al (2005) Reduced hippocampal volumes and memory loss in patients with early-and late-onset depression. Br J Psych 186:197–202

Hildebrandt MA, Hoffmann C, Sherrill-Mix SA, Keilbaugh SA, Hamady M, Chen YY et al. (2009) High-fat diet determines the composition of the murine gut microbiome independent-lyofobesity. Gastroenterology 137:1716–1724.e1-2

Hinwood M, Morandini J, Day TA, Walker FR (2012) Evidence that microglia mediate the neurobiological effects of chronic psychological stress on the medial prefrontal cortex. Cereb Cortex 22:1442–1454

Hinwood M, Tynan RJ, Charnley JL, Beynon SB, Day TA, Walker FR (2013) Chronic stress induced remodeling of the prefrontal cortex: structural re-organization of microglia and the inhibitory effect of minocycline. Cereb Cortex 23:1784–1797

Hirose Y, Imai Y, Nakajima K, Takemoto N, Toya S, Kohsaka S (1994) Glial conditioned medium alters the expression of amyloid precursor protein in SH-SY5Y neuroblastoma cells. Biochem Biophys Res Commun 198:504–509

Hock C, Golombowski S, Müller-Spahn F, Naser W, Beyreuther K, Mönning U et al (1998) Cerebrospinal fluid levels of amyloid precursor protein and amyloid beta-peptide in Alzheimer's disease and major depression—inverse correlation with dementia severity. Eur Neurol 39:111–118

Hof PR (1997) Morphology and neurochemical characteristics of the vulnerable neurons in brain aging and Alzheimer's disease. Eur Neurol 37:71–81

Hollon SD, Jarrett RB, Nierenberg AA, Thase ME, Trivedi M, Rush AJ (2005) Psychotherapy and medication in the treatment of adult and geriatric depression: which monotherapy or combined treatment? J Clin Psych 66:455–468

Holmes A, Wellman CL (2009) Stress-induced prefrontal reorganization and executive dysfunction in rodents. Neurosci Biobehav Rev 33:773–783

Hoozemans JJ, Rozemuller AJ, van Haastert ES, Eikelenboom P, van Gool WA (2011) Neuroinflammation in Alzheimer's disease wanes with age. J Neuroinflammation 8:171

Hosoi T, Okuma Y, Nomura Y (2000) Electrical stimulation of afferent vagus nerve induces IL-1beta expression in the brain and activates HPA axis. Am J Physiol Regul Integr Comp Physiol 279:R141–R147

Hsieh MH, McQuoid DR, Levy RM, Payne ME, MacFall JR, Steffens DC (2002) Hippocampal volume and antidepressant response in geriatric depression. Int J Geriatr Psych 17:519–525

Iannotti RJ, Wang J (2013) Patterns of physical activity, sedentary behavior, and diet in US adolescents. J Adolesc Health 53:280–286

Jack CR Jr, Petersen RC, Xu YC, O'Brien PC, Smith GE, Ivnik RJ et al (1999) Prediction of AD with MRI based hippocampal volume in mild cognitive impairment. Neurology 52:1397–1403

Jack CR Jr, Petersen RC, Xu Y, O'Brien PC, Smith GE, Ivnik RJ et al (2000a) Rates of hippocampal atrophy correlate with change in clinical status in aging and AD. Neurology 55:484–489

Jack CR Jr, Petersen RC, Farber NB, Rubin EH, Newcomer JW, Kinscherf DA et al (2000b) Increased neocortical neurofibrillary tangle density in subjects with Alzheimer disease and psychosis. Arch Gen Psych 57:1165–1173

Jack CR Jr, Dickson DW, Parisi JE, Xu YC, Cha RH, O'Brien PC et al (2002) Antemortem MRI findings correlate with hippocampal neuropathology in typical aging and dementia. Neurology 58:750–757

Jacobson TK, Howe MD, Schmidt B, Hinman JR, Escabi MA, Markus EJ (2013) Hippocampal theta, gamma, and theta{-}- gamma coupling: Effects of aging, environmental change, and cholinergic activation. J Neurophysiol 2013 Jan 9. (Epub ahead of print)

Janssen J, Hulshoff Pol HE, de Leeuw FE, Schnack HG, Lampe IK, Kok RM, et al. (2007) Hippocampal volume and subcortical white matter lesions in late life depression: comparison of early and late onset depression. J Neurol Neurosurg Psych 78:638–440

Jarrett RB, Minhajuddin A, Gershenfeld H, Friedman ES, Thase ME (2013) Preventing depressive relapse and recurrence in higher-risk cognitive therapy responders: a randomized trial of continuation phase cognitive therapy, fluoxetine, or matched Pill Placebo. JAMA Psych. 2013 Sep 4. doi:10.1001/jamapsychiatry.2013.1969. (Epub ahead of print)

Jennings JM, Jacoby LL (1997) An opposition procedure for detecting age-related deficits in recollection: telling effects of repetition. Psychol Aging 12:352–361

Jeong YH, Park CH, Yoo J, Shin KY, Ahn SM, Kim HS et al (2006) Chronic stress accelerates learning and memory impairments and increases amyloid deposition in APPV717I-CT100 transgenic mice, an Alzheimer's disease model. FASEB J 20:729–731

Jernigan TL, Archibald SL, Fennema-Notestine C, Gamst AC, Stout JC, Bonner J et al (2001) Effects of age on tissues and regions of the cerebrum and cerebellum. Neurobiol Aging 22:581–594

Jespersen SN, Kroenke CD, Østergaard L, Ackerman JJ, Yablonskiy DA (2007) Modeling dendrite density from magnetic resonance diffusion measurements. Neuroimage 34:1473–1486

Jumpertz R, Le DS, Turnbaugh PJ, Trinidad C, Bogardus C, Gordon JI et al (2011) Energy-balance studies reveal associations between gut microbes, caloric load, and nutrient absorption in humans. Am J Clin Nutr 94:58–65

Kadokura A, Yamazaki T, Lemere CA, Takatama M, Okamoto K (2009) Regional distribution of TDP-43 inclusions in Alzheimer disease (AD) brains: their relation to AD common pathology. Neuropathology 29:566–573

Kahn MS, Kranjac D, Alonzo CA, Haase JH, Cedillos RO, McLinden KA et al (2012) Prolonged elevation in hippocampal Aβ and cognitive deficits following repeated endotoxin exposure in the mouse. Behav Brain Res 229:176–184

Kasselman LJ, Kintner J, Sideris A, Pasnikowski E, Krellman JW, Shah S et al (2007) Dexamethasome treatment and ICAM-1 deficiency impair VEGF-induced angiogenesis in adult brain. J Vasc Res 44:283–291

Kempton MJ, Salvador Z, Munafò MR, Geddes JR, Simmons A, Frangou S et al (2011) Structural neuroimaging studies in major depressive disorder. Meta-analysis and comparison with bipolar disorder. Arch Gen Psych 68:675–690

Kendler KS, Karkowski LM, Prescott CA (1999) Causal relationship between stressful life events and the onset of major depression. Am J Psych 156:837–841

Kennedy SH, Evans KR, Krüger S, Mayberg HS, Meyer JH, McCann S et al (2001) Changes in regional brain glucose metabolism measured with positron emission tomography after paroxetine treatment of major depression. Am J Psych 158:899–905

Kessler RC (1997) The effects of stressful life events on depression. Annu Rev Psychol 48:191–214

Kim JJ, Song EY, Kosten TA (2006) Stress effects in the hippocampus: synaptic plasticity and memory. Stress 9:1–11

Kim JJ, Lee HJ, Welday AC, Song E, Cho J, Sharp PE et al (2007) Stress-induced alterations in hippocampal plasticity, place cells, and spatial memory. Proc Natl Acad Sci USA 104:18297–18302

Kluck RM, Wetzel EB, Green DR, Newmeyer DD (1997) The release of cytochrome c from mitochondria: a primary site for Bcl-2 regulation of apoptosis. Science 275:1132–1136

Kopelman PG (2000) Obesity as a medical problem. Nature 404:635–643

Kramer JH, Mungas D, Reed BR, Wetzel ME, Burnett MM, Miller BL et al (2007) Longitudinal MRI and cognitive change in healthy elderly. Neuropsychology 21:412–418

Kong D, Zheng T, Zhang M, Wang D, Du S, Li X et al (2013) Static mechanical stress induces apoptosis in rat endplate chondrocytes through MAPK and mitochondria-dependent caspase activation signaling pathways. PLoS One 8(7):e69403

Konturek PC, Brzozowski T, Konturek SJ (2011) Stress and the gut: pathophysiology, clinical consequences, diagnostic approach and treatment options. J Physiol Pharmacol 62:591–599

Krstic D, Madhusudan A, Doehner J, Vogel P, Notter T, Imhof C et al (2012) Systemic immune challenges trigger and drive Alzheimer-like neuropathology in mice. J Neuroinflammation 9:151

Lawson MA, McCusker RH, Kelley KW (2013) Interleukin-1 beta converting enzyme is necessary for development of depression-like behavior following intracerebroventricular administration of lipopolysaccharide to mice. J Neuroinflammation 10:54

Lee HY, Tae WS, Yoon HK, Lee BT, Paik JW, Son KR et al (2011) Demonstration of decreased gray matter concentration in the midbrain encompassing the dorsal raphe nucleus and the limbic subcortical regions in major depressive disorder: An optimized voxel-based morphometry study. J Affect Disord 133:128–136

Lee JW, Lee YK, Yuk DY, Choi DY, Ban SB, Oh KW et al (2008) Neuro-inflammation induced by lipopolysaccharide causes cognitive impairment through enhancement of beta-amyloid generation. J Neuroinflammation 5:37

Lee SY, Kang JS, Song GY, Myung CS (2006) Stress induces the expression of heterotrimeric G protein beta subunits and the phosphorylation of PKB/Akt and ERK1/2 in rat brain. Neurosci Res 56:180–192

Lee YJ, Han SB, Nam SY, Oh KW, Hong JT (2010) Inflammation and Alzheimer's disease. Arch Pharm Res 33:1539–1556

Liu X, Kim CN, Yang J, Jemmerson R, Wang X (1996) Induction of apoptotic program in cell-free extracts: requirement for dATP and cytochrome c. Cell 86:147–157

Liu X, Wu Z, Hayashi Y, Nakanishi H (2012) Age-dependent neuroinflammatory responses and deficits in long-term potentiation in the hippocampus during systemic inflammation. Neuroscience 216:133–142

Lowy MT, Gault L, Yamamoto BK (1993) Adrenalectomy attenuates stress-induced elevations in extracellular glutamate concentrations in the hippocampus. J Neurochem 61:1957–1960

MacMaster FP, Kusumakar V (2004) Hippocampal volume in early onset depression. BMC Med 29(2):2

MacQueen G, Frodl T (2011) The hippocampus in major depression: evidence for the convergence of the bench and bedside in psychiatric research? Mol Psych 16:252–264

MacQueen GM, Campbell S, McEwen BS, Macdonald K, Amano S, Joffe RT et al (2003) Course of illness, hippocampal function, and hippocampal volume in major depression. Proc Natl Acad Sci USA 100:1387–1392

Maes M (2008) The cytokine hypothesis of depression: inflammation, oxidative & nitrosative stress (IO&NS) and leaky gut as new targets for adjunctive treatments in depression. Neuro Endocrinol Lett 29:287–291

Maes M, Kubera M, Obuchowiczwa E, Goehler L, Brzeszcz J (2011) Depression's multiple comorbidities explained by (neuro)inflammatory and oxidative & nitrosative stress pathways. Neuro Endocrinol Lett 32:7–24

Magariños AM, McEwen BS (1995) Stress-induced atrophy of apical dendrites of hippocampal CA3c neurons: involvement of glucocorticoid secretion and excitatory amino acid receptors. Neuroscience 69:89–98

Magarinos AM, McEwen BS, Flugge G, Fuchs E (1996) Chronic psychosocial stress causes apical dendritic atrophy of hippocampal CA3 pyramidal neurons in subordinate tree shrews. J Neurosci 16:3534–3540

Mancini M, Nicholson DW, Roy S, Thornberry NA, Peterson EP, Casciola-Rosen LA et al (1998) The caspase-3 precursor has a cytosolic and mitochondrial distribution: implications for apoptotic signaling. J Cell Biol 140:1485–1495

Marras CE, Chiesa V, De Benedictis B, Franzini A, Rizzi M, Villani F et al (2013) Vagus nerve stimulation in refractory epilepsy: new indications and outcome assessment. Epilepsy Behav 28:374–378

Mayer JL, Klumpers L, Maslam S, de Kloet ER, Joëls M, Lucassen PJ (2006) Brief treatment with the glucocorticoid receptor antagonist mifepristone normalises the corticosterone-induced reduction of adult hippocampal neurogenesis. J Neuroendocrinol 18(8):629–631

McEwen BS (1999) Stress and hippocampal plasticity. Ann Rev Neurosci 22:105–122

McEwen BS (2002) Protective and damaging effects of stress mediators: the good and bad sides of the response to stress. Metabolism 51:2–4

McKinnon MC, Yucel K, Nazarov A, MacQueen GM (2009) A meta-analysis examining clinical predictors of hippocampal volume in patients with major depressive disorder. J Psych Neurosci 34:41–54

Méndez P, Pazienti A, Szabó G, Bacci A (2012) Direct alteration of a specific inhibitory circuit of the hippocampus by antidepressants. J Neurosci 32:16616–16628

Michaud M, Balardy L, Moulis G, Gaudin C, Peyrot C, Vellas B, Cesari M, et al (2013) Proinflammatory cytokines, aging, and age-related diseases. J Am Med Dir Assoc. 2013 Jun 20. pii: S1525-8610(13)00280-6. doi:10.1016/j.jamda.2013.05.009. (Epub ahead of print)

Mika A, Mazur GJ, Hoffman AN, Talboom JS, Bimonte-Nelson HA, Sanabria F et al (2012) Chronic stress impairs prefrontal cortex-dependent response inhibition and spatial working memory. Behav Neurosci 126:605–619

Mönnikes H, Tebbe JJ, Hildebrandt M, Arck P, Osmanoglou E, Rose M et al (2001) Role of stress in functional gastrointestinal disorders. Evidence for stress-induced alterations in gastrointestinal motility and sensitivity. Dig Dis 19:201–211

Mooijaart SP, Sattar N, Trompet S, Lucke J, Stott DJ, Ford I et al (2013) Circulating interleukin-6 concentration and cognitive decline in old age: the PROSPER study. J Intern Med 274:77–85

Moore AH, Wu M, Shaftel SS, Graham KA, O'Banion MK (2009) Sustained expression of interleukin- 1beta in mouse hippocampus impairs spatial memory. Neuroscience 164:1484–1495

Moreno H, Wu WE, Lee T, Brickman A, Mayeux R, Brown TR et al (2007) Imaging the abeta-related neurotoxicity of Alzheimer disease. Arch Neurol 64:1467–1477

Mormino EC, Kluth JT, Madison CM, Rabinovici GD, Baker SL, Miller BL et al (2009) Episodic memory loss is related to hippocampal-mediated beta-amyloid deposition in elderly subjects. Brain 132:1310–1323

Mu Q, Xie J, Wen Z, Weng Y, Shuyun ZA (1999) quantitative MR study of the hippocampal formation, the amygdala, and the temporal horn of the lateral ventricle in healthy subjects 40 to 90 years of age. AJNR Am J Neuroradiol 20:207–211

Mueller SG, Schuff N, Raptentsetsang S, Elman J, Weiner MW (2008) Selective effect of Apo e4 on CA3 and dentate in normal aging and Alzheimer's disease using high resolution MRI at 4 T. Neuroimage 42:42–48

Muller D, Wang C, Skibo G, Toni N, Cremer H, Calaora V et al (1996) PSA-NCAM is required for activity-induced synaptic plasticity. Neuron 17:413–422

Mulugeta S, Maguire JA, Newitt JL, Russo SJ, Kotorashvili A, Beers MF (2007) Misfolded BRICHOS SP-C mutant proteins induce apoptosis via caspase-4 and cytochrome c-related mechanisms. Am J Physiol Lung Cell Mol Physiol 293:L720–L729

Mutlu E, Keshavarzian A, Engen P, Forsyth CB, Sikaroodi M, Gillevet P (2009) Intestinal dysbiosis: a possible mechanism of alcohol-induced endotoxemia and alcoholic steatohepatitis in rats. Alcohol Clin Exp Res 33:1836–1846

Mutlu EA, Gillevet PM, Rangwala H, Sikaroodi M, Naqvi A, Engen PA et al (2012) Colonic microbiome is altered in alcoholism. Am J Physiol Gastrointest Liver Physiol 302:G966–G978

Nair A, Vadodaria KC, Banerjee SB, Benekareddy M, Dias BG, Duman RS et al (2007) Stressor-specific regulation of distinct brain-derived neurotrophic factor transcripts and cyclic AMP response element-binding protein expression in the postnatal and adult rat hippocampus. Neuropsychopharmacology 32:1504–1519

Namekawa Y, Baba H, Maeshima H, Nakano Y, Satomura E, Takebayashi N et al (2013) Heterogeneity of elderly depression: increased risk of Alzheimer's disease and Aβ protein metabolism. Prog Neuropsychopharmacol Biol Psych 43:203–208

Nasman B, Olsson T, Viitanen M, Carlstrom K (1995) A subtle disturbance in the feedback regulation of the hypothalamic-pituitary-adrenal axis in the early phase of Alzheimer's disease. Psychoneuroendocrinology 20:211–220

Neal MD, Leaphart C, Levy R, Prince J, Billiar TR, Watkins S et al (2006) Enterocyte TLR4 mediates phagocytosis and translocation of bacteria across the intestinal barrier. J Immunol 176:3070–3079

Neish AS (2009) Microbes in gastrointestinal health and disease. Gastroenterology 136:65–80

Newman MC, Kaszniak AW (2000) Spatial memory and aging: Performance on a human analog of the Morris water maze. Aging Neuropsychol Cogn 7:86–93

Nibuya M, Morinobu S, Duman RS (1995) Regulation of BDNF and TrkB mRNA in rat brain by chronic electroconvulsive seizure and antidepressant drug treatments. J Neuosci 15:7539–7547

Nibuya M, Takahashi M, Russell DS, Duman RS (1999) Repeated stress increases catalytic TrkB mRNA in rat hippocampus. Neurosci Lett 267:81–84

Nihonmatsu-Kikuchi N, Hayashi Y, Yu XJ, Tatebayashi Y (2013) Depression and Alzheimer's disease: novel postmortem brain studies reveal a possible common mechanism. J Alzheimers Dis. 2013 Jul 19. (Epub ahead of print)

Nousen EK, Franco JG, Sullivan EL (2013) Unraveling the mechanisms responsible for the comorbidity between metabolic syndrome and mental health disorders. Neuroendocrinology. 2013 Sep 21. (Epub ahead of print)

Nunes SO, Vargas HO, Prado E, Barbosa DS, de Melo LP, Moylan S et al (2013) The shared role of oxidative stress and inflammation in major depressive disorder and nicotine dependence. Neurosci Biobehav Rev 37:1336–1345

Parachikova A, Agadjanyan MG, Cribbs DH, Blurton-Jones M, Perreau V, Rogers J et al (2007) Inflammatory changes parallel the early stages of Alzheimer disease. Neurobiol Aging 28:1821–1833

Pardon MC, Roberts ME, Marsden CA, Bianchi M, Latif ML, Duxon MS et al (2005) Social threat and novel cage stress-induced sustained extracellular-regulated kinase 1/2 (ERK 1/2) phosphorylation but differential modulation of brain-derived neurotrophic factor (BDNF) expression in the hippocampus of NMRI mice. Neuroscience 132:561–574

Park AJ, Collins J, Blennerhassett PA, Ghia JE, Verdu EF, Bercik P et al (2013) Altered colonic function and microbiota profile in a mouse model of chronic depression. Neurogastroenterol Motil 25:733–e575

Penner MR, Roth TL, Chawla MK, Hoang LT, Roth ED, Lubin FD et al (2010) Age-related changes in Arc transcription and DNA methylation within the hippocampus. Neurobiol Aging 32:2198–2210

Pedersen BK, Bruunsgaard H, Ostrowski K, Krabbe K, Hansen H, Krzywkowski K et al (2000) Cytokines in aging and exercise. Int J Sports Med 21:S4–S9

Peila R, Rodriguez BL, Launer LJ (2002) Type 2 diabetes, APOE gene, and the risk for dementia and related pathologies: the Honolulu-Asia aging study. Diabetes 51:1256–1262

Perry VH, Cunningham C, Holmes C (2007) Systemic infections and inflammation affect chronic neurodegeneration. Nat Rev Immunol 7:161–167

Persson J, Pudas S, Lind J, Kauppi K, Nilsson LG, Nyberg L (2012) Longitudinal structure-function correlates in elderly reveal MTL dysfunction with cognitive decline. Cereb Cortex 22:2297–2304

Peterson AL, Murchison C, Zabetian C, Leverenz J, Watson GS, Montine T et al. (2013) Memory, mood, and Vitamin D in persons with Parkinson's disease. J Parkinsons Dis. 2013 Sep 30. (Epub ahead of print)

Pittenger C, Sanacora G, Krystal JH (2007) The NMDA receptor as a therapeutic target in major depressive disorder. CNS Neurol Disord Drug Targets 6:101–115

Player MJ, Taylor JL, Weickert CS, Alonzo A, Sachdev P, Martin D et al (2013) Neuroplasticity in depressed individuals compared with healthy controls. Neuropsychopharmacology 38:2101–2108

Price JL, Ko AI, Wade MJ, Tsou SK, McKeel DW, Morris JC (2001) Neuron number in the entorhinal cortex and CA1 in preclinical Alzheimer disease. Arch Neurol 58:1395–1402

Rabinovici GD, Jagust WJ (2009) Amyloid imaging in aging and dementia: testing the amyloid hypothesis in vivo. Behav Neurol 21:117–128

Rasmusson AM, Shi L, Duman R (2002) Downregulation of BDNF mRNA in the hippocampal dentate gyrus after re-exposure to cues previously associated with footshock. Neuropsychopharmacology 27:133–142

Raz N, Gunning-Dixon F, Head D, Rodrigue KM, Williamson A, Acker JD (2004) Aging, sexual dimorphism, and hemispheric asymmetry of the cerebral cortex: replicability of regional differences in volume. Neurobiol Aging 25:377–396

Reichenberg A, Yirmiya R, Schuld A, Kraus T, Haack M, Morag A et al (2001) Cytokine-associated emotional and cognitive disturbances in humans. Arch Gen Psych 58:445–452

Reitz C, Brickman AM, Brown TR, Manly J, DeCarli C, Small SA et al (2009) Linking hippocampal structure and function to memory performance in an aging population. Arch Neurol 66:1385–1392

Revest JM, Di Blasi F, Kitchener P, Rougé-Pont F, Desmedt A, Turiault M et al (2005) The MAPK pathway and Egr-1 mediate stress-related behavioral effects of glucocorticoids. Nat Neurosci 8:664–672

Reynolds CF III, Frank E, Perel JM, Imber SD, Cornes C, Miller MD et al (1999) Nortriptyline and interpersonal psychotherapy as maintenance therapies for recurrent major depression: a randomized controlled trial in patients older than 59 years. JAMA 281:39–45

Richard E, Reitz C, Honig LH, Schupf N, Tang MX, Manly JJ et al (2013) Late-life depression, mild cognitive impairment, and dementia. JAMA Neurol 70:374–382

Robitsek RJ, Fortin NJ, Koh MT, Gallagher M, Eichenbaum H (2008) Cognitive aging: a common decline of episodic recollection and spatial memory in rats. J Neurosci 28:8945–8954

Rodríguez-Martínez E, Martínez F, Espinosa-García MT, Maldonado P, Rivas-Arancibia S (2013) Mitochondrial dysfunction in the hippocampus of rats caused by chronic oxidative stress. Neuroscience. 2013 Aug 27. pii: S0306-4522(13)00696-9. doi:10.1016/j.neuroscience.2013. 08.018. (Epub ahead of print)

Rohn TT, Head E, Su JH, Anderson AJ, Bahr BA, Cotman CW et al (2001) Correlation between caspase activation and neurofibrillary tangle formation in Alzheimer's disease. Am J Pathol 158:189–198

Roozendaal B, Griffith QK, Buranday J, De Quervain DJ, McGaugh JL (2003) The hippocampus mediates glucocorticoid-induced impairment of spatial memory retrieval: dependence on the basolateral amygdala. Proc Natl Acad Sci USA 100:1328–1333

Rössler M, Zarski R, Bohl J, Ohm TG (2002) Stage-dependent and sector-specific neuronal loss in hippocampus during Alzheimer's disease. Acta Neuropathol 103:363–369

Rothman SM, Herdener N, Camandola S, Texel SJ, Mughal MR, Cong WN et al (2012) 3xTgAD mice exhibit altered behavior and elevated Aβ after chronic mild social stress. Neurobiol Aging 33:830.e1–830.e12

Row BW (2007) Intermittent hypoxia and cognitive function: implications from chronic animal models. Adv Exp Med Biol 618:51–67

Rusch BD, Abercrombie HC, Oakes TR, Schaefer SM, Davidson RJ (2001) Hippocampal morphometry in depressed patients and control subjects: relations to anxiety symptoms. Biol Psych 50:960–964

Russo-Neustadt A, Ha T, Ramirez R, Kesslak JP (2001) Physical activity-antidepressant treatment combination: impact on brain-derived neurotrophic factor and behavior in an animal model. Behav Brain Res 120:87–95

Sapolsky RM, Uno H, Rebert CS, Finch CE (1990) Hippocampal damage associated with prolonged glucocorticoid exposure in primates. J Neurosci 10:2897–2902

Sapolsky RM (2000) Glucocorticoids and hippocampal atrophy in neuropsychiatric disorders. Arch Gen Psych 57:925–935

Sapolsky RM (2003) Stress and plasticity in the limbic system. Neurochem Res 28:1735–1742

Sarkar D, Fisher PB (2006) Molecular mechanisms of aging-associated inflammation. Cancer Lett 236:3–23

Sastre M, Dewachter I, Landreth GE, Willson TM, Klockgether T, van Leuven F et al (2003) Modulate immunostimulated processing of amyloid precursor protein through regulation of beta-secretase. J Neurosci 23:9796–9804

Saylam C, Ucerler H, Kitis O, Ozand E, Gönül AS (2006) Reduced hippocampal volume in drug-free depressed patients. Surg Radiol Anat 28:82–87

Scahill RI, Frost C, Jenkins R, Whitwell JL, Rossor MN, Fox NC (2003) A longitudinal study of brain volume changes in normal aging using serial registered magnetic resonance imaging. Arch Neurol 60:989–994

Schaaf MJ, de Kloet ER, Vreugdenhil E (2000) Corticosterone effects on BDNF expression in the hippocampus. Implications for memory formation. Stress 3:201–208

Schram MT, Euser SM, de Craen AJ, Witteman JC, Frölich M, Hofman A et al (2007) Systemic markers of inflammation and cognitive decline in old age. J Am Geriatr Soc 55:708–716

Schuff N, Amend DL, Knowlton R, Norman D, Fein G, Weiner MW (1999) Age-related metabolite changes and volume loss in the hippocampus by magnetic resonance spectroscopy and imaging. Neurobiol Aging 20:279–285

Selye H (1936) Syndrome produced by diverse nocuous agents. Nature 138:32

Shaw KN, Commins S, O'Mara SM (2001) Lipopolysaccharide causes deficits in spatial learning in the water maze but notin BDNF expression in the rat dentate gyrus. Behav Brain Res 124:47–54

Sheline YI, Wang PW, Gado MH, Csernansky JG, Vannier MW (1996) Hippocampal atrophy in recurrent major depression. Proc Natl Acad Sci USA 93:3908–3913

Sheline YI, Sanghavi M, Mintun MA, Gado MH (1999) Depression duration but not age predicts hippocampal volume loss in medically healthy women with recurrent major depression. J Neurosci 19:5034–5043

Shen J, Obin MS, Zhao L (2013) The gut microbiota, obesity and insulin resistance. Mol Aspects Med 34:39–58

Shors TJ, Seib TB, Levine S, Thompson RF (1989) Inescapable versus escapable shock modulates long-term potentiation in the rat hippocampus. Science 244:224–226

Small SA (2001) Age-related memory decline: current concepts and future directions. Arch Neurol (Chicago) 58:360–364

Small SA, Tsai WY, DeLaPaz R, Mayeux R, Stern Y (2002) Imaging hippocampal function across the human life span: is memory decline normal or not? Ann Neurol 51:290–295

Small SA, Chawla MK, Buonocore M, Rapp PR, Barnes CA (2004) From the cover: imaging correlates of brain function in monkeys and rats isolates a hippocampal subregion differentially vulnerable to aging. Proc Natl Acad Sci USA 101:7181–7186

Smith MA, Makino S, Kvetnansky R, Post RM (1995) Stress and glucocorticoids affect the expression of brain-derived neurotrophic factor and neurotrophin-3 mRNAs in the hippocampus. J Neurosci 15:1768–1777

Smits JA, Minhajuddin A, Thase ME, Jarrett RB (2012) Outcomes of acute phase cognitive therapy in outpatients with anxious versus nonanxious depression. Psychother Psychosom 81:153–160

Sparkman NL, Martin LA, Calvert WS, Boehm GW (2005) Effects of intraperitoneal lipopolysaccharide on Morris maze performance in year-old and 2-month-old female C57BL/6J mice. Behav Brain Res 159:145–151

Stranahan AM, Haberman RP, Gallagher M (2010) Cognitive decline is associated with reduced reelin expression in the entorhinal cortex of aged rats. Cereb Cortex 21:392–400

Su JH, Kesslak JP, Head E, Cotman CW (2002) Caspase-cleaved amyloid precursor protein and activated caspase-3 are co-localized in the granules of granulovacuolar degeneration in Alzheimer's disease and Down's syndrome brain. Acta Neuropathol (Berl) 104:1–6

Swanwick GR, Kirby M, Bruce I, Buggy F, Coen RF, Coakley D et al (1988) Hypothalamic-pituitary-adrenal axis dysfunction in Alzheimer's disease: lack of association between longitudinal and cross-sectional findings. Am J Psychiatry 155:286–289

Taylor WD, Steffens DC, Payne ME, MacFall JR, Marchuk DA, Svenson IK et al (2005) Influence of serotonin transporter promoter region polymorphisms on hippocampal volumes in late-life depression. Arch Gen Psych 62:537–544

Thomas RM, Hotsenpiller G, Peterson DA (2007) Acute psychosocial stress reduces cell survival in adult hippocampal neurogenesis without altering proliferation. J Neurosci 27:2734–2743

Thornberry NA, Rano TA, Peterson EP, Rasper DM, Timkey T, Garcia-Calvo M et al (1997) A combinatorial approach defines specificities of members of the caspase family and granzyme B. Functional relationships established for key mediators of apoptosis. J Biol Chem 272:17907–17911

Tiihonen K, Ouwehand AC, Rautonen N (2010) Effect of overweight on gastrointestinal microbiology and immunology: correlation with blood biomarkers. Br J Nutr 103:1070–1078

Tran L, Greenwood-Van Meerveld B (2013) Age-associated remodeling of the intestinal epithelial barrier. J Gerontol A Biol Sci Med Sci 68:1045–1056

Tran TT, Srivareerat M, Alhaider IA, Alkadhi KA (2011) Chronic psychosocial stress enhances long-term depression in a subthreshold amyloid-beta rat model of Alzheimer's disease. J Neurochem 119:408–416

Turrin NP, Gayle D, Ilyin SE, Flynn MC, Langhans W, Schwartz GJ et al (2001) Pro-inflammatory and anti-inflammatory cytokine mRNA induction in the periphery and brain following intraperitoneal administration of bacterial lipopolysaccharide. Brain Res Bull 54:443–453

Umegaki H (2012) Neurocognitive dysfunction in old diabetics: management and treatment. Adv Exp Med Biol 771:465–470

Uno H, Tarara R, Else JG, Suleman MA, Sapolsky RM (1989) Hippocampal damage associated with prolonged and fatal stress in primates. J Neurosci 9:1705–1711

Vakili K, Pillay SS, Lafer B, Fava M, Renshaw PF, Bonello-Cintron CM et al (2000) Hippocampal volume in primary unipolar major depression: a magnetic resonance imaging study. Biol Psych 47:1087–1090

van Heesch F, Prins J, Konsman JP, Westphal KG, Olivier B, Kraneveld AD et al (2013) Lipopolysaccharide-induced anhedonia is abolished in male serotonin transporter knockout rats: an intracranial self-stimulation study. Brain Behav Immun 29:98–103

Vander Velt AJ, McAllister RJ (1962) Psychiatric illness in hospitalized clergy: alcoholism. Q J Stud Alcohol 23:124–130

Vander Weele CM, Saenz C, Yao J, Correia SS, Goosens KA (2013) Restoration of hippocampal growth hormone reverses stress-induced hippocampal impairment. Front Behav Neurosci 7:66

Varela-Nallar L, Aranguiz FC, Abbott AC, Slater PG, Inestrosa NC (2010) Adult hippocampal neurogenesis in aging and Alzheimer's disease. Birth Defects Res C Embryo Today 90:284–296

Venero C, Borrell J (1999) Rapid glucocorticoid effects on excitatory amino acid levels in the hippocampus: a microdialysis study in freely moving rats. Eur J Neurosci 11:2465–2473

Videbech P, Ravnkilde B, Pedersen TH, Hartvig H, Egander A, Clemmensen K et al (2002) The Danish PET/depression project: clinical symptoms and cerebral blood flow.a regions-of-interest analysis. Acta Psych Scand 106:35–44

Videbech P, Ravnkilde B (2004) Hippocampal volume and depression: a meta-analysis of MRI studies. Am J Psych 161:1957–1966

Vindigni SM, Broussard EK, Surawicz CM (2013) Alteration of the intestinal microbiome: fecal microbiota transplant and probiotics for Clostridium difficile and beyond. Expert Rev Gastroenterol Hepatol 7:615–628

Walhovd KB, Fjell AM, Reinvang I, Lundervold A, Dale AM, Eilertsen DE et al (2005) Effects of age on volumes of cortex, white matter and subcortical structures. Neurobiol Aging 26:1261–1270

Walhovd KB, Westlye LT, Amlien I, Espeseth T, Reinvang I, Raz N et al (2011) Consistent neuroanatomical age-related volume differences across multiple samples. Neurobiol Aging 32:916–932

Wang B, Van Veldhoven PP, Brees C, Rubio N, Nordgren M, Apanasets O et al. (2013a) Mitochondria are targets for peroxisome-derived oxidative stress in cultured mammalian cells. Free Radic Biol Med. 2013 Aug 27. pii: S0891-5849(13)00576-5. doi:10.1016/j.freeradbiomed.2013.08.173. (Epub ahead of print)

Wang J, Sereika SM, Styn MA, Burke LE (2013b) Factors associated with health-related quality of life among overweight or obese adults. J Clin Nurs 22:2172–2182

West MJ (1993) Regionally specific loss of neurons in the aging human hippocampus. Neurobiol Aging 14:287–293

West MJ, Coleman PD, Flood DG, Troncoso JC (1994) Differences in the pattern of hippocampal neuronal loss in normal ageing and Alzheimer's disease. Lancet 344:769–772

West MJ, Kawas CH, Martin LJ, Troncoso JC (2000) The CA1 region of the human hippocampus is a hot spot in Alzheimer's disease. Ann NY Acad Sci 908:255–259

WHO (2009) Obesity and overweight. World Health Organization. Available at: http://www.who.int/dietphysicalactivity/publications/facts/obesity/en/

Willner P (2005) Chronic mild stress (CMS) revisited: consistency and behavioural-neurobiological concordance in the effects of CMS. Neuropsychobiology 52:90–10

Wilson AC, Dugger BN, Dickson DW, Wang DS (2011) TDP-43 in aging and Alzheimer's disease—a review. Int J Clin Exp Pathol 4:147–155

Wilson IA, Gallagher M, Eichenbaum H, Tanila H (2006) Neurocognitive aging: prior memories hinder new hippocampal encoding. Trends Neurosci 29:662–670

Wilson RS, Schneider JA, Bienias JL, Arnold SE, Evans DA, Bennett DA (2003) Depressive symptoms, clinical AD, and cortical plaques and tangles in older persons. Neurology 61:1102–1107

Wilson RS, Barnes LL, Bennett DA, Li Y, Bienias JL, Mendes de Leon CF, et al (2005) Proneness to psychological distress and risk of Alzheimer disease in a biracial community. Neurology 64:380–382

Wilson RS, Schneider JA, Boyle PA, Arnold SE, Tang Y, Bennett DA (2007) Chronic distress and incidence of mild cognitive impairment. Neurology 68:2085–2092

Wilson RS, Lei Y, Trojanowski JQ, Chen E, Boyle PA, Bennett DA, et al (2013) TDP-43 pathology, cognitive decline, and dementia in old age. JAMA Neurol. Published online September 30, 2013. doi:10.1001/jamaneurol.2013.3961

Wium-Andersen MK, Orsted DD, Nielsen SF, Nordestgaard BG (2012) Elevated C-reactive protein levels, psychological distress, and depression in 73 131 individuals. Arch Gen Psych 24:1–9

Woodmansey EJ (2007) Intestinal bacteria and ageing. J Appl Microbiol 102:1178–1186

Yaffe K, Laffan AM, Harrison SL, Redline S, Spira AP, Ensrud KE et al (2011) Sleep-disordered breathing, hypoxia, and risk of mild cognitive impairment and dementia in older women. JAMA 306:613–619

Yang CH, Huang CC, Hsu KS (2004) Behavioral stress modifies hippocampal plasticity through corticosterone-induced sustained extracellular signal-regulate kinase/mitogen-activated protein kinase activation. J Neurosci 24:11029–11034

Yang DS, Kumar A, Stavrides P, Peterson J, Peterhoff CM, Pawlik M et al (2008) Neuronal apoptosis and autophagy cross talk in aging PS/APP mice, a model of Alzheimer's disease. Am J Pathol 173:665–681

Yassa MA, Muftuler LT, Stark CE (2010) Ultrahigh-resolution microstructural diffusion tensor imaging reveals perforant path degradation in aged humans in vivo. Proc Natl Acad Sci USA 107:12687–12691

Yassa MA, Mattfeld AT, Stark SM, Stark CEL (2011) Age-related memory deficits linked to circuit- specific disruptions in the hippocampus. Proc Natl Acad Sci USA 108:8873–8878

Yun J, Koike H, Ibi D, Toth E, Mizoguchi H, Nitta A et al (2010) Chronic restraint stress impairs neurogenesis and hippocampus-dependent fear memory in mice: possible involvement of a brain-specific transcription factor Npas4. J Neurochem 114:1840–1851

Zakzanis KK, Leach L, Kaplan E (1998) On the nature and pattern of neurocognitive function in major depressive disorder. Neuropsychiatry Neuropsychol Behav Neurol 11:111–119

Zhang K, Xu H, Cao L, Li K, Huang Q (2013) Interleukin-1β inhibits the differentiation of hippocampal neural precursor cells into serotonergic neurons. Brain Res 1490:193–201

Zhang R, Miller RG, Gascon R, Champion S, Katz J, Lancero M et al (2009a) Circulating endotoxin and systemic immune activation in sporadic amyotrophic lateral sclerosis (sALS). J Neuroimmunol 206:121–124

Zhang S, Xu Z, Gao Y, Wu Y, Li Z, Liu H et al (2012) Bidirectional crosstalk between stress-induced gastric ulcer and depression under chronic stress. PLoS One 7:e51148

Zhang Y, Popovic ZB, Bibevski S, Fakhry I, Sica DA, Van Wagoner DR et al (2009b) Chronic vagus nerve stimulation improves autonomic control and attenuates systemic inflammation and heart failure progression in a canine high-rate pacing model. Circ Heart Fail 2:692–699

Zhou H, Liang H, Li ZF, Xiang H, Liu W, Li JG (2013) Vagus nerve stimulation attenuates intestinal epithelial tight junctions disruption in endotoxemic mice through α7 nicotinic acetylcholine receptors. Shock 40:144–151

Zoetendal EG, Vaughan EE, de Vos WM (2006) A microbial world within us. Mol Microbiol 59:1639–1650

Zschocke J, Bayatti N, Clement AM, Witan H, Figiel M, Engele J et al (2005) Differential promotion of glutamate transporter expression and function by glucocorticoids in astrocytes from various brain regions. J Biol Chem 280:34924–34932

Stress and Psychological Resiliency

Alan L. Peterson, Tabatha H. Blount and Donald D. McGeary

Abstract Over the past decade, there has been an enormous increase in research and scientific publications targeting psychological resiliency. However, compared to the research on the neurobiology of stress, resiliency research is in its relative infancy. Much of the resiliency research has focused on theoretical models and the conceptualization of psychological resiliency. Resiliency research has been limited by (1) the broad use of the term resiliency; (2) the lack of standardized definitions of resiliency; (3) a primary focus on descriptive, assessment, and measurement studies; (4) relatively few randomized controlled trials to evaluate the efficacy of resiliency enhancement programs; and (5) methodological challenges inherent in conducting applied resiliency research. Although many recent programs have been initiated in attempts to enhance psychological resiliency in targeted populations, such as military personnel, relatively few randomized controlled trials have been conducted. Translational research, prospective longitudinal cohort studies, and clinical intervention trials are needed to better understand the behavioral neurobiology of stress and psychological resiliency.

Keywords Stress · Psychological resiliency · Hardiness

A. L. Peterson (✉) · T. H. Blount · D. D. McGeary
Division of Behavioral Medicine, Department of Psychiatry,
The University of Texas Health Science Center at San Antonio,
7550 IH 10 West, Suite 1325, San Antonio, TX 78229, USA
e-mail: petersona3@uthscsa.edu

T. H. Blount
e-mail: blountt@uthscsa.edu

D. D. McGeary
e-mail: mcgeary@uthscsa.edu

Contents

1	An Overview of Psychological Resiliency	298
2	Definitions of Psychological Resiliency	300
3	Models and Conceptualizations of Psychological Resiliency	300
	3.1 The Trait Conceptualization of Resiliency	301
	3.2 The Outcome Conceptualization of Resiliency	301
	3.3 Military Demand-Resource Model of Resiliency	302
	3.4 The Blister-Callus Model of Psychological Resilience	303
4	The Assessment and Measurement of Psychological Resiliency	303
	4.1 Connor–Davidson Resilience Scale	304
	4.2 Resilience Scale for Adults	304
	4.3 Brief Resilience Scale	304
5	Resiliency Research	304
	5.1 Military Resiliency Programs	305
6	Future Directions	309
References		310

1 An Overview of Psychological Resiliency

Over the past decade, there has been a tremendous increase in research and scientific publications targeting psychological resiliency (McGeary 2011; Peterson et al. 2009). The majority of this increased interest and research has been a result of the ongoing military conflicts in Iraq and Afghanistan. Over the past 12 years, approximately 2.5 million United States military personnel have been deployed to the Middle East (Institute of Medicine 2012). Exposure to traumatic events is almost universal among deployed military service members. The majority of these military personnel have been exposed to multiple forms of combat-related or deployment-related stress such as blast explosions accompanied by mutilating injuries and the death of others, as well as significant risk of personal injury or death. It has been estimated that 5–17 % of service members returning from deployments to Iraq and Afghanistan are at significant risk for combat-related posttraumatic stress disorder (PTSD; Gates et al. 2012; Hoge et al. 2004; Peterson et al. 2011; Tanielian and Jaycox 2008). The prevalence estimates of PTSD among service members and veterans vary widely based on the population assessed, the screening methodology, and the time frame of assessment.

Considering that almost all military personnel are exposed to some form of combat-related trauma during deployments, those who develop PTSD reflect only a small percentage of military service members who experience trauma. To state the converse, 83–95 % of military personnel deployed to a war zone and repeatedly exposed to extreme levels of stress and trauma appear to be relatively resilient. Some of the questions being asked by investigators are, "What type of individuals appear to be resilient under stress? What factors influence those who develop stress-related disorders versus those who remain resilient despite exposure to extreme stress? Are there personality, genetic, behavioral, or neurobiological differences? Are these differences related to nature or nurture? Can these differences be detected through behavioral or biological assessments? Can resiliency be enhanced through behavioral or biological interventions?"

As reviewed by McGeary (2011), resilience has become an increasingly popular psychological construct to study. The first formal description of resiliency appeared in the 1970s, and it has been increasingly studied in the medical and psychological research literature since that time. A keyword MEDLINE search for the terms "resiliency" or "resilience" or "hardiness" in 2011 returned over 5,000 citations (McGeary 2011). The number of annual scientific manuscript citation counts related to resiliency in PsychINFO increased over a 10-year period from 130 in 2000 to 679 in 2010 (Britt et al. 2013).

The concept of resiliency has been examined in a number of high-risk groups, including the following: vulnerable children and adolescents (e.g., Evans et al. 2010; Gomez and McLaren 2006; Jaffee et al. 2007); older adults (e.g., Windle et al. 2008); military medical personnel (Maguen et al. 2008), police officers (e.g., Paton et al. 2008); medical professionals (Gillespie et al. 2007); individuals with psychological disorders (McLaren et al. 2007); and individuals exposed to trauma (e.g., Klasen et al. 2010). Moreover, the concept of resilience serves as the keystone for many prevention programs and has been used to inform social and public policy recommendations (e.g., Jenson 2007; Kaminsky et al. 2007). While the present literature provides interesting and valuable individual findings, a theoretical structure that can unify the literature is wanting. Thus far, increased empirical attention has resulted in multiple, often inconsistent, conceptualizations of resilience and has not lead to improved understanding of the latent construct. While a complete review of the literature is beyond the scope of this chapter, some of the current conceptualizations of psychological resilience are outlined below.

Some of the first studies of psychological resilience were conducted by psychologists interested in vulnerable children raised in high-risk environments (e.g., Werner and Smith 1992). Over the past decade, interest in psychological resilience has been expanded into a wide range of at-risk groups including older adults, individuals in high-risk professions (e.g., military, law enforcement, emergency medical personnel), individuals with psychological disorders, and individuals with trauma exposure (Evans et al. 2010; Gillespie et al. 2007; Gomez and McLaren, 2006; Jaffee et al. 2007; Klasen et al. 2010; Maguen et al. 2008; McLaren et al. 2007; Paton et al. 2008; Windle et al. 2008). The published literature on resiliency has been used to inform social and public policy (Jenson 2007; Kaminsky et al. 2007). Unfortunately, there is considerable variation in the definition and conceptualization of psychological resilience across studies. This hinders the synthesis of the scientific literature and limits the translation of research findings into applied settings.

2 Definitions of Psychological Resiliency

According to the Merriam-Webster online dictionary (2014), the basic definition of resilience is *"the ability to become strong, healthy, or successful again after something bad happens"* and *"the ability of something to return to its original shape after it has been pulled, stretched, pressed, or bent."* The Merriam-Webster medical definition of resilience is *"the capability of a strained body to recover its size and shape after deformation caused especially by compressive stress"* and *"an ability to recover from or adjust easily to misfortune or change."*

In 2011, the RAND Center for Military Health Policy Research published a comprehensive monograph titled *Promoting Psychological Resilience in the U.S. Military* (Meredith et al. 2011). Part of the RAND report examined the definitions of resilience. Altogether, they found over 100 individual definitions of psychological resilience across 270 publications. The definitions of resilience were classified into three broad categories: basic, adaptation, and growth. According to the RAND report (Meredith et al. 2011, p. 20), basic definitions involve the idea that psychological resilience is a "process or capacity that develops over time." The adaption definitions of resiliency "incorporate the concept of bouncing back" after exposure to stress or trauma. The growth definitions of resiliency "involve growth after experiencing adversity or trauma." The definitional differences uncovered in the RAND report are consistent with observations from other researchers (e.g., Agiabi and Wilson 2005; McGeary 2011). After the completion of their rigorous review of the definitional literature on resiliency, the RAND reviewers (Meredith et al. 2011, p. 3) selected the definition by Jenson and Fraser (2005) as the one thought to best describe the construct of resiliency: *"Resilience is the capacity to adapt successfully in the presence of risk and adversity."*

3 Models and Conceptualizations of Psychological Resiliency

Models of resilience attempt to explain the dynamic relations among factors that diminish or enhance psychological resilience. The level of complexity varies considerably in these models (Bates et al. 2010). The compensatory model, risk-protective model, protective–protective model, and challenge model of psychological resilience are relatively simple models that examine the interactions between risk factors and protective factors. These models were originally described by developmental psychologists in the 1980s and 1990s to describe the trajectories and outcomes of high-risk, disadvantaged youth (e.g., Garmezy et al.1984; Rutter 1985).

3.1 The Trait Conceptualization of Resiliency

The most straightforward conceptualizations define resilience as individual traits such as high self-esteem, intelligence, determination, strong coping skills, or hardiness (Kobassa et al. 1982; Metzl and Morrell 2008). Hardiness has been shown to be a moderator of stress (Pengilly and Dowd 2000) and is conceptualized to be a trait that includes three components: control, commitment, and challenge. When confronted with a stressful life event, hardy individuals (1) believe they can control or influence the outcome, (2) are committed or dedicated to resolve the situation, and (3) see stressful life events as challenges and believe that one can learn and grow from both positive and negative life experiences.

The simplicity of conceptualizing resiliency as a trait allows for greater ease in conducting research. Moreover, assuming the trait may be amenable to training or other interventions, this conceptualization has implications for developing resiliency enhancement programs for at-risk individuals or populations. Despite these potential advantages, there are significant limitations with this conceptualization. First, individual traits are most often considered to be innate characteristics that are relatively stable over time and difficult to change. Studies that rely on this formulation generally identify the construct as a correlate or predictor variable within the analyses and not as the criterion variable. Consequently, the results do not provide the information necessary to advance the field's understanding of resilience. Furthermore, the conceptualization of resilience as a trait fails to capture the complex processes involved in resilience as well as the context or system in which resilience occurs within any given individual (Gillespie et al. 2007; Waller 2001). Consequently, while these unidimensional, nondynamic definitions can help answer specific questions associated with individual studies, they fail to provide a satisfactory model of resilience that can be utilized in a broader context, such as promoting resilience through education and training programs.

3.2 The Outcome Conceptualization of Resiliency

Resilience has also been conceptualized as either the presence of a positive outcome (e.g., academic success) or the absence of a negative outcome (e.g., lack of psychological symptoms) following adversity or trauma (as reviewed by Metzl and Morrell 2008). For example, in the case of war, resilience may be inferred if a service member exposed to combat-related trauma does not develop acute stress disorder or posttraumatic stress symptoms. As with a trait conceptualization, this approach generates straightforward research designs and provides valuable information on one important facet of resilience. However, similar to a trait conceptualization of psychological resilience, outcome conceptualizations also fail to delineate the dynamic processes that promote or diminish resilience, thereby providing only a limited snapshot of the construct (Gillespie et al. 2007; Luthar et al. 2000).

The examination of posttrauma/adversity outcomes alone is insufficient for theory building or program development.

Efforts have been made to describe the interaction between protective and risk factors that contribute to positive or negative outcome in high-risk individuals. The majority of this work was spearheaded in the 1980s and 1990s by developmental researchers interested in the trajectories and outcomes of high-risk, disadvantaged youth (e.g., Garmezy et al.1984; Rutter 1985). As discussed by Zimmerman and Arunkumar (1994), four main theoretical frameworks of resilience emerged from this literature: the compensatory model, the risk-protective model, the protective–protective model, and the challenge model. According to the compensatory model, protective factors have a direct effect on outcome and combine in an additive manner to determine outcome. In other words, higher levels of a protective factor are associated with better outcomes. The risk-protective and protective–protective models emphasize the interaction between risk and protective factors in the prediction of outcome. In the risk-protective model, the protective factor moderates the relation between risk factor and outcome, whereas, in the protective–protective model, a second protective factor moderates the relation between an initial protective factor and the outcome. In the challenge model, a curvilinear relation occurs between the risk factor and outcome such that lower and higher levels of the risk factor are associated with poor outcome, whereas a moderate level of the risk factor is associated with more positive outcomes.

Primarily, these models have been applied to developmental concerns such as adolescent aggression and victimization (e.g., Hollister-Wagner et al. 2001). For example, in a recent study on adolescent victimization, Marsh et al. (2009) found support for the challenge model for both male and female adolescents and for the compensatory and protective–protective models for adolescent males. However, research has yet to clarify whether these models apply to military personnel in combat or deployed settings. Moreover, these models may not fully capture the systemic and cultural factors that can impact resilience on an individual level.

3.3 Military Demand-Resource Model of Resiliency

Recent progress has been made to incorporate evidence-based factors into models of resiliency. For example, Bates and colleagues (Bates et al. 2010; Bowles and Bates 2010) proposed the Military Demand-Resource Model as a comprehensive and integrative model of psychological fitness. In this model, psychological fitness is described as "the integration and optimization of mental, emotional, and behavioral abilities and capacities to optimize performance and strengthen the resilience of warfighters" (Bates et al. 2010, p. 21). According to this model, the effects of military demands are mediated by internal resources such as awareness, beliefs, and engagement, as well as external resources such as leadership, unit members, families, training, and support programs that enhance or undermine resilience. The Military Demand-Resource Model posits that resource loss is more impactful than resource gain. It is proposed that resource investment helps mitigate resource loss, whereas the resource environment can facilitate resource development.

Bates and colleagues highlight the importance of physical fitness training as part of the military culture and acknowledge the importance of providing sufficient challenge to push skill development without exceeding resources (Bates et al. 2010; Bowles and Bates 2010). They propose that psychological fitness can be developed using training principles similar to physical fitness training. A limitation of the Military Demand-Resource Model is its complexity, which may leave some to be confused when trying to conceptualize how best to implement the model in order to enhance military resiliency.

3.4 The Blister-Callus Model of Psychological Resilience

Similar to the physical calluses that form from blisters following physical exertion, the Blister-Callus Model of Psychological Resiliency (Blount et al. 2012) posits that psychological resilience develops from repeated and gradually increasing periods of psychological and physical stress, followed by periods of recovery. The Blister-Callus Model does not disregard other potential contributors to psychological resilience, such as genetics, personality factors, or other innate attributes or characteristics of an individual. However, the model does posit that the primary contributing factor to psychological resiliency is life events involving exposure to psychological and physical stress. It also supports physical conditioning programs such as those that occur as part of basic military training. For most individuals, repeated and gradually increasing levels of physical training can result in increased levels of physical fitness. It is believed that the same concept of repeated episodes of gradually increasing levels of psychological stress followed by periods of recovery is the key to enhancing psychological resiliency.

4 The Assessment and Measurement of Psychological Resiliency

The evaluation of interventions and policies designed to promote resilience requires reliable and valid approaches to assessment and measurement. Windle et al. (2011) conducted a methodological review of resilience measurement scales developed for use in general and clinical populations. The authors used eight electronic abstract databases to identify published journal articles where resilience was a key focus or was assessed. As a result of this review, 15 resilience measures were identified. Overall, measures with the best psychometric properties were the Connor–Davidson Resilience Scale (Connor and Davidson 2003), the Resilience Scale for Adults (Friborg et al. 2003, 2005) and the Brief Resilience Scale (Smith et al. 2008).

4.1 Connor–Davidson Resilience Scale

The Connor–Davidson Resilience Scale (Connor and Davidson 2003) was developed for clinical practice as a measure of stress coping ability. It includes five factors (personal competence, trust/tolerance/strengthening effects of stress, acceptance of change and secure relationships, control, and spiritual influences). The scale contains 25 items, and respondents indicate their degree of endorsement on five-point scales ranging from 0 ("not true at all") through 4 ("true nearly all the time") of items such as "Having to cope with stress makes me stronger."

4.2 Resilience Scale for Adults

The Resilience Scale for Adults (Friborg et al. 2003, 2005) is a 37-item self-report measure developed to examine intrapersonal and interpersonal protective factors that are presumed to facilitate adaptation to psychosocial adversities. It includes five factors: personal competence, social competence, family coherence, social support, and personal structure. The measure can be used in clinical and health psychology as an assessment tool of protective factors important to prevent maladjustment and psychological disorders.

4.3 Brief Resilience Scale

The Brief Resilience Scale (Smith et al. 2008) is a six-item, self-report scale of resiliency designed as an outcome measure to assess the ability to bounce back or recover from stress. The authors note that most other measures of resilience have focused on examining the resources and protective factors that might facilitate a resilient outcome. The Brief Resilience Scale was developed to have a specific focus on recovering or bouncing back after stress exposure.

5 Resiliency Research

As noted earlier, the RAND Center for Military Health Policy Research published a comprehensive monograph in 2011 titled *Promoting Psychological Resilience in the U.S. Military* (Meredith et al. 2011). This review of the evidence-informed scientific literature is arguably the most comprehensive literature review to date on psychological resilience. Eleven subject-matter experts were contracted to complete this review, and 270 manuscripts published over a 10-year period (2000–2009) were identified for inclusion. The primary focus of RAND was to conduct a systematic review of the scientific literature on psychological resilience to (1) identify evidence-informed factors that promote psychological resilience, and (2) assess the strength of the evidence base associated with each factor. Overall, there was generally very little rigorous research available across the different resilience factors. Only 11 publications reported results from randomized controlled trials, the gold-standard design for intervention research.

The review and synthesis yielded 20 evidence-informed factors associated with resilience. The resilience factors were categorized according to whether they operated at the individual, family, organization (or unit), or community level. This framework was used to distinguish intrinsic factors that promote resilience within an individual from external or environmental resilience factors (e.g., family, organization, or community). Table 1 provides a summary of evidence-informed factors that promote resilience according to the RAND report.

5.1 Military Resiliency Programs

Battlemind Training Battlemind is an early intervention resiliency enhancement program developed at the Walter Reed Army Institute of Research (WRAIR) to target increasing resiliency in service members returning home from combat deployments. The basic concept, as described in training materials by the WRAIR Land Study Team (2006), is that Battlemind skills help service members survive in combat, but they may cause problems if they are not adapted when service members get home. The program uses military terminology to normalize and promote the benefit of these approaches while deployed and how they must be adapted to effectively survive on the home front. Using the acronym BATTLE-MIND, the program utilizes a cognitive and skills-building approach to teach 10 specific concepts (WRAIR Land Study Team 2006, p. 2).

Buddies (cohesion) versus Withdrawal
Accountability versus Controlling
Targeted Aggression versus Inappropriate Aggression
Tactical Awareness versus Hypervigilance
Lethally Armed versus "Locked and Loaded" at home
Emotional Control versus Anger/Detachment
Mission Operational Security (OPSEC) versus Secretiveness
Individual Responsibility versus Guilt
Nondefensive (combat) Driving versus Aggressive Driving
Discipline and Ordering versus Conflict

By actively normalizing the difficulties that can occur during the reintegration period, the program also attempts to decrease the stigma associated with seeking out treatment for combat-related problems. An example of one of these concepts (Buddies versus Withdrawal) is provided by the authors to help clarify the Battlemind concept:

Table 1 RAND Center for Military Health Policy Research summary of evidence-informed factors that promote psychological resilience

Individual level resilience factors	Definition
Positive coping	The process of managing taxing circumstances, expending effort to solve personal and interpersonal problems, and seeking help to reduce or tolerate stress or conflict, including active/pragmatic, problem-focused, and spiritual approaches to coping
Positive affect	Feeling enthusiastic, active, and alert, including having positive emotions, optimism, a sense of humor (ability to have humor under stress or when challenged), hope, and flexibility about change
Positive thinking	Information processing, applying knowledge, and changing preferences through restructuring, positive reframing, making sense out of a situation, flexibility, reappraisal, refocusing, having positive outcome expectations, a positive outlook, and psychological preparation
Realism	Realistic mastery of the possible/having realistic outcome expectations, self-esteem/self-worth, confidence, self-efficacy, perceived control/acceptance of what is beyond control or cannot be changed
Behavioral control	The process of monitoring, evaluating, and modifying reactions to accomplish a goal (i.e., self-regulation, self-management, self-enhancement)
Physical fitness	Bodily ability to function efficiently and effectively in life domains
Altruism	Selfless concern for the welfare of others, motivation to help without reward
Family level resilience factors	*Definition*
Emotional ties	Emotional bonding among family members, including shared recreation and leisure time
Communication	The exchange of thoughts, opinions, or information, including problem solving and relationship management
Support	Perceiving that comfort is available from (and can be provided to) others, including emotional, tangible, instrumental, informational, and spiritual support
Closeness	Love, intimacy, and attachment
Nurturing	Parenting skills
Adaptability	Ease of adapting to changes associated with military life, including flexible roles within the family
Unit level resilience factors	*Definition*
Positive command climate	Facilitating and fostering intraunit interaction, building pride/support for the mission, leadership, positive role modeling, and implementing institutional policies
Teamwork	Work coordination among team members, including flexibility
Cohesion	Team ability to perform combined actions; bonding together of members to sustain commitment to each other and the mission

(continued)

Table 1 (continued)

Community level resilience factors	Definition
Belongingness	Integration, friendships; group membership, including participation in spiritual/faith-based organizations, protocols, ceremonies, social services, schools, and so on; and implementing institutional policies
Cohesion	The bonds that bring people together in the community, including shared values and interpersonal belonging
Connectedness	The quality and number of connections with other people in the community; includes connections with a place or people of that place; aspects include commitment, structure, roles, responsibility, and communication
Collective efficacy	Group members' perceptions of the ability of the group to work together

Adapted from Meredith et al. 2011

> *While deployed, no one understands your experience except your buddies who were there with you. Your life depended on your trust in your buddies. After returning home, you may prefer to be with your battle buddies rather than with spouse, family, or other friends. You may assume only those who were there with you in combat can understand you or are interested in you. You may avoid speaking about yourself to friends and family. Combat results in bonds with fellow Soldiers that will last a lifetime. Back home, your friends and family have changed. Re-establishing these bonds takes time and work. Renew relationships at home. Spend individual time with each of your loved ones. Balance time spent with buddies and family. Provide and accept support from them* (Adapted from WRAIR Land Study Team 2006).

The extant research indicates that Battlemind training is associated with high acceptability and with improved mental health outcomes in US soldiers returning from combat deployments. For example, Alder and colleagues (2009) examined early intervention training in 2,297 soldiers following a 12-month deployment. Participants were randomized by platoons into either stress education, Battlemind debriefing, small group Battlemind, or large group Battlemind. In comparison with the standard stress education, participants in the Battlemind training groups had fewer depressive symptoms. Battlemind briefing was also superior to stress education in decreasing PTSD symptoms and sleep problems. Although these differences are statistically significant at a population level, the actual differences from a clinical perspective are relatively small. In Canada, Battlemind training has been incorporated into Third-Location Decompression programs (Zamorski et al. 2012). However, some findings suggest that the Battlemind training may have limited benefits for non-US military personnel (Mulligan et al. 2012).

Comprehensive Soldier Fitness Program The Comprehensive Soldier Fitness (CSF) program represents the largest, most ambitious, and most controversial attempt at resiliency training ever devised or implemented in the United States military. Described by General George Casey as "a prevention program that seeks to enhance psychological resilience among all members of the Army community,"

CSF is a multi-module resiliency program designed to bolster and enhance various proposed dimensions of resiliency among US Army Soldiers and their families (Casey 2011, p. 1). Based on a positive psychology paradigm (Seligman 2011), CSF modules are designed to improve the health and performance of Soldiers and, more notably, to decrease the incidence of PTSD after deployment through comprehensive assessment using the online Soldier Fitness Tracker (Fravell et al. 2011) and training modules delivered through Master Resilience Trainers (Lester et al. 2011). Lester and colleagues report some of the initial research efforts designed to assess and describe the effectiveness of CSF. These efforts have included a longitudinal assessment of training effectiveness, studies of the impact of CSF module training on more specific resilience variables (e.g., physical, psychological), and an examination of the influence of CSF on socioeconomic outcomes (e.g., career progression and retention of active duty).

Research on Comprehensive Soldier Fitness, while ambitious and potentially fruitful, has not been without some controversy. Some have questioned the value and ethics of the CSF research program, expressing concern about the large expenditures required to establish and run CSF (without a priori pilot studies to better describe the potential impact of this work), as well as ethical concerns about large-scale CSF research efforts across the Army Soldier population (Eidelson et al. 2011). Additionally, Smith (2013) has expressed some concern that the positive psychology focus of CSF (which has shown promise in child and adolescent populations) may actually work against resiliency aims for deployed service members confronting trauma. Many have criticized the lack of published outcomes from the proposed CSF research described by Lester and colleagues (Steenkamp et al. 2013), and a 2014 PSYCINFO search using the term "Comprehensive Soldier Fitness" returns only 14 publications, none of which describe CSF outcomes. The true value of this ambitious program will only be realized once CSF program outcomes begin to appear in the extant peer-reviewed research. Unfortunately, because CSF is not being implemented as a randomized controlled trial, the ability to evaluate its potential efficacy will be limited.

Total Force Fitness Total Force Fitness is an expanded version of the biopsychosocial model (Engel 1977; Peterson et al. 2014) including behavioral, social, physical, environmental, medical/dental, spiritual, nutritional, and psychological factors related to resiliency. In 2010, Admiral Mike Mullen, then Chairman of the Joint Chiefs of Staff, described his vision of "Total Force Fitness" for the United States military (Mullen 2010). Characterized as "dynamic" and focusing on physical and mental "readiness," Total Force Fitness is described by Admiral Mullen as an effort to maintain peak physical and mental readiness to perform despite the constantly fluctuating landscape of challenges that confront military service members. The eight components of Total Force Fitness are outlined in Table 2 (Jonas et al. 2010). Colonel Beverly Land (US Army Medical Corps) has further explained Total Force Fitness as an ongoing attempt to maximize the health and well-being of military service members, achievable only through defining specific "end states" of expected functioning and designing training that will help service members achieve those end states (Land 2010). As is

Table 2 Total Force Fitness model of psychological resilience

Dimension	Description
Behavioral	Substance abuse
	Hygiene
	Risk mitigation
Social	Family cohesion
	Social support
	Task cohesion
	Social cohesion
Physical	Strength
	Endurance
	Power
	Flexibility
	Mobility
Environmental	Heat/cold
	Altitude
	Noise
	Air quality
Medical/dental	Immunizations
	Screening
	Prevention
Spiritual	Perspective
	Core values
	Identity, meaning, and purpose
	Ethical foundation
	Embracing diversity
Nutritional	Food quality
	Nutrient requirements
	Food choices
Psychological	Coping
	Awareness
	Beliefs/appraisals
	Decision-making
	Engagement

Adapted from Jonas et al. 2010

the case for Comprehensive Soldier Fitness, Total Force Fitness was developed with a specific outcomes assessment plan (Walter et al. 2010), though there have been no notable peer-reviewed publications detailing the outcomes of Total Force Fitness as of the writing of this chapter. Indeed, a 2014 PSYCINFO search using the term "Total Force Fitness" returned only 10 manuscripts, none of which detailed outcomes of Total Force Fitness implementation efforts.

6 Future Directions

Research on psychological resiliency has increased exponentially over the past decade. The majority of the research was stimulated by concerns about military personnel who have been exposed to extreme stress and trauma during deployments to Iraq and Afghanistan. Unfortunately, resiliency research has lagged behind research on the neurobiology of stress, primarily because of the lack of standardized definitions and randomized controlled trials. Future translational research should include operational definitions of resiliency, prospective longitudinal cohort studies, and randomized clinical intervention trials.

References

Adler AB, Bliese PD, McGurk D et al (2009) Battlemind debriefing and battlemind training as early interventions with soldiers returning from Iraq: randomization by platoon. J Consult Clin Psychol 77(5):928–940. doi:10.1037/a0016877

Agaibi CE, Wilson JP (2005) Trauma, PTSD, and resilience: a review of the literature. Trauma Violence Abuse 6(3):195–216. doi:10.1177/1524838005277438

Bates MJ, Bowles S, Hammermeister J et al (2010) Psychological fitness. Mil Med 175(8S):21–38. doi:10.7205/MILMED-D-10-00073

Blount T, Peterson AL, McGeary D et al (2012) The blister-callus model of psychological resilience. Panel discussion presented at the annual convention of the association for behavioral and cognitive therapies, National Harbor, MD, 15–18 November 2012

Bowles SJ, Bates MJ (2010) Military organizations and programs contributing to resilience building. Mil Med 175(6):382–385. doi:10.7205/MILMED-D-10-00099

Britt TW, Sinclair RR, McFadden AC (2013) Introduction: the meaning and importance of military resilience. In: Sinclair RR, Britt TW (eds) Building psychological resilience in military personnel: theory and practice. American Psychological Association, Washington, pp 3–17

Casey GW Jr (2011) Comprehensive soldier fitness: a vision for psychological resilience in the US Army. Am Psychol 66(1):1–3. doi:10.1037/a0021930

Connor KM, Davidson JRT (2003) Development of a new resilience scale: the Connor–Davidson Resilience Scale (CD-RISC). Depress Anxiety 18(2):76–82

Eidelson R, Pilisuk M, Soldz S (2011) The dark side of comprehensive soldier fitness. Am Psychol 66(7):643–644. doi:10.1037/10025272

Engel GL (1977) The need for a new medical model: a challenge for biomedicine. Sci 196:129–136. doi:10.1126/science.847460

Evans WP, Marsh SC, Weigel DJ (2010) Promoting adolescent sense of coherence: testing models of risk, protection, and resiliency. J Community Appl Soc Psychol 20(1):30–43. doi:10.1002/casp.1002

Fravell M, Nasser K, Cornum R (2011) The soldier fitness tracker: global delivery of comprehensive soldier fitness. Am Psychol 66(1):73–76. doi:10.1037/a0021632

Friborg O, Hjemdal O, Rosenvinge JH et al (2003) A new rating scale for adult resilience: What are the central protective resources behind healthy adjustment? Int J Methods Psychiatr Res 12(2):65–76. doi:10.1002/mpr.143

Friborg O, Barlaug D, Martinussen M et al (2005) Resilience in relation to personality and intelligence. Int J Methods Psychiatr Res 14(1):29–42. doi:10.1002/mpr.15

Garmezy N, Masten AS, Tellegen A (1984) The study of stress and competence in children: a building block for developmental psychopathology. Child Dev 55(1):97–111. doi:10.2307/1129837

Gates MA, Holowka DW, Vasterling JJ et al (2012) Posttraumatic stress disorder in veterans and military personnel: epidemiology, screening, and case recognition. Psychol Serv 9(4):361–382. doi:10.1037/a0027649

Gillespie BM, Chaboyer W, Wallis M et al (2007) Resilience in the operating room: developing and testing of a resilience model. J Adv Nurs 59(4):427–438. doi:10.1111/j.1365-2648.2007.04340.x

Gomez R, McLaren S (2006) The association of avoidance coping style, and perceived mother and father support with anxiety/depression among late adolescents: applicability of resiliency models. Pers Individ Dif 40(6):1165–1176. doi:10.1016/j.paid.2005.11.009

Hoge CW, Castro CA, Messer SC et al (2004) Combat duty in Iraq and Afghanistan, mental health problems, and barriers to care. N Eng J Med 351(1):13–22. doi:10.1056/NEJMoa040603

Hollister-Wagner GH, Foshee VA, Jackson C (2001) Adolescent aggression: models of resiliency. J Appl Soc Psychol 31(3):445–466. doi:10.1111/j.1559-1816.2001.tb02050.x

Institute of Medicine (2012) Treatment for posttraumatic stress disorder in military and veteran populations: initial assessment. The National Academies Press, Washington, DC

Jaffee SR, Caspi A, Moffitt TE et al (2007) Individual, family, and neighborhood factors distinguish resilient from non-resilient maltreated children: a cumulative stressors model. Child Abuse Negl 31(3):231–253. doi:10.1016/j.chiabu.2006.03.011

Jenson JM (2007) Research, advocacy, and social policy: lessons from the risk and resilience model. Soc Work Res 31(1):3–5. doi:10.1093/swr/31.1.3

Jenson JM, Fraser MW (2005) A risk and resilience framework for child, youth, and family policy. In: Jenson and Fraser (eds) Social policy for children and families: a risk and resilience perspective. Sage Publications, Thousand Oaks

Jonas WB, O'Connor FG, Deuster P et al (2010) Why total force fitness? Mil Med 175(8S):6–13. doi:10.7205/MILMED-D-10-00280

Kaminsky M, McCabe OL, Langlieb AM et al (2007) An evidence-informed model of human resistance, resilience, and recovery: the Johns Hopkins' outcome-driven paradigm for disaster mental health services. Brief Treat Crisis Interv 7(1):1–11. doi:10.1093/brief-treatment/mhl015

Klasen F, Oettingen G, Daniels J et al (2010) Posttraumatic resilience in former Ugandan child soldiers. Child Devel 81(4):1096–1113. doi:10.1111/j.1467-8624.2010.01456.x

Kobassa SC, Maddi SR, Kahn S (1982) Hardiness and health: a prospective study. J Pers Soc Psychol 42(1):168–177. doi:10.1037/0022-3514.42.1.168

Land (2010) Current department of defense guidance for total force fitness. Mil Med 175(8):3–5. doi:10.7205/MILMED-D-10-00138

Lester PB, McBride S, Bliese PD, Adler AB (2011) Bringing science to bear: an empirical assessment of the comprehensive soldier fitness program. Am Psychol 66(1):77–81. doi:10.1037/a0022083

Luthar SS, Cicchetti D, Becker B (2000) The construct of resilience: a critical evaluation and guidelines for future work. Child Dev 71(3):543–562. doi:10.1111/1467-8624.00164

Maguen S, Turcotte DM, Peterson AL et al (2008) Description of risk and resilience factors among military medical personnel before deployment to Iraq. Mil Med 173(1):1–9

Marsh SC, Evans WP, Weigel DJ (2009) Exploring models of resiliency by gender in relation to adolescent victimization. Vict Offender 4(3):230–248. doi:10.1080/15564880903048487

McGeary D (2011) Making sense of resilience. Mil Med 176(6):603–604. doi:10.7205/MILMED-D-10-00480

McLaren S, Gomez R, Bailey M et al (2007) The association of depression and sense of belonging with suicidal ideation among older adults: applicability of resiliency models. Suicide Life Threat Behav 37(1):89–102. doi:10.1521/suli.2007.37.1.89

Meredith LS, Sherbourne CD, Gaillot S et al (2011) Promoting psychological resilience in the US military. RAND Corporation, Santa Monica

Merriam-Webster (2014) Resilience. http://www.merriam-webster.com/dictionary/resilience?show=0&t=1391419988. Accessed 3 February 2014

Metzl E, Morrell M (2008) The role of creativity in models of resilience: theoretical exploration and practical applications. J Creativity Ment Health 3(3):303–318. doi:10.1080/15401380802385228

Mullen M (2010) On Total Force Fitness in war and peace. Mil Med 175(8):1–2. doi:10.7205/MILMED-D-10-00246

Mulligan K, Fear N, Jones N et al (2012) Postdeployment battlemind training for the UK armed forces: a cluster randomized controlled trial. J Consult Clin Psychol 80(3):331–341. doi:10.1037/a0027664

Paton D, Violanti JM, Johnston P et al (2008) Stress shield: a model of police resiliency. Int J Emerg Ment Health 10(2):95–108

Pengilly JW, Dowd ET (2000) Hardiness and social support as moderators of stress. J Clin Psychol 56(6):813–820. doi:10.1002/(SICI)1097-4679(200006)56:6<813:AID-JCLP10>3.0.CO;2-Q

Peterson AL, Cigrang JA, Isler WC (2009) Future directions: trauma, resilience, and recovery research. In: Freeman SM, Moore B, Freeman A (eds) Living and surviving in harm's way: a psychological treatment handbook for pre-and post-deployment of military personnel. Taylor & Francis, Routledge, pp 467–493

Peterson AL, Luethcke CA, Borah EV et al (2011) Assessment and treatment of combat-related PTSD in returning war veterans. J Clin Psychol Med Settings 18(2):164–175. doi:10.1007/s10880-011-9238-3

Peterson AL, Goodie JL, Andrasik F (2014) Introduction and overview of biopsychosocial assessment in clinical health psychology. In: Andrasik F, Goodie JL, Peterson AL (eds) Biopsychosocial assessment in clinical health psychology: a handbook. Guilford, New York, pp 1–6

Rutter M (1985) Resilience in the face of adversity: protective factors and resistance to psychiatric disorder. Br J Psychiatry 147(6):598–611. doi:10.1192/bjp.147.6.598

Seligman ME (2011) Building resilience. Harv Bus Rev 89(4):100–106

Smith SL (2013) Could Comprehensive Soldier Fitness have iatrogenic consequences? A commentary. J Health Behav Hlth Serv Res 40(2):242–246. doi:10.1007/s11414-012-9302-2

Smith BW, Dalen J, Wiggins K et al (2008) The brief resilience scale: assessing the ability to bounce back. Int J Behav Med 15(3):194–200. doi:10.1080/10705500802222972

Steenkamp MM, Nash WP, Litz BT (2013) Post-traumatic stress disorder: review of the Comprehensive Soldier Fitness program. Am J Prev Med 44(5):507–512. doi:10.1016/j.amepre.2013.01.013

Tanielian T, Jaycox LH (eds) (2008) Invisible wounds of war: psychological and cognitive injuries, their consequences, and services to assist recovery. RAND Corporation, Santa Monica

Waller MA (2001) Resilience in ecosystemic context: evolution of the concept. Am J Orthopsychiatry 71(3):290–297. doi:10.1037/0002-9432.71.3.290

Walter JA, Coulter I, Hilton L et al (2010) Program evaluation of total force fitness programs in the military. Mil Med 175(8S):103–109. doi:10.7205/MILMED-D-10-00279

Walter Reed Army Institute of Research Land Study Team (2006) Battlemind training I: transitioning from combat to home. Available at http://www.ptsd.ne.gov/pdfs/WRAIR-battlemind-training-Brochure.pdf. Accessed 6 February 2014

Werner EE, Smith RS (1992) Overcoming the odds: high risk children from birth to adulthood. Cornell University Press, Ithaca

Windle G, Markland DA, Woods RT (2008) Examination of a theoretical model of psychological resilience in older age. Aging Ment Health 12(3):285–292. doi:10.1080/13607860802120763

Windle G, Bennett KM, Noyes J (2011) A methodological review of resilience measurement scales. Health Qual Life Outcomes 9:8. doi:10.1186/1477-7525-9-8

Zamorski M, Guest K, Bailey S et al (2012) Beyond battlemind: evaluation of a new mental health training program for Canadian forces personnel participating in third-location decompression. Mil Med 177(11):1245–1253. doi:10.7205/MILMED-D-12-00064

Zimmerman MA, Arunkumar R (1994) Resiliency research: implications for schools and policy. Soc Res Child Dev 8(4):1–19

Index

A

Addiction, 238–243, 245–248, 250, 251, 253, 257
Adolescent stress, 241
Adrenocorticotrophic hormone, 14
Aging, 265–269, 271, 275, 276
Alzheimer's disease, 265, 266
Animal models, 130, 131, 136, 138, 141, 142
    acute stress, 134
    chronic corticosterone, 134
    chronic mild stress, 131
    chronic single stress, 133
    chronic social stress, 132
    chronic stress, 138
    depression, 130
    postnatal, 130, 135
    prenatal, 130, 135
    repeated restraint, 137
    social defeat, 136
    stress-induced, 130, 131, 136, 138
    social isolation, 136
Antidepressants, 33–35
Anxiety disorder, 49, 74, 82, 162, 165, 167, 173, 191–193, 200, 202, 205, 209
Arginine vasopressin, 67, 68, 72
Assessment and measurement of psychological resiliency, 303
Awakening cortisol response, 16

B

Battlemind Training, 306
Behaviour, 68
Bipolar disorder, 16
Blister-Callus Model of Psychological Resilience, 303
Brain derived neurotrophic factor (BDNF), 242, 250, 252

C

Childhood adversity, 19
Childhood trauma, 18
Cognitive-Behavioral Therapy (CBT), 255–257
Cognitive decline, 265, 276, 277, 280
Comprehensive Soldier Fitness Program, 307
Corticotropin-releasing factor, 14, 67, 68
Cortisol, 14–20, 29, 30, 69, 73, 93, 95, 100, 101, 107–110, 114, 115, 126, 128, 129, 192, 193, 197–202, 204–209, 243, 244, 249, 251, 252, 255
CAMP response element-binding (CREB), 242, 250

D

Depression, 265–267, 272–274, 277–277, 279, 280
    cortisol, 128, 129
    dexamethasone, 128
    HPA-axis, 128, 129
    resistance, 142
    stress, 127, 129
    susceptibility, 142, 143
    vulnerability, 15, 25, 29, 32, 36, 37, 127–130, 142, 143
Dexamethasone, 14
Dopaminergic, 47, 51–53

E

Early developmental stress, 238, 240
Early life experiences, 18
Early life stressors, 246
Endocannabinoid, 57–59
Environmental enrichment, 253
Epigenetic, 217, 226, 227

Epigenetic programming, 19
Exercise, 3, 8, 9

F
Fear extinction, 161, 174–176, 179, 180
FKBP5, 19
Fractalkine, 6

G
Gender, 251, 252, 257
Glucocorticoid receptor, 14
Glucocorticoids, 46, 47, 52, 58, 59
Glutamatergic, 47–49, 57
Gut dysbiosis, 276, 277, 280

H
Hair cortisol, 17
Hardiness, 299, 301
Hippocampus, 26–29, 31, 32, 36, 38
Hypothalamic–pituitary–adrenal (HPA) axis, 14, 29, 30, 33–35, 37, 221–223, 226, 227, 240, 243–245, 248–250, 252, 255
   ACTH, 126
   corticotropin-releasing factor, 125
   CRF, 125, 126
   estrogens, 126
   glucocorticoids, 126
   growth hormon, 126
   progesterone, 126
   prolactin, 126
   testosterone, 126
   thyroid hormones, 126

I
Initiation, 257

M
Microglia, 1, 3, 4, 6, 7, 9
Military Demand-Resource Model of Resiliency, 302
Military Resiliency Programs, 306
Mineralocorticoid receptor, 14

N
Neurogenesis, 25–29, 32–34, 36–38, 140, 141
Neurodegeneration, 140, 141

Neuroinflammation, 265, 277, 279, 280
Neurotrophin, 141
Noradrenergic, 47, 49, 51

O
Outcome Conceptualization of Resiliency, 301

P
Panic, 165, 167, 191, 193, 197–202, 204, 206, 209
Peritrauma, 20
Phobias, 165, 179, 191, 193, 197, 205–209
Pituitary, 15
Pituitary volume, 16
Post-traumatic stress disorder (PTSD), 17, 161–169, 172, 173, 175, 176, 178, 179, 181
Prolactin, 16
Psychological resiliency, 297, 300, 303, 309
Psychosis, 16, 217–223, 225, 227
Psychosocial stressors, 15
Psychotic, 224

R
Reward pathways, 242, 245, 250
Reward pathway systems, 257
Reward system, 247
Rodent models
   chronic mild stress, 137
   resistant, 138
   susceptible, 137, 138

S
Serotonergic, 54–56
Stress, 28, 30–34, 37, 38, 67–73, 124, 139, 140, 266–268, 269–277, 280, 297, 298, 300, 301, 303, 304, 306, 307, 309
   anxiety, 93–96, 100
   behaviour, 93, 95–99, 114, 115
   depression, 93–96, 100, 102, 105, 107, 124
   glucocorticoids, 125
   maternal, 140
   postnatal, 139
   prenatal, 139
   sympathomimetic, 26, 27, 125
Stress resistance, 1, 8
Stress response, 246, 257
Stress response system, 243–245, 247, 248

Stress-vulnerability, 218, 222
Substance use initiation, 239, 243
System, 242

T
Total Force Fitness, 308
Trait Conceptualization of Resiliency, 301

Trauma exposure, 17
Trauma, 221, 225, 227
Traumatic, 226
Treatment, 245, 251–258

V
Vasopressin, 14